ANIMAL PHYSIOLOGY

ANIMAL PHYSIOLOGY

By

KAVITA JUNEJA

ANMOL PUBLICATIONS PVT. LTD.
NEW DELHI - 110 002 (INDIA)

ANMOL PUBLICATIONS PVT. LTD.
4374/4B, Ansari Road, Daryaganj
New Delhi - 110 002
Ph.: 3261597, 3278000
Visit us: www.anmolbooks.com

Animal Physiology

First Edition, 2002

ISBN 81-261-1167-4

PRINTED IN INDIA

Published by J.L. Kumar for Anmol Publications Pvt. Ltd., New Delhi - 110 002 and Printed at Mehra Offset Press, Delhi.

CONTENTS

Preface

This book is an attempt to make the students of under graduate level understand the current state of animal body at the level of its functional features. Emphasis in this book is mainly on the functions of its component organs and organ systems under its natural environment. Phenomena at subcellular to tissue levels are discussed to provide perspective at organs to whole body level.

The subject of physiology being exhaustive, the author has treated only those topics in each system which have been included in the course outlines of Animal Physiology designed by the Indian Universities for their under-graduate and graduate classes in the subject of Zoology.

The phenomena of life processes cannot be explained without going into the details of biochemical events that occur at molecular level; the first three chapters of this book are devoted to the structure, chemical properties and biological significance of macromolecules of physiological importance to prepare the students to understand the part of the material which follows in the subsequent 12 chapters. It has been author's endeavour to include up-to-date information in simple language. Efforts have been made to produce a text that will be useful, informative and stimulating for the students opting for Animal Physiology as one of the core subject in animal sciences and other interested readers.

Though, the author has taken special care to present a current account, yet she is fully aware about her limitations, and the readers may come across the mistakes of various types. For all types of mistakes she extends her due apology.

The author has freely consulted other standard books and research papers while preparing the manuscript, so the author claims no originality of the work.

The author expresses her thanks to all friends and colleagues whose continuous inspirations have invited her to bring out this text.

The author is also thankful to Shri J.L. Kumar of M/s Anmol Publications Pvt. Ltd. for bringing out a handsome print of this book in a comparatively short time.

Constructive criticisms and suggestions for improvement of the book will thankfully acknowledged.

Kavita Juneja

1

The Cell

The cell is a unit of biological activity delimited by a semipermeable membrane and capable of self-reproduction in a suitable non-living medium. The body of all living organisms (bacteria, blue green algae, plants and animals) except viruses has cellular organization and may contain one or many cells. The organisms with only one cell in their body are called acellular organisms (e.g., bacteria, blue green algae, some algae, protozoan, etc.). The organisms having many cells in their body are called multicellular organisms (e.g., most plants and animals). Any cellular organism may contain only one type of cell Prokaryotic cells, Eukaryotic cells.

It was *Robert Hooke* who first of all in 1665 observed under a microscope certain honey comb like structures in cork, and he applied the term 'cell' (L., *cella;* small room) for the same. Previously it was believed that there is only one component of the cell which is cell wall. But in 1831, one more peculiar structure was observed by *Robert Brown.* He gave the name 'nucleus'. Later on the 1835 the word 'Sarcode' was proposed for the jelly like material present inside the cell by *Dujardin.* In 1840 *Purkinje* replaced the word sarcode by Protoplasm which is now used universally.

Prokaryotic Cells

The prokaryotic (Gr., *pro*-primitive, *Karyon*-nucleus) cells are the most primitive cells from morphological point of view. They occur in bacteria and blue green algae. Prokaryotes are small, single cells organism, usually less than a micrometer (abbreviated μm; 1000 μm = 1 millimeter, abbreviated mm) are generally not longer than 3 μm. Their structural organization is simple than that of the eukaryotes. They

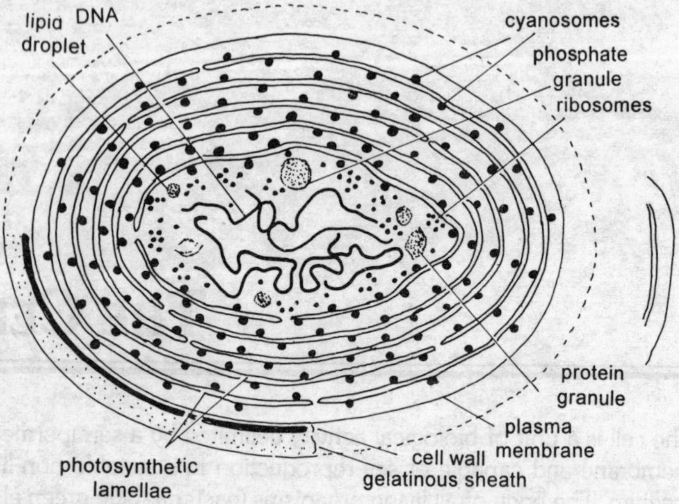

Fig. 1.1. A prokaryotic cell of Escherichia coli.

do not have a distinct nucleus, and they are lacking in many membranous structures found in the more complex cell types. The hereditary information is contained in a single circular structure that lacks many of the molecular components of the eukaryotic hereditary structure called the chromosome. In addition, the prokaryotes seem to lack any kind of special apparatus for cell division that is why they divide amitotically or by fission.

The prokaryotes are characterized by:

(i) the absence of a membrane around the nuclear material.
(ii) the absence of clearly defined membrane-limited organelles like mitochondria, chloroplast, Golgi complex and lysosomes.
(iii) the genetic material is located on a single chromosome which consists of a circular double strand of DNA.
(iv) the basic protein—histones, which are one of the most important constituents of chromosomes of eukaryotic cells, are absent in prokaryotic chromosomes.
(v) the absence of nucleolus and mitotic apparatus.
(vi) the cell wall is non-cellulosic, being formed of carbohydrates and amino acids.
(vii) the plasma membrane which lies below the cells wall is produced into the cytoplasm and acts as the mitochondrial membrane carrying respiratory enzymes, and

(viii) the cytoplasm neither exhibits streaming nor the amoeboid movement.

The prokaryotes include bacteria, blue green algae, spirochaete, *Mycoplasma* or pleuropneumonia like organism (PPLO).

Bacterial Cell

The bacteria are microscopic, cellular, achlorophillous, asexually reproducing prokaryotes which occur in fresh and salt water, in soil and in plants and animal. They lead either a saprophytic or parasitic mode of existence. The saprophytic species of bacteria are of great economic significance for man. Some parasitic species of bacteria are pathogenic (disease producing) to plants, animals and man.

Bacteria are very small having an average diameter of 1.25μ.

The smallest bacterium is *Dialister pneumosintes* (0.15 to 0.3μ) and largest bacterium is *Spirillum volutans* (13-15μ) in length.

One of the best studied bacteria is *Escherichia coli*. It is about 2 μm long and 1μm wide and contains several thousand different kinds of molecules. Structurally, *E. coli* is rod like and has a typical cell wall and underlying cell membrane. The membrane contains the special regions called *mesosomes,* which apparently have a role in cellular respiration and cell division. There may be one or several "nucleoid" areas, which contain a single, circular molecule of hereditary information (DNA).

The rest of the bacterial cytoplasm is full of structures called ribosomes, which differ slightly from the ribosomes of eukaryotic cells. These structures are the sites of protein synthesis. In addition, recent studies have revealed the presence of small circular DNA molecules in the cytoplasm called *plasmids.* These structures are very important in genetic engineering.

Mycoplasma

The *Mycoplasma* (previously called PPLO, pleuropneumonia like organisms) are small (0.2 to 0.3 μm) and more simply organized than bacteria. The hereditary material (DNA) is contained in a non-membrane-bound, nucleus like region, and it may exist either as strands or as a circular molecule as in the bacteria. However, there is only one tenth the amount of genetic material in Mycoplasma as there is in bacteria. Similarly, the Mycoplasma contains 1/50 to 1/100 the number of bacterial ribosomes per cell. A delimiting cell membrane exists, but no cell walls have been detected. A variety of other cytoplasmic inclusions, such as vacuoles and granules, have been detected, but their function is not known. As in other prokaryotes, there are no intracellular membranous structures.

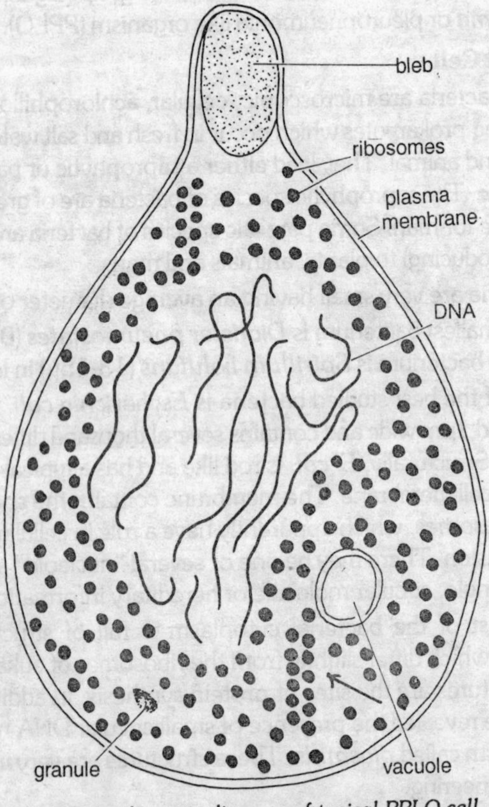

Fig. 1.2. A schematic diagram of typical PPLO cell.

Mycoplasma has been known to cause certain respiratory diseases in animals and man. They also occur as routine contaminants in eukaryote tissue culture and currently are of much concern to researchers who use these culture procedures in their studies.

Blue-Green Algae

The blue-green algae are the third major group of prokaryotic organisms. The cellular organization is more comples than that of the other prokaryotes, primarily because of the existence of photosynthetic machinery. The blue-green algae contain chlorophyll as well as other more unusual pigments called *phycobilins*. The chlorophylls are bound

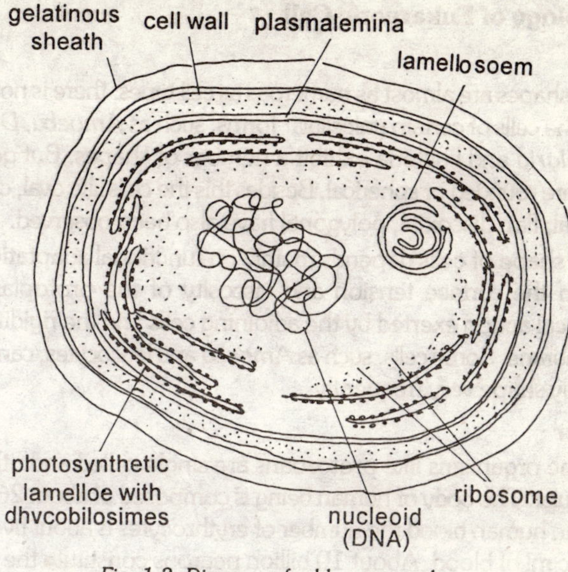

Fig. 1.3. Diagram of a blue green alga.

to membranes in flattened rows called lamellae, which give the algal cytoplasm a rather organized and structured appearance. The phycobilins appear to be located in granular structures called *cyanosomes,* which apparently are attached to the outer lamellar membrane surface. Like the other prokaryotes, the blue-green algae do not have a membrane-bound nucleus or other membrane-delimited organelles. The genetic material exists as free strands of DNA, and the ribosomes are free in the cytoplasm as in the bacteria and *Mycoplasma.* There appears to be other cytoplasmic granules of a protein and phosphate nature, and the entire cell is surrounded by a cell wall with an underlying plasma membrane.

EUKARYOTIC CELLS

Eukaryotic cells may be acellular organisms, such as protozoans and acellular algae, or they may be cells that make up the tissues and organs of multicellular organisms. Though the eukaryotic cells have different shape, size, and physiology but all the cells are typically composed of plasma membrane, cytoplasm and its organelles, viz., mitochondria, endoplasmic reticulum, ribosomes, Golgi complex, etc., and a true nucleus.

Morphology of Eukaryotic Cell

Shape

Cell shapes are almost as numerous as cell types: there is not typical shape. The cells of certain unicellular forms, such as *Amoeba, Diatoms, Acetabularia* and bacteria exhibit a number of shapes. But generally the cells are rounded or spherical. Besides this the cells like oval, cuboidal, cylindrical, flat, discoidal, polygonal have also been observed.

The shape of cells depends mainly on functional adaptations and partly on the surface tension and viscosity of the protoplasm, the mechanical action exerted by the adjoining cells and the rigidity of the cell membrane. Some cells, such as *Amoeba* and leucocytes, can change their body shape very frequently.

Number

Some organisms like protozoans are single celled and others are multicellular. The body of human being is composed of about 26 trillions of cells. In human blood the number of erythrocytes is about five million per cubic ml of blood. About 10 billion neurons constitute the nervous system in human being.

Size

The size of different cells ranges within broad limits. Some plant and animal cells are visible to the naked eye, such as eggs of certain birds have a diameter of several centimeters. But the great majority of cells are visible only under microscope. The smallest living cells are found among bacteria and viruses where the size ranges 0.1μ to 1μ (1μ = 0.01 mm). Organisms like Pleuropneumonia-like organisms so called "elementary bodies" have been observed with diameters of 100 mμ. These appear to be a resting from of the bacteria which may grow into bodies of 250 mμ in diameter during the active metabolism of the organism. It would seem that 200-250 mμ in diameter is the lower limit for the size of inactive, living cells. This lower limit may be set by the minimum number and size of the component necessary for independent cellular existence.

Usually the vast majority of cells lie in the range of .5 to 20μ in diameter. The length of acellular diatoms is about 100μ. And the size of *Amoeba proteus* is 1 m.m. (1000μ) in length. The size of human R.B.C. is 7-8μ in diameter.

The long nerve is also an example of the extremes in cell sizes, the range of which is quite remarkable in the living world.

Factors Restricting Cell Size

In general, the size of the cell is correlated with its function. In no way it is associated with the size of the organism. For instance, a microscopic cell of bread mould can be several metres long, whereas cells in the body of whale or elephant may be as small as 10μ. Two major factors seem to restrict cell size.

1. The nucelocytoplasmic ratio

The nucleus controls activities of the cell by protein synthesis through RNA synthesis. In a large cell cytoplasm requires more proteins and consequently more RNA. The DNA content of a cell is constant, which can control the activities of the cell upto a specific size, beyond which proper control is not possible. For this reason a cell after attaining a specific size stops growing.

2. The surface area

As shown in fig, the small cells have more surface area per unit volume and with an increase in size, this ratio between surface area and volume gradually decreases. However, a large cell naturally requires more nutrients and oxygen for its metabolism, which diffuse into the cell through its surface. It means, the surface area of a cell affects its ability to exchange materials with its environment and puts a check on cell volume. For this reason, the cells that are metabolically more active are small in size.

Similarly, large cells face problem in the diffusion of excretory products from cell cytoplasm to the exterior.

Structure of Fixed Cells

Examination of the living cell is limited to light microscopy and is based mainly on the differences in refractive index of the different cell components. Sometimes the use of stains that act on the living organism (vital staining) facilitates observation of the living cell. However, more important in the morphologic study of the cell are methods of fixation, by which cell death results in such a way that physiologic structure and chemical composition are preserved as much as possible.

The complexity of the structural organization in a cell of the higher plants and animals is most impressive. Although there are great differences between the primitive forms of life, and the higher plant and animal cells, the similarities between primitive and advanced cells are also notable. Eukaryotic cells are characterized by a true nucleus with a nuclear membrane or envelope which divides the cell into two main compartments: nucleus and cytoplasm. The cytoplasm is turn is limited by the plasma membrane. In a plant cell the plasma membrane is covered and protected on the outside by a thicker cell wall through which there are tunnels, the *plasmodesmata,* by which the cell intercommunicates with neighbouring cells via fine cell process. In animal cells, parts of the plasma membrane are covered by a thin layer of material, which is

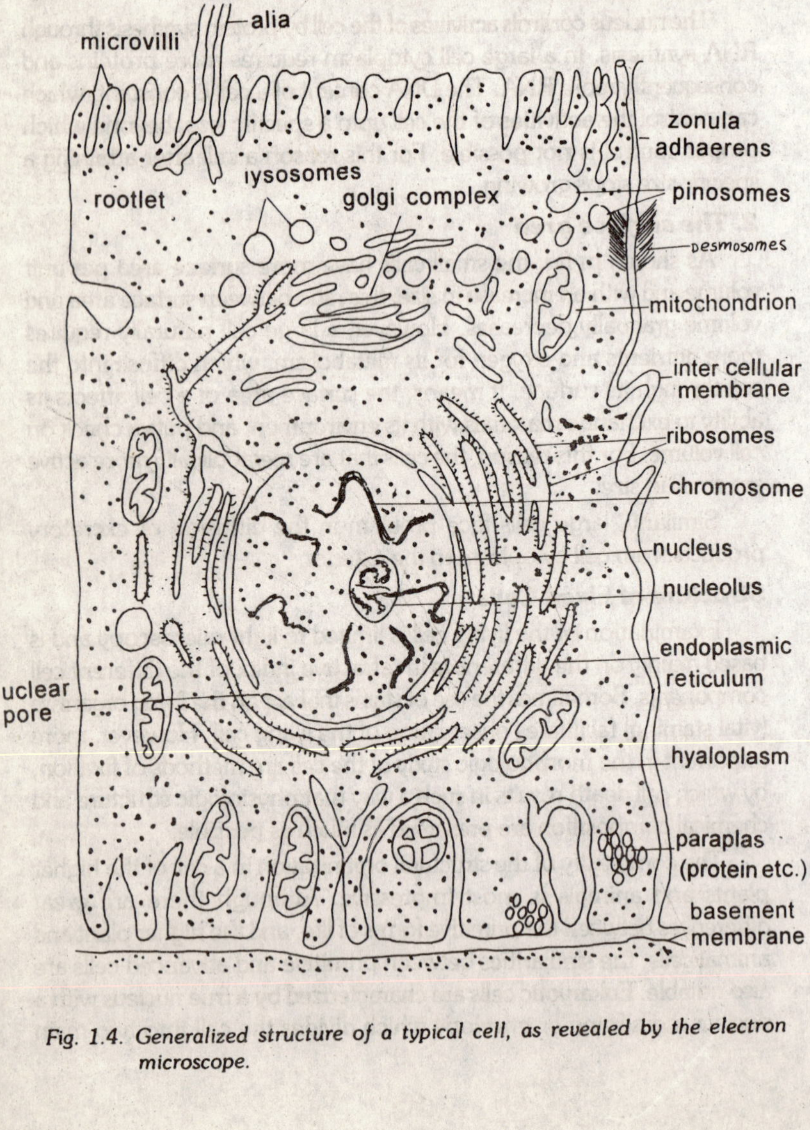

Fig. 1.4. Generalized structure of a typical cell, as revealed by the electron microscope.

generally described as the extraneous coat of the plasma membrane. The co-called *basement membranes* correspond to this extraneous coat.

In every stained cell usually one nucleus is present. It is central organelle of cell and is the main site of hereditary material. The nucleus is surrounded by a nuclear membrane but in bacteria and blue green algae it is without nuclear membrane. Within the nucleus a single nucleolus and network of chromatin is present. These two materials remain suspended in the nucleoplasm.

A small cytoplasmic organoid 'centriole' is present in the majority of animal cells and in some cells of the lower plants. The position is generally fixed for each type of cell. Generally the cell center or centriole occupies an axial position behind the nucleus. In some cells, it has the tendency to occupy the geometrical center i.e., in the leucocytes it occupies this position within the nucleus which is horseshoe shaped. The morphology of centriole changes with the functional state of the cell.

It is cytoplasm of the cell, there are several organelles out of them, only mitochondria and Golgi bodies are visible under light microscope. Mitochondria or chondriosomes are scattered in the cytoplasm of all kinds of cells. Their shape in the fixed cells depends upon the fixation and handling of the tissues before preparation they may appear as long rods, short rods or as small granules. Golgi material appears in its very prominent form in nervous and secretory cells in animals. Plastids are cytoplasmic organelles of plant cells may or may not contain pigments. In green plants chlorophyll is present in these organelles and thus termed as chloroplasts.

Electron Microscopic Structure of a Typical Cell

Under ordinary light microscope only few cell organelles like mitochondria, Golgi complex, chloroplasts and nucleus are visible. However, under electron microscope, several other cytoplasmic organelles such as endoplasmic reticulum, ribosomes, lysosomes, nuclear membrane, plasma membrane, etc. are seen.

Some important cellular components are being described here under the following heads:

Plasma Membrane

The outermost boundary of the animal cell is called the 'plasma membrane'. It is invisible under a light microscope but under electron microscope, it appears to be composed of two dense layers which are called the outer dense layer and inner dense layer. Both these layers are about 20Å in thickness, and are separated from each other by a less dense area of about 35Å in thickness. By this way the total thickness of plasma membrane is about 75Å. Before the advance of electron microscope people used to think of the plasma membrane has been stretched tightly over the cell. But now we know that a single plasma membrane can have as many as 3000 microvilli. Another peculiarity of the structure of the membrane is that, it remains connected with the endoplasmic reticulum.

Plasma membrane is lipo-proteinaceus in composition.

Functions

(i) Gives an identity to the cell.
(ii) Protects the cell organelles.
(iii) Helps in digestion through phagocytosis and pinocytosis.
(iv) Being permeable to water, water can pass through it.
(v) Help in the transport of ions.
(vi) Helps in the formation of cytoplasmic organelles.

Mitochondria

The term mitochondria was, for the first time used by *Benda* (1902) to designate the filamentous and granules of cell cytoplasm. Most of the knowledge related to mitochondrial fine structure is due to the studies of *Palade* (1952, 55, 56) and *Sjostrand* (1955). Due to these studies, now mitochondria are defined as the double walled structures of cell organelles within which the membranous structures called *cristae* are present. The outer chamber of mitochondrion is made up of two dense layers and a less dense layer having the total thickness of 140 Å to 180Å while the outer and inner dense layers vary in their measurements from 35Å to 60Å in thickness depending upon the physiological state of the cell. The less dense layer is 40-47Å in thickness.

The matrix of the mitochondria may have some fine opaque granules of various size. *David E. Green* (1964) observed in a very highly magnified electron microscope that the inside of the inner membrane and the outside of outer membrane are covered with very small particles. They measure about 90-100Å in diameter. The particles on the outer membrane differ from those on the inner membrane. The former generally appear as simple spheres packed closely together which give to the surface of the mitochondrion, a rough or pimpled appearance. The particle on the inside of the inner membrane has the more complex configuration. In some cells, but not all, this particle is seen to have a base, a stalk and a spherical head. Mitochondria represent the power generating centres of the cell. They can bring about oxidation of the intermediates of carbohydrates, fat and amino acid metabolism in the presence of O_2 releasing energy. The released energy is trapped in the form of adenosine triphosphate (ATP).

Golgi Complex

Golgi comples have probably been first observed earlier in 1867 by *La. Vallete St. George,* but any how credit goes to *Camillo Golgi* (1898) who described it as internal reticulate apparatus' and after it, was named *Golgi apparatus.* It is variously called Golgi element, Golgi complex, Dalton's complex etc. Each Golgi complex is composed of many lamella, tubules, vesicles and vacuoles. The membranes of Golgi complex are of lipoproteins and these are supposed to be originated from the membranes of endoplasmic reticulum. The function of the Golgi complex is storage of proteins and enzymes which are synthesized by ribosomes and transported by endoplasmic reticulum to them. Further the Golgi complex has most important secretory function. It secretes many secretory granules and lysosomes. In plant cells the Golgi complex is known as dictyosome and secretes necessary materials for cell wall formation during cell division.

Lysosomes

The lysosomes (Gr. *lyso;* digestive+*some;* body) are also found scattered in the cytoplasm. By gently centrifuging lysosomes can be separated from other cell organelles. They are usually spherical but may be irregular in shape, covered with a membrane of lipoprotein. Inside structure is always variable. Lysosomes contain hydrolysing enzymes, such as acid phosphatase, acid ribonuclease, acid deoxyribonuclease and cathespins. But they do not have oxidative enzymes. This property distinguishes them from the mitochondria. Lysosomes are polymorphic in nature and are commonly known as suicide bags and are responsible for postmortem changes in cells.

Endoplasmic Reticulum

The term "Microsome" was introduced to the modern cytology by *Claude* (1943) to represent one of the submicroscopic cellular components isolated by centrifugation. These are now known to be the broken parts of the endoplasmic reticulum after the studied of *Kollman* (1953). *Porter* (1945) was the 1st man to describe their electron microscopic structure in cultured cells. He described them as lace-like reticulum. These are membrane bounded sacs in the form of double membrane or cisternae. The term cisternae was introduced by *Sjostrand*

(1953) to describe long or elongated rod like parts of the endoplasmic reticulum measuring 50-200Å. *Wiess* (1953), described the vesicles of the endoplasmic reticulum which usually has a diameter of 25-500Å while *Bradfield* (1913) came across of another type of endoplasmic reticulum termed them as tubules, they measures from 50 to 100Å in diameter.

On the basis of presence or absence of ribosomes or RNP particles the endoplasmic reticulum (E.R) is distinguished in two varieties. The *rough walled endoplasmic reticulum,* and *smooth walled endoplasmic reticulum.*

Rough Walled Endoplasmic Reticulum

Rough walled endoplasmic reticulum (R.E.R.) is that variety which bears the ribosomes at the external surface of the cisternae. The distribution of the ribosomes can be circular, spiral, or rosette type. These particles may be induced to leave the surface of the cisternae without damaging the structure of the cisternae or the capacity to associate again with the particles. Usually the rough walled endoplasmic reticulum is richly distributed in those cells which are engaged in the synthesis of the protein.

Smooth Walled Endoplasmic Reticulum

The other large division of the system owes its identity in parts of the absence of the particles and therefore commonly referred to as the smooth surfaced a granular form. Like the R.E.R., the S.E.R. shows a characteristic morphology which is tubular rather than cisternae. The tubules have the diameter nearly of 50-100 mμ. As studied by *Fawcett* (1960) S.E.R. is richly distributed in those cells which are engaged in the synthesis of steroids.

Continuity between the S.E.R. of a particular cell type possess a characteristic pattern of structure. Thus it is typical for rat liver cells to show groups of eight or ten slender profiles in parallel array.

Plastids

Plastids are present only in plant cells (not in animal cells). These are double-membrane structures relatively larger than the mitochondria. Some of the plastids are colourless and store starch (Leucoplasts). Other contain various pigments and impart different colours (red, brown, yellow, etc.) to different parts of the plant. These are collectively called chromoplasts. Those containing green pigment—cholorophyll, are called choloroplasts. These contain several enzymes in their matrix necessary for photosynthesis. The chloroplasts capture light energy of the sun and convert it into the chemical energy—ATP, which is used in the biosynthesis of glucose from CO_2 and H_2O.

Ribosomes

Many minute, spherical structures known as *ribosomes.* They remain attached with the membrane of endoplasmic reticulum and forms granular or rough type of E.R. The ribosomes are originated in the nucleolus and consist of mainly the ribonucleic acids (RNA) and proteins. Each ribosome is composed of two structural units, small subunit known as 40S subunit and a larger subunit known as 60S subunit. The ribosomes remain attached with the membranes of endoplasmic reticulum by the 60S subunits. The 40S subunits occur on the larger sub unit and form a cap-like structure. The ribosomes consists of 3 types of RNA's known as ribosomal RNAs or rRNA, viz., 5S, 18S and 28S rRNAs. The 28S and 5S rRNAs occur in large (60S) subunit, while 18S rRNA occurs in the smaller ribosomal subunit. The ribosomes also occur freely in the cytoplasm. They are the sites of protein synthesis.

Peroxisomes

These are also bounded by a single membrane. These contain enzymes for the breakdown of hydrogen peroxide to form water and oxygen. These are protective organelles of the cell.

Cilia and Flagella

The cells of many unicellular organisms and ciliated epithelium of multicellular organisms consists of some hair-like cytoplasmic projections outside the surface of the cell. These are known as cilia or flagella. They help in locomotion of the cell. The cilia and flagella consists of nine outer fibrils around the two large central fibrils. Each outer fibrils consists of two microtubules. The two microtubules of each peripheral pair or *doublet* remain morphologically distinct. The central fibrils are separate but remain enveloped in a common *central sheath.* Cilia and flagella are envolved in locomotion, feeding, cleansing etc.

Vacuoles

These structures are fluid-filled "bubbles" found in the cytoplasm. Vacuoles are bounded by a membrane, called *tonoplast,* which is structurally similar to, if not identical with, a cell-membrane. Inside the vacuoles, food materials or wastes are found.

Vacuoles are consipicuous in plant-cells but they are found less in animal cells. Vacuoles are absent in actively dividing cells. In a young plant-cell, there are many vacuoles but, when a cell matures, they unite to form a single, large, central vacuole.

Guilliermond (1941) was of the opinion that vacuoles are colloidal with a strong imbibing capacity. Vacuoles arise *de novo* i.e., spontaneously formed from pre-existing vacuoles.

In many protozoan, vacuoles are formed at the bottom of the gullet. In these vacuoles, food-materials, such as bacteria, are collected. Some vacuoles are formed by cell-drinking. This type of vacuoles are called *pinosomes*. Food-vacuoles formed by evagination of cell-membrane surrounding food-particles with some fluid, are called *phagosomes*. In animals cells, the vacuoles are formed by a lipoprotein membrane that helps in storage and transport of materials and in the maintenance of internal pressure of the cell.

Vacuoles are not a living part of the protoplasm but are a storing place of reserve materials and waste products.

Crystals and oil droplets

In certain plant cells the food or waste material is found deposited in the form cf crystals. The oil droplets are round in both plant and animal cells, which appear as tiny glistening white spheres. These serve as a reserve supply of energy rich fuel.

Nucleus

It is the most vital part of cell which is invariably present in all the cells of higher animals and plants. Normally one nucleus is present in a cell but more than one may be present in certain cells.

Structurally it is enveloped by a porous *nuclear membrane* that separates it from the surrounding cytoplasm. It has the characteristic permeability. During cell division, it breaks up and soon after the mitosis it is reconstituted. The inside of nucleus is filled with *nuclear sap* or *karyolymph*. In nuclear sap remain embedded other nuclear constituents. The *chromatin* is the Feulgen positive material in the interphase nucleus which during division becomes the definite chromosomes. In short it represents the chromosomes substance. The sex-chromatin are the specific heterochromatic bodies, found generally near the periphery of the nucleus. In nucleus usually one sex chromatin body is found but more than one can also be observed. *Nucleolus* is relatively large, generally spherical ball-like body. It has a high concentration of RNA and thus is an important structure involved in protein synthesis. Its number in a nucleus vary in different cells and depend on either the species or the sets of chromosomes.

The nucleus co-ordinates many cell activities. DNA molecules are the source of coded instructions, which pass to RNA molecule which then act as messenger, moving out of the nucleus to one of the various structures of the cytoplasm. It also controls the cell reproduction. Here too, DNA is the source of coded instructions.

The Cytosol, Hyaloplasm or Ground Substance

The material that remains suspended under conditions sufficient to sediment even the ribosomes is referred to by biochemists as the *soluble fraction* or *cytosol*. This fraction usually contains some material lost from the organelles in the course of homogenization and centrifugation. However, it also contains other molecules, including some enzymes, that are not known to be part of the intracellular structures indentifiable in the microscope. Such molecules are usually thought to derive from the *hyaloplasm* (also called ground substance, cell matrix, or cell sap), which was originally defined by microscopists as the apparently structureless medium that occupies the space between the visible organelles. As will be evident in subsequent chapters, with the improved techniques less and less of the cell appears structureless. Most investigators continue to believe that there does exist in the cytoplasm a "soluble phase"—a solution containing inorganic ions and small molecules, materials in transit from one cell region to another and dissolved or suspended enzymes that function as individual molecules or perhaps or parts of small groups of molecules rather than as components of large, microscopically recognizable organelles. But there is a lively debate going on over proposals that many components once though to be suspended in this solution are actually bound, perhaps loosely, in structured arrangements that do not survive the disruption involved in cell fractionation.

Table 1.1. The Relative Volumes Occupied by the Major Intracellular Compartments in a Typical liver cell (Hepatocyte)

Intracellular Compartment	*Percent of Total Cell volume*	*Approximate Number per Cell*
Cytosol	54	1
Mitochondria	22	1700
Rough endoplasmic reticulum cisternae	9	1
Smooth endoplasmic reticulum cisternae plus Golgi cisternae	6	
Nucleus	6	1
Peroxisomes	1	400
Lysosomes	1	300

Table 1.2. Comparison of Prokaryotic and Eukaryotic Organisms

	Prokaryotes	*Eukaryotes*
Organisms	bacteria and cyanobacteria	protists, fungi, plants, and animals
Cell Size	generally 1 to 10 μm in linear dimension	generally 10 to 100 μm in linear dimension
Metabolism	anaerobic or aerobic	aerobic
Organelles	few or none	nucleus, mitochondria chloroplasts, endoplasmic reticulum, etc.
DNA	circular DNA in cytoplasm	very long DNA containing many non-coding regions; organized into chromosomes and bounded by nuclear envelope
RNA and protein	RNA and protein in same compartment	RNA synthesized and processed in nucleus; proteins synthesized in cytoplasm.
Cytoplasm	no cytoskeleton, cytoplasmic streaming, endocytosis, or exocytosis	cytoskeleton composed of protein filaments; cytopl-asmic streaming; endocytosis and exocytosis
Cell division	by binary fission	by mitosis (or meiosis)
Cellular organization	mainly acellular	mainly multicellular, with differentiation of cells.

Table 1.3. Comparison of plant and animal cells

Plant Cell	*Animal Cell*
1. Generally larger than the animal cell.	1. Usually smaller than the plant cell.
2. Plant cells are surrounded by a rigid cellulose wall.	2. Cellulose wall is not found in animal cell.
3. Majority of the mature plant cells have a large central sap vacuole.	3. Vacuoles in animal cells are small.
4. Plastids, especially chloroplasts are found usually in cells of green plants.	4. Chloroplast is not found in animal cell except in some Protozoa.

5. In the cell division, the division of cytoplasm of plant cells take by formation of a partition called place cell plate.	5. Animal cells divide by constriction during cell division.

Table 1.4. Functions of Cellular Organelles

Cell Organelles	*Functions*
Cell or Plasma membrane	Serves as a differentially permeable membrane through which extra-cellular substances may be selectively sampled and cell products may be liberated. Gives external form. Provides mechanical support and thus protects the cell.
Cell Wall (in plants)	Thick cellulose wall around the cell membrane gives strength and rigidity of the cell.
Cytoplasm	Provides greatly expanded surface. Endoplasmic recticulum-area for various biochemical reactions which normally take place at or across membrane surfaces.
Ribosomes	Sites of protein synthesis and known as "Protein factories on the cell"
Mitochondria	Oxidation of food substances, changes potential energy into a form of energy to be used by cell. Energy production (Krebs' cycle, electron transport chain, beta oxidation of fatty acids etc.). Power house of cell.
Centrioles (usually not seen in plant cells)	Form poles for the cell division. Capable of replication usually.
Golgi body or apparatus	Produce cellular secretions known as dictyosomes in plants.
Lysosome (in animals)	Produce intracellular digestive enzymes which help in disposal of bacteria and other foreign bodies. If ruptured, may cause cell destruction.

Plastids (in plants)	Serve for storage of starch, pigments and other cellular products. Photosynthesis takes place in chloroplasts.
Vacuoles	Store excess water, waste products, soluble pigments etc.
Hyaloplasm	Contains enzymes for glycolysis and structural materials, viz., sugars, amino acids, water, vitamins, nucleotides etc.
Cilia and flagella	Serve to move the animals or to move the material.
Nucleus	Provides selective continuity.
Nuclear membrane	between nuclear and cytoplasmic materials.
Nucleoplasm	Contains materials, for building DNA and messenger molecules which serve as intermediates between nucleus and cytoplasm.
Nucleolus	May synthesize ribosomes. Disappears during cellular replication. Function uncertain.
Chromosomes	Bearers of hereditary instructions. Regulation of cellular processes (evident only during nuclear division). Control the nucleus and metabolism.

2

PROTOPLASM

Protoplasm is a highly complex mixture of some elements and compounds found in the bodies of living beings. The protoplasm is variously knonw as the *living matter, living substance* or *physical basis of life*. It is the basic fundamental substance exhibiting all the vital processes of the cell. The protoplasm was first observed by *Corti* in 1772. In 1835 *Dujardin* a Frenchman described it as a soft and gumy substance and named it as *scarcode* i.e. *flesh*. *Purkinje*, a Bohemian physiologist (1839), was the first biologist, who gave the name *protoplasm* to this living substance. *Hugo von Mohl*, a German botanist, in 1839, also suggested the name *protoplasm* for the granular and viscous substance found in plants similar to that found in the animals. The above-mentioned German botanist popularized the word protoplasm as a name given to the living matter found in plants and animals. Protoplasm is the most complex and interesting substance. It is not to be thought of a chemical compound but rather as very complex organized system. Protoplasm varies somewhat in its nature from cell to cell and from organism to organism, but basically it must be the same, as evidenced by its common manifestations of metabolism, growth, reproduction and by some other peculiarities.

PHYSICAL NATURE OF PROTOPLASM

Different workers have proposed different theories to explain the physical nature of protoplasm as given under the following heads:

1. Granular Theory

This theory was propounded by *Altmann* in 1893. According to this theory, protoplasm consists of numerous tiny granules as shown in *Amoeba*. *Henle, Maggi*, etc., considered these protoplasmic granules *plastidules. Altmann* reconginzed them as "elementary organisms," or *bioplasts* (or cytoplasts).

2. Alveolar Theory

The alveolar nature of protoplasm was suggested by *Butschilli* in 1892. According to him, protoplasm consists of many suspended droplets or alveoli or minute bubbles, resembling to foam of emulsion.

3. Fibrillar Theory

This theory was put forward by *Flemming*. According to him, protoplasm consists of fibres embedded in the inner mass of matrix. The fibrillae are called mitome or spongioplasm and ground substance is temed paramitome or hyaloplasm.

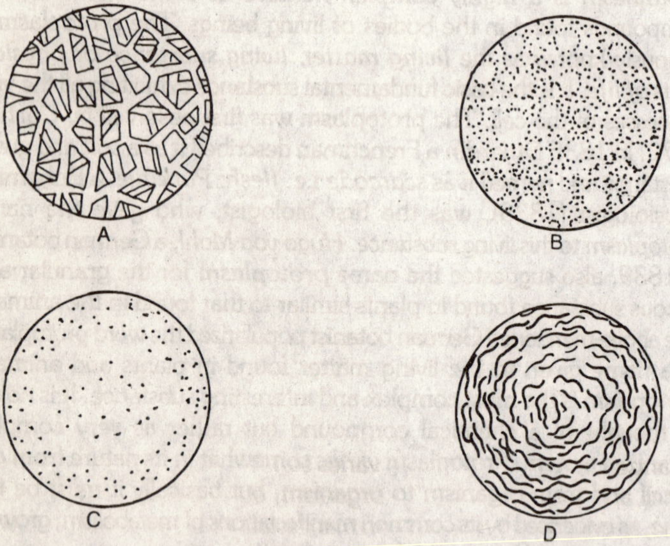

Fig. 2.1. Physical appearance of protoplasm—A. Reticular, B. Alveolar, C. Granular, D. Fibrillar.

4. Reticulate Theory

This theory was proposed by *Klein, Cornoy* etc., suggesting that the protoplasm consists of a network or reticulum of fibres in the ground substance.

5. Colloidal Theory

Proposed by *Wilson* in 1925. According to this theory the protoplasm having any of the aforesaid appearances is always a fluid-colloidal system having various chemical inclusions in *gel phase.* A gel is a semi-solid condition which presents jelly-like appearance. Here the molecules are held together by vrious bonds depends upon the imature and strength. By absorbing water the gel changes to more liquid-like phase. This is known as *sol* and the process is termed as *solation*. The sol can stream easily and by losing water again changes into gel-state. Both these states of colloidal matrix are interchangeable according to the various physiological, mechanical and biochemical activities of the cell. This represents phase reversal in the colloidal system. These sol-gel conditions of colloidal system are the basis for the mechanical behaviour of protolplasm.

PROPERTIES OF PROTOPLASM

1. Cohesiveness

The various particles or molecules of protoplasm are adhered with each other by forces, such as Van der Waal's bonds, that hold long chains of molecules together. These Van der Waal's bonds are weak and non-specific forces between non-polar groups of atoms. This property varies with the strength of these forces.

2. Contractility

This property is significant in various stomatal operations in plants. The contractility of protoplasm is important for the absorption and removal of water as they generally occur in protoplasm.

3. Electrical Charge

Protein molecules repel each other because their overall charge is similar. All molecules are either positively charged or negatively charged. However, if the molecules approach one another close enough so that valency forces can act, then they may be attracted to each other.

4. Precipitation

Addition of certain amount of electrolyte in a colloidal suspension causes its dispersed particles to colloide, aggregate and finally to precipitate as suspension. For example, the addition of HCl to a colloidal system of arsenic sulphide causes precipitation.

5. Viscosity

The viscosity of the ground substance of the cell varies greatly. It may be as low as that of water, or may be very high in the gelating cytoplasm of pseudopodia of *Amoeba*.

6. Streaming Movement or Cyclosis

The protoplasm exhibits various sorts of streaming movements inside the cell. These have been studied in *Amoeba, Paramecium* etc. The movement involves only the localized portions with no visible changes in the protoplasm. No complete explanation of this has been given as yet. But it is seen that its rate depends upon the rate of cell metabolism. It is due to the fact that the energy is supplied by respiration.

7. Amoeboid Movement

The amoeboid movement as exhibited by *Amoeba* and other protozoans involves the movement of entire protoplasm of the cell, where the cytoplasm moves as one mass carrying the various inclusion with it. It is due to the continuous change of gel to sol and sol to gel.

8. Brownian Movement

It is characterized by the zigzag motion of suspended colloidal particles, occurring due to the bombardment of one particle or molecule by other. This type of movement of particles was first of all observed by *Robert Brown* in 1827 in the colloidal solution and hence such movements are known as Brownian movements. The higher the temperature, more rapid the movement and thus viscosity of cell is decreased. This means that high viscosity indicates a more gel-like state of protoplasm and low viscosity, a more sol-like condition.

9. Tyndall Effect

Colloidal particles of protoplasm have the property of scattering light. When a beam of light is passed through a colloidal solution it becomes visible. This is the Tyndall effect. A colloidal solution of proteins in water shows a typical Tyndall cone.

10. Adsorption

The tendency of particles, molecules or ions to adhere to the surface of certain solids or liquids is known as adsorption and is exhibited by the particles of colloidal system. The phenomenon helps the matrix to form protein boundaries.

11. Biological Properties

Protoplasm has all the biological properties of a living organism. It is capable of nutrition, respiration, excertion, metabolism, growth and reproduction. It has the property of irritability, e.g., it responds to stimuli like heat, light and chemicals. It also has the property of conductivity, i.e., of conducting impulses produced by stimuli.

Chemical Nature of Protoplasm

For the detailed study the chemical components of the cell can be classified as inorganic (mineral salts) and organic (proteins, carbohydrates, nucleic acids, lipids and so forth) substance. Although the most prominent constitutent of protoplasm is water—the substance whcih gives protoplasm its characteristic structure is protein. Lipids are important in all membranes and carbohydrates serve as nutrient stores. The protoplasm of a plant or animal cell contains 75 to 85% water, 10-20% protein, 2-3% lipids, 1% carbohydrates and 1% inorganic material. The following Table 2.1. gives approximate active protoplasm.

Table 2.1. Chemical Analysis of Protoplasm

Substance	*Percent*	*Average molecules weight*	*Number of molecues in relation to protein*
Water	85	18	180
Protein	10	36000	1
DNA	0.4	10	—
RNA	0.7	4.0×10^4	—
Lipid	2	700	10
Other Organic matter	0.4	250	20
Inorganic substances	1.5	55	100

Water

Water is a substance that is justifiably called the *fluid of life*. It comprises over 90% of the chemical content of most organisms. In human body about 55% of the water (20-22 litres) is intracellular water and remains confined to the cells and the rest is found in extracellular fluids such as blood, tissue fluid and lymph. Water participates directly or indirectly in all metabolic reactions. The biologically active conformations of macromolecules and arrangement of phospholipids in the lipid bilayer of membranes are dependent on water. Water helps to keep minerals ionised in body fluids. It ionises itself to provide hydrogen ion (H^+) concentration to body fluids. It also helps in maintaining the constancy of the internal environment of an organism.

Water is a remarkable compound with unique properties that result from its molecular configuration and hydrogen bonding. Some important properties of water are as follows.

1. Polarity

The water molecule is composed of two hydrogen atoms covalently bonded to one side of an oxygen atom. Since the mean angle (105⁰) between the hydrogen atoms is not rigid, water can absorb large quantities of heat and be subject to other physical stresses without breakage of the bonds. Water is a polar molecule and as with other polar molecules it has a surface charge. Water is a dipolar substance in that the hydorgen pole is positively charged and the other pole is negatively charged due to electrophilic (electron attracting) properties of oxygen. The polar nature of the water molecule causes salts to be held in solution through charge interaction (ionization; salt dissolved in water exist in the form of positive and negative charges).

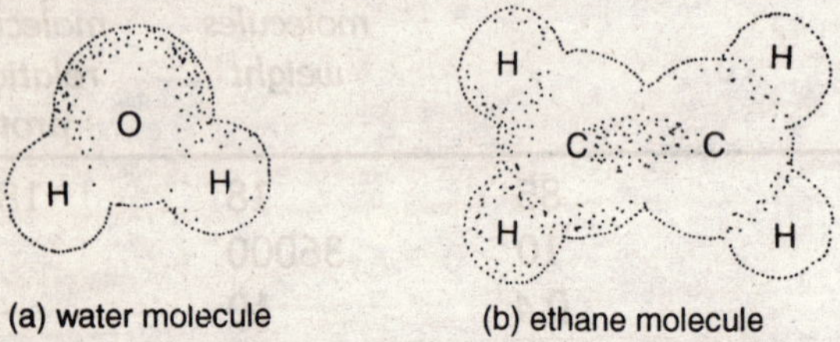

Fig. 2.2. Two water molecules showing polarity and formation of a hydrogen bond between them.

2. Cohesion and adhesion

Because of the asymmetrical distribution water molecules associate with each other (cohesion) and 'wet' other substances (adhesion). The attraction of the positive hydrogen atom of one water molecule for the negative oxygen atom of another water molecule results in a *hydrogen bond*. Hydrogen bonding of water produces a dipolar molecule and favours the formation of a lattice-like structure that enables the packing of many atoms into a small area and stabilizes the molecular structure of water. Fluidity of water is maintained by very rapid formation and dissociation of hydrogen bonds between water molecules.

3. Latent heat and specific heat

The presence of hydrogen bonds is directly responsible for the high heat of fusion, high specific heat, and high heat of vaporization of water. The energy required to break hydrogen bonds for melting of ice, heating of water, and evaporation of water is considerably greater than the energy needed to overcome the Van der Waal's forces that are normally found in the weak association of molecules of ethane, ether and benzene. Due to high specific heat, water affords protection against sudden temperature changes in living organisms. High latent heat of evaporation of water causes elimination of excess heat through evaporation of sweat. This maintains a constant body temperature.

4. Density of freezing properties

The density of water decreases below 4^{0}C and ice therefore tends to float. It is the only substance whose solid form is less dense than its liquid form. The fact that water below 4^{0}C tends to rise helps to maintain circulation in large bodies of water. This may result in nutrient cycling and colonisation of water to greater depths.

5. Solvent action

A property of water that is very important to the living cell is its solvent action. Because it forms a solution with a vast array of compounds, water is sometimes referred to as the *universal solvent.* The solvent action of water is an effect of its ability to form hydrogen bonds due to the asymmetrical distribution of its charges. In solution with water, compounds like sugars, alcohols, and amino acids—which contain oxygen atoms, hydroxyl (–OH) groups, amino (–NH_2) groups—form hydrogen bonds with the molecules of water.

The solvent action of water is of tremendous importance for the living system. The essential elements necessary for normal growth, the compounds necessary for energy transfer and storage, and the components of structural compounds all require water as a translocation and reaction medium. Indeed, physiological processes in dilute solutions and suspensions; and the reactions are, therefore, under control of the physical and chemical laws that govern the activities of dilute solutions and suspensions.

Inorganic Compounds

A number of inorganic salts occur both in free as well as in the ionised state. The elements that occur in quantity in protoplasm are oxygen, carbon, hydrogen and nitrogen. Always present but in much smaller amounts are potassium, phosphorus, calcium, sulphur, magnesium and iron. These ten are generally referred to as *essential elements.* Chlorine and sodium are necessary components of most animal protoplasm but are apparently not essential for plant protoplasm. Copper, boron, iodine, maganese, zinc and several other elements which are present in all protoplasm but only in minute quantities, are called *trace elements.* This term must be used with care, because these elements are still important even though they are present only in very small quantities. The essential elements found in human protoplasm are listed below:

Table 2.2. Showing Percentage by Weight to the Element in Protoplasm

Elements	*Symbol*	*Percent*
Oxygen	O	65.04
Carbon	C	18.24
Hydrogen	H	10.05
Nitrogen	N	3.15
Potassium	K	1.60
Phosphorus	P	0.84
Calcium	Ca	0.25
Sulphur	S	0.20
Magnesium	Mg	0.04
Iron	Fe	0.01
Chlorine	Cl	0.27
Sodium	Na	0.26

The elements are almost always present in protoplasm in the form of chemical compounds rather than elements. Many of these compounds inorganic salts, whch are usually in solution. These salts have numerous functions. They serve as source of elements to be built into other compounds, and some act as *buffer* in maintaining the proper acidity or alkalinity in the protoplasm. Because salts in solutions are electrical conductors, they also function in some way in connection with the electrical properties of protoplasm, however, this is not well understood at present.

ORGANIC COMPOUNDS

Proteins

Protein is an indispensable constituent of the diet because it is the only source of the amino acids. Amino acids are needed to built up new tissue during he period of growth; to maintain the structure of every tissue cell including its content of protein—containing enzyme systems; to provide raw material for the manufacturing of digestive enzymes and certain hormones; and to maintain the normal concentrations of plasma proteins and haemoglobin.

The proteins are macromolecules of very high molecular weight. The polymer molecules of protein are formed by linking together of a number of amino acids. About 20 odd different kinds of amino acids are known. Each amino acid has an amino (NH_2) and a carboxyl (COOH) group. A bond is formed between the amino group of one amino acid with the carboxyl group of another amino acid. The bond is called peptide bond. With the help of such peptide bonds a variety of amino acids can form a long chain molecule called polypeptide.

The nutritional value of a protein depends upon its amino acid composition. Synthesis of amino acids from organic keto-acids is of common occurrence in bacteria and plants. Animals can synthesize only about half of the naturally occurring amino acids (20). Thus there are two categories of amino acids:

1. *Essential amino acids*

The amino acids which cannot synthesized by an organisms i.e. they mut be obtained from the dietary proteins are called essential amino acids. These are *arginine, histidine, isoleucine, leucine, lysine, methconine, theronine, phenylalanine, tryptophan* and *valine.*

2. *Non-essential amino acids*

Although these amino acids are required by the animal as they are found in the protein of the tissues, but they can apparently be synthesized from alpha-keto acids, by the process of amination. These amino acids are *tyrosine, serine, alanine, asparagine, proline, hydroxyproline, asparatic acid, glycine, glautamine, glutamic acid* and *cystine.*

People who eat diets lacking only one of the essential amino acids are unable to synthesize normal body proteins and become protein deficient, despite having a normal nitrogen intake.

Most animal proteins, are *complete or first class proteins;* that is they contain all the necessary amino acids in appropriate proportions for ultilization by man as eggs, meat, kidney, fish, liver, poultry, milk etc.

Plant proteins are *incomplete* or *second class proteins,* since they lack certain essential amino acids. as. cereals, nuts, legumes etc. Thus people living on a strict vegetarian diet may become protein deficient.

Associated with protein deficiency is *Kwashiorker,* which affects millions of children in developing countries. It must commonly occurs in infants after weaning and particularly when given an inadequate and predominantly carbohydrate diet. Many of the children also have an inadequate energy intake and the disorder is known as *protein-energy malnutrition* (PEM). The affected child stops growing and loses weight. The skin and hair may become depigmented, and oedema associated with low plasma levels of albumin often develop. Untreated Kwashiorker is often fatal.

Structure of Protein

The critical determination of biological function of a protein is its conformation, which is defined as the three-dimensional arrangement of the atoms of a molecule. Four basic structural levels are assigned to proteins. These are:

1. Primary Structure

Arrangement of amino acid in a polypeptids chain is the primary structure of a protein. Amino acid sequence of the smallest protein known, that is of insulin, was studied by *F. Sanger* (1954). Insulin consists of two chains, A and B, joined together by two disulplide bonds (A chain conains 21 and B chain 30 amino acids). Normal human haemoglobin consists of four polypeptide chains (2α and 2β) to be held together by non-covalent forces. Each α-chain is composed of 141 amino acids, and each β chains 146 amino acids.

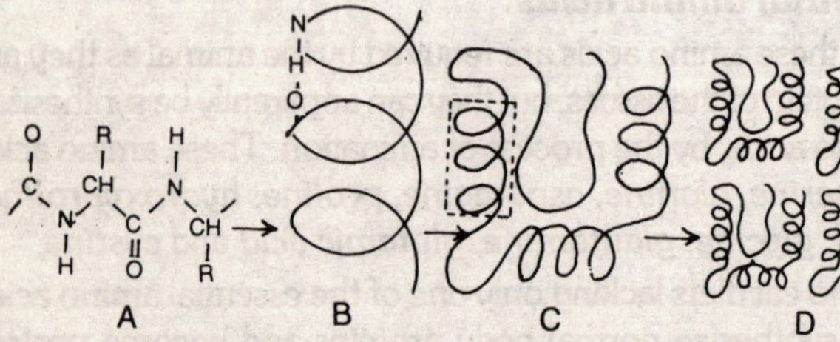

Fig. 2.3. Diagrammatic representation of different levels of protein structure. A—Primary. B—Secondary. C—Tertiary. D—Quaternary.

Fig. 2.4. Structure of hemoglobin; the molecule consists of four chains: two α-chains and two β-chains.

2. Secondary Structure

The polypeptides are held together to give rise a definite shape (helical pattern) of the protein molecules by hydrogen bonds. Hydrogen bonds can be broken by many physical and chemical treatments, for example, by excessive heat, pressure, *pH*, electricity, heavy metals and other agents that changes the environment of protein.

The most common type of secondary structure is α-helix. This is formed by the bending of polypeptide chain to form hydrogen bonds in the same chain. The bending of the chain is very regular and chain assumes the shape of a helix.

Pauling and *Corey* have considered an alternate type of secondary structure. In this case the polypeptide chains are lead linearly either parallel or antiparallel with respect to one another. This is called β-structure of β-helix.

3. Tertiary Structure

The arrangement and inter-relationship of the twisted chains of protein into specific loops and bends in called the tertiary structure. Such a structure enables the proteins to form specific layers, crystals or fibres. The tertiary structures is maintained by hydrogen bonds disulphide bonds, Van der Waals interaction, hydrophobic interactions and ionic bonds. The tertiary structure is important and found in globular proteins. If this structure is disrupted, the biological activity of protein would be lost.

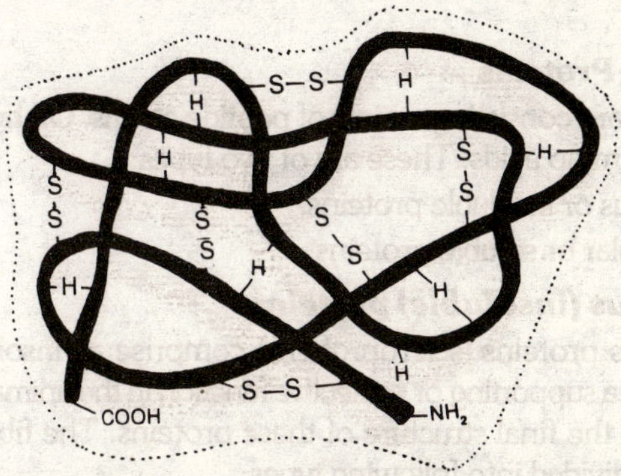

Fig. 2.5. Tertiary structure of a hypothetical protein molecule; note the extensive folding of the secondary structure which imparts a globular shape to the protein.

4. Quaternary Structure

This defines the degree of polymerisation of a protein unit. This structure is found in only oligomeric proteins. They are composed of subunit peptide chains linked together by any or all of the forces that can act between amino acid side chains. Haemoglobin is a fine example of quaternary structure and was studied by *Kendrew* and *Perutz* (1963).

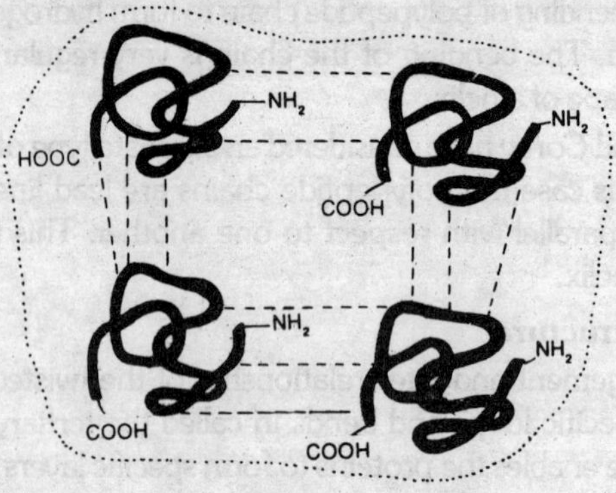

Fig. 2.6. Quaternary level of protein structure; note a diagrammatic representation showing four sub-units which are polymerised in the functional protein molecule.

Classification of Protein

The proteins are mainly classified into two—simple and conjugated proteins.

1. Simple Proteins

These are consisting mainly of peptide chains. On hydrolyses they yield only amino acids. These are of two types:

(A) Fibrous or insoluble proteins.

(B) Globular or soluble proteins.

(A) Fibrous (insoluble) proteins

Fibrous proteins (scleroproteins) comprise all insoluble proteins which have a supporting or protective function in the animals. Secondary structure is the final structure of these proteins. The fibrous proteins can be subdivided into following types:

(i) *Collagens.* Approximately 30% of the total protein in mammalian body is collagen. They are the principle proteins of skin, tendon and bones and are resistant to peptic and tryptic digestion. They contain large amounts of hydroxyproline. They can be converted into soluble gelatin by boiling with water, dilute acids or alkalies.

(ii) *Elastins*. They are present in ligament, arteries and other elastic tissue. They cannot be converted into gelatin.

(iii) *Keratins*. They are present in animal skin, nails, hairs, horns, hooves, claws, feathers etc. they are derived from ectoderm.

(iv) *Myosin*. The major protein of muscles is also a fibrous protein.

(B) Globular or Soluble proteins

These are soluble simple proteins with more or less definite molecular weight. These are speherical or ovoid in shape. They are generally soluble in water. This group includes enzymes, oxygen carrying proteins (globulins), hormones like oxytocin, vassopressin, insulin, glucogon etc.

(i) *Albumins*. They are soluble in water, dilute acids and bases. They coagulate when heated. Common examples are albumins of egg white, blood serum and lactalbumin.

(ii) *Globulins*. They are insoluble in water but soluble in dilute salt solution. They are also coagulated when heated. Examples are serum globulin, fibrinogen, ovoglobulin in egg-yolk etc.

(iii) *Glutelins*. These are insoluble in water but soluble in dilute acids or base solution. They are commonly found in various plant seeds. Examples are wheat, cereals etc.

(iv) *Protamins*. These are basic proteins. They are very soluble, small, stable proteins and cannot be coagulated by heating. They are exceptionally rich in arginine amino acid. They are found in the ripe sperm cells of certain fishes.

(v) *Histones*. These are water-soluble basic proteins. They coagulate on heating. These are found associated with nucleic acids in nucleoproteins. They are obtainable from the thymus, spleen and nucleated erythrocytes of birds.

2. Conjugated Proteins

The conjugated proteins differ from the simple proteins in that they consist of proteins combined wiht some non-protein substances. The term *prosthetic group* is generally used to designate such substances (non-proteinous). They are classified according to the nature of prosthetic group as follows:

(i) *Nucleoprotins*. These are formed by the combination of nucleic acids with protein.

(ii) *Mucoproteins*. They contain carbohydrate bound to protein molecules. The protein components of *mucoids* are combined with large amount (more than 4%) of carbohydrate. Examples are ovomucoid-b from egg white; orosomucoid from blood serum.

(iii) *Glycoproteins.* The carbohydrate percentage is less than 4% in glycoproteins. The carbonhdrate in these is usually the mucopolysaccharids. Mucin of saliva, chorionic gonadotropins and some other hormones of pituitary such as follicle stimulating hormone and luteinizing hormones are glycoproteins.

(iv) *Lipoproteins.* They are compounds of lipid and protein. The protein part is water soluble and lipid unsoluble. Examples are cholesterol, phosphatidylglycerides, lipovitelline of egg-yolk and serum lipoporteins.

(v) *Chromoproteins.* These are the protiens which have a matelloprophyrin or some similar substnace. Examples are haemoglobin which contains a basic protein globin with an iron porphyrin as Heme. Haemocyanin, haemoerythrin, erythrocuorins are other examples.

(vi) *Phosphoproteins.* In thse the protein molecule is linked to phosphoric acid when treated with dilute sodium hydroxide yield inorganic phosphate. Examples are casein of milk, vitellin of egg-yolk and pepsin.

(vii) *Metalloproteins.* These proteins contain metals as an inherent part of their molecules. They include tyrosinase (Cu), carbonic anhydrase (Zn), arginase (Mn or Mg) etc.

(viii) *Flavoproteins.* The prosthetic group of flavin component remains permanently attached to the protein. Flavoproteins act as coenzymes which catalyze oxidation-reduction reactions.

Derived Proteins

These include the products obtained from proteins by the action of heat and other physical agents or by hydrolysis.

1. Denatured and Coagulated proteins

Upon treatment with certain chemial agents or heat treatment the proteins become insoluble on account of some sort of intramolecular rearrangement and it is said to have denatured and when it gets separated in the form of precipitate, it is said to have undergone coagulation.

2. Peptides

These include the various fragments of the proteins which are broken off during hydrolysis of the big protein molecules either the presence of the acids or enzymes. The larger fragments had previously been named as the proteoses and the smaller fragments as peptones. This terminology is being abandoned now.

Properties of Proteins

1. ***Colour and Taste.*** Proteins are colourless and usually tasteless. These are homogenous and crystalline.
2. ***Molecular weight.*** The proteins generally have large molecular weights ranging between 5 x 10^3 and 1 x 10^6. It might be noted that the values of molecular weights of many proteins lie close to or multiples of 35,000 and 70,000.
3. ***Colloidal nature.*** Because of their giant size the proteins exhibit many colloial propertis such as (a) their diffusion rates are extremely slow, (b) they may produce considerable light— scattering in solution, thus resulting in visible turbidity (Tyndall effect).
4. ***Denaturation.*** In refers to the changes in the properties of a protein. The process of denaturation is followed by coagulation. Denaturation may be brought by a variety of agents. The physical agents, include mechanical action, heat treatment, cooling and freezing operations, ultraviolet rays etc. The chemical agents include X-rays, acetone, alcohol (solvents), salicyclates etc.
5. ***Amphoteric nature.*** The proteins are amphoteric i.e., they act as acids and alkalies both. These migrates in an electric field and the direction of migration depends upon the net charge possessed by the molecule. The net charge is influenced by the *pH* value. Each portion has a fixed value of isoelectric point (pl) at which it will move in an electric field.
6. ***Solubility.*** It is influenced by *pH* solubility is lowest at isoelectric point and increases with increasing acidity or alkalinity.
7. ***Hydrolysis.*** The proteins are hydrolyzed by a variety of hydrolytic agents.

(a) *By acidic agents.* Proteins, upon hydrolysis which come HCl (6-12 N) at 100-110°C for 6-20 hrs yields amino acids in the form of their hydrochlorides.

(b) *By alkaline agents.* Proteins may be hydrolyzed with 2N NaOH. This process is less used as it is highly disadvantageous.

(c) *By proteolytic enzymes.* Under relatively mild conditions of temperature and acidity, certain proteolytc enzymes such as pepsin and trypsin hydrolyze the proteins. Enzyme hydrolysis is used for the tryptophan.

8. ***Reactions involving—COOH group***

(a) *Reaction with alkalies* (salt formation). The carboxylic group of amino acids release a H^+ ion and forms carboxylae (Coo^-) ions.

These may be neutralized by cations like Na^+ and Ca^{++} to form salts.

(b) *Reaction with alcohols (esterification).* With alcohols, corresponding esters are produced.

$$\underset{\underset{NH2}{|}}{\overset{\overset{H}{|}}{R{-}C{-}COOH}} + HOC_2H_5 \xrightarrow[\text{catalyst}]{\text{Acid}} \underset{\underset{NH_2}{|}}{\overset{\overset{H}{|}}{R{-}C{-}COOC_2H_5}} + H_2O$$

Ethyl ester of AA.

(c) *Reaction with amines.* Forms amides.

$$\underset{\underset{NH_2}{|}}{R{-}CH{-}COOH} + HHN\text{-}R \longrightarrow \underset{\underset{NH_2}{|}}{R{-}CH{-}CO{-}NH{-}R} + H_2O$$

9. *Reaction involving NH2 groups*

(a) *Reaction with mineral acids* (Salt formation). When AA or proteins are treated with HCl the acid salts are formed.

(b) *Reaction with formaldehyde.* With formaldehyde, he hydroxy methyl derivatives are formed which are insoluble in water and resistant to the attack of microorganisms.

(c) *Reaction with Nitrous acid (Van Slyke reaction).* The amino acids react with HNO_2 to liberate N_2 gas and to produce α-hydroxy acid.

(d) *Reaction with fluro-dinitrobenzene (FDNB) or Sanger's reagent.* In mildly alkaline solution, 1-fluro - 2, 4-dinitrobenzene reats with a-amino acids to produce yellow coloured derivative, DNB-amino acid.

10. *Reaction involving both COOH and NH_2 groups* (a) Reaction with triketohydrindene hydrate (*Ninhydrin reaction*). Ninhydrin in a powerful oxidizing agent causes oxidative decarboxylation of a-amino acids producing CO_2, NH_3 and an aldehyde. The reduced ninhydrin then reacts with the liberated NH_3 forming blue-coloured Ruheman's complex.

11. ***Reactions involving—R group or side chain***

(a) *Biuret test.* Compounds containing peptide bonds produce a purple colour when treats with Biuret reagent (.2% alkaline $CuSO_4$). A substance Biuret is formed. Dipeptides do not respond to this reaction.

(b) *Xenthoproteic test.* Yellow colour develops when proteins are boiled with conc. HNO_3 due to presence of benzene ring. This reaction is due to nitration of phenyl rings. This is the test for tyrosine, tryptophan, phenylalanine.

(c) *Million's test.* When proteins are heated with $HgNO_3$ in HNO_3, a red colour develops. This reaction is specific for tyrosine. Tryptophan also responds to this reaction.

(d) *Hoplin's—Cole test* (or *Glyoxylic acid test*). Violet ring develops on addition of Conc. H_2SO_4 at the junction of protein and glyoxylic acid solution. This test is specific for tryptophan.

(e) *Folin's test.* Blue colour develops with phosphomolybdotungstic acid in alkaline soln-due to phenol group. Specific test for tyrosine.

Biological Importance of Proteins

Proteins constitute a large part of he cell-structure and are present in all the tissue. Many proteins have special physiological functions.

1. *Membrane proteins.* The integral proteins include translocases, the peripheral proteins include cytochrome—C and monamine oxidase.
2. *Enzymes.* The enzymes are biocatalysts to influence the rats of a chemical reaction. All the enzymes are proteins.
3. *Hormones.* Some of the messangers of our body, the hromones, are proteins.
4. *Blood proteins.* 6 major plasma components are albumen, α-1 globulin, α-2 globulin, β-globulin, γ-globilin and fibrinogen. All are proteins.

CARBOHYDRATES

The class of substances, known as carbohydrates, is comprises of a large number of relatively heterogenous compounds. These are synthesized form carbon dioxide and water in chlorophyll bearing cells during the process of photosynthesis. They are especial constituents of plants (cellulose, starch etc.) but also occur and serve important functions in animals. They serve as the chief source of energy in the food of many animals. Their energy vlaue is 4 cal per gram and this energy is provided to the various synthetic needs of a cell. In carbohydrate carbon, hydrogen and oxygen are generally found in 1:2:1 ratio.

Structure

Structurally the carbohydrate are the polyhydric alcohols of carbon possessing active aldehyde or ketonic groups, which on hydrolysis yield to such products.

Classification

The carbohydrates may be classified according to their complexity. They are classified into following groups:

1. Monosaccharides
2. Oligosaccharides
3. Polysaccharides.

Monosaccharides

Monosaccharides are those sugars which cannot be hydrolysed into a simple form. Their genral formula is $(CH_2O)n$. The value of n ranges from 3-7. If the monosaccharide has an aldehyde group—CHO it is called an aldose e.g. glucose and if a keto group > C = O is present it is called a ketose e.g. fructose.

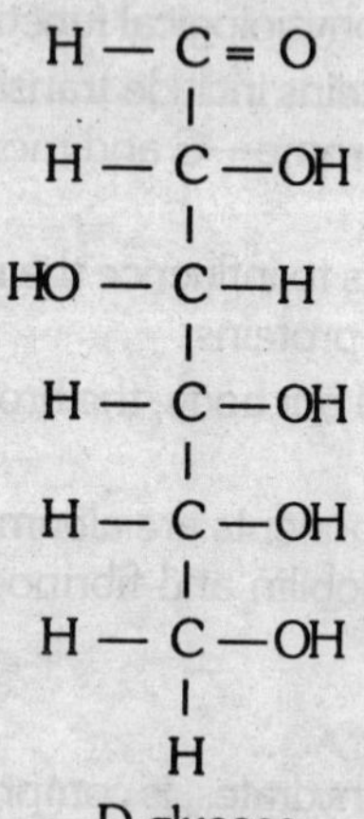

D-glucose

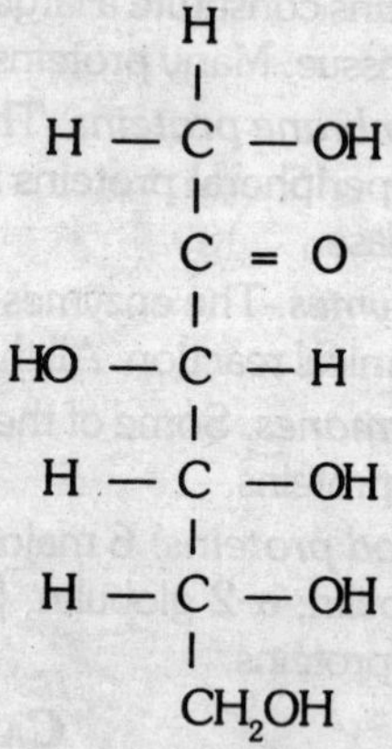

D-fructose

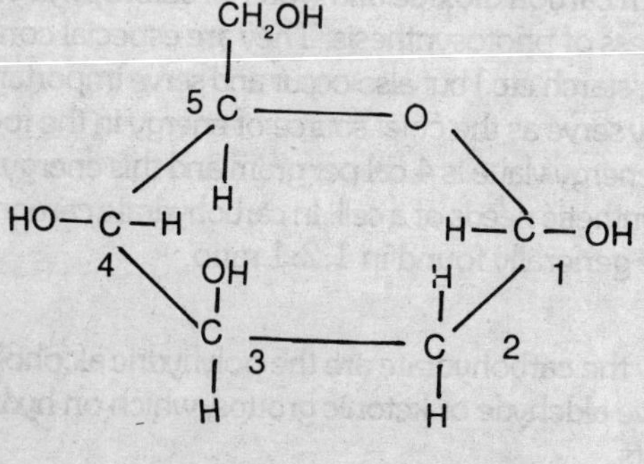

Ring structure of glucose

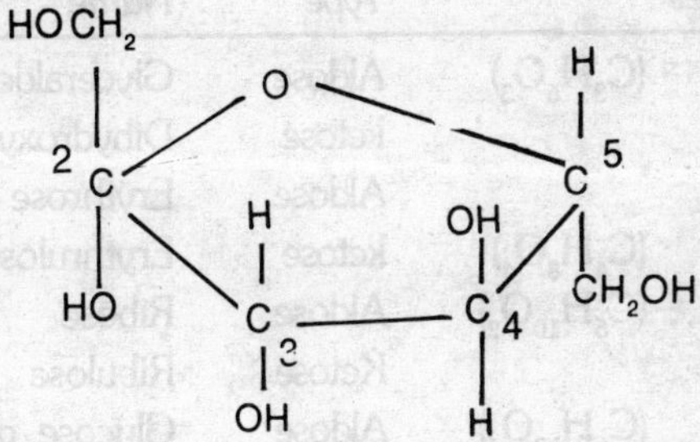

Fructose

Monosacchrides include trioses, tetroses, pentoses, hexoses, heptoses. This classification is based on the basis of carbon atoms present in their molecules. The recent trend in the classification of monosaccharides is a combination of both the system. The aldeosugars and ketosesugars are classified as below. The suffix ('- *ore*) denotes that the sugar belongs to the aldosugar grop and the suffix (-*ulose*) denotes the ketosugas.

Glyceraldehyde occurs in two forms D-glyceraldehyde and L-glyceraldelyde. In D-glyceraldelyde the hydroxyl group (OH) on right side. In L-glyceraldelyde the position of the hydroxyl group is on the left on the asymmetric carbon atom.

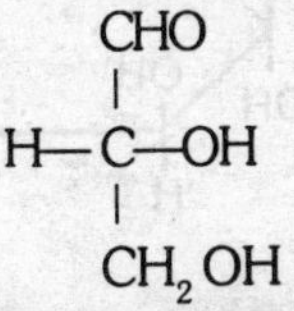

D-glyceraldehyde

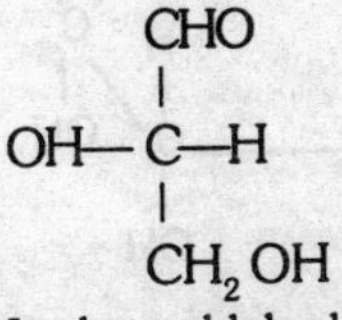

L-glyceraldehyde

It has now been well-establsihed that in the crystalline form, the monosaccharides containing 5 or more carbon atoms exist in the *tautomeric* ring forms. Even in solutions, these sugars mostly occur in the ring forms. The free sugars mostly contain six membered ring in which five members are carbon atoms and the remaining one member is an atom of oxygen. The six membered ring is known as *pyranose ring*. In adition, there is evidence of the existence of five membered rign in the sugar in which four members are carbon atoms and the remaining one member is the atom of oxygen. The five membered rign is known as *furanose rang*. Trioses and tetroses do not possess ring structure. The free pentose sugars are largely found in pyranose forms but in glycosides and nucleic acids these exist in the pyranose form.

Monosaccharides		*Type*	*Name*
1. Triose	($C_3H_6O_3$)	Aldose	Glyceraldehyde
		ketose	Dihydroxy acetone
		Aldose	Erythrose
2. Tetrose	($C_4H_8O_4$)	ketose	Erythrulose
3. Pentose	($C_5H_{10}O_2$)	Aldose	Ribose
		Ketose	Ribulose
4. Hexoses	($C_6H_{12}O_6$)	Aldose	Glucose, glactose
		Ketose	Fructose
		Aldose	Persenlose
5. Heptose	($C_7H_{14}O_7$)	Ketose	Sedoheptulose

In the pyranose type of hexoses, carbon at 'I' position is asymmetrical. If the hydroxyl group at position 'I' is *cis* to the hydroxyl group at position 2, it is known as α-form and if *trans* it is known as β-form. The α-form canreadily pass into β-form when the sugar is brought into solution.

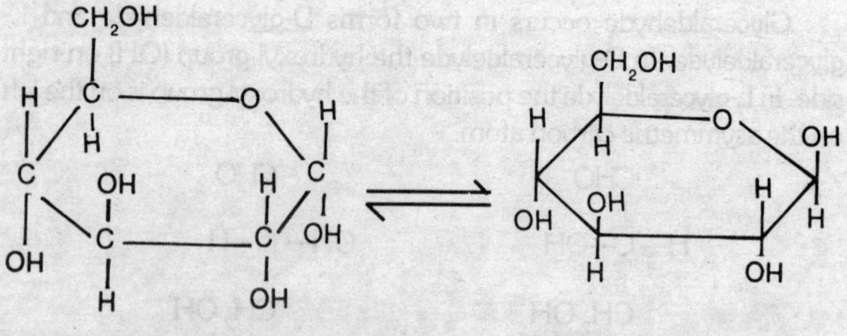

α-Form (glucose) β-Form (glucose)

Properties of Monosaccharides

(a) Physical

The monosaccharides are sweet testing, colourless solids. They are soluble in water, partially in alcohols and insoluble in other.

When a polarized light is passed through a solution of these carbohydrates, the plane of light is rotated to either right or left side.

They contain asymmetric carbon atom hence exist in different isometric forms. The degree of optical rotation may change due to interconversion of isomeres. Fresh solution of glucose gives an optical rotation of +112^0 which change to + 52.7^0 on standing. The change in optical rotation is called *mutarotation*.

(b) Chemical

(i) *Oxidation.* They can easily be oxidized by oxidizing angets. Glucose yields gluconic acid after oxidation with Tollen's reagent (ammonical Ag_2O) or Fehlings solution (alkaline $CuSO_4$). The reduction of Tollen's reagent yields silver as polishing on the surface of tube whereas with the Fehling solution red ppt are obtained.

When strong oxidizing agent is used like conc HNO3, gluconic acid, is ultimately oxidizes to discarboxylic saccharic acid.

Glucuronic acid is obtained in animal body on slow oxidation of glucose. Glucuronic acid combines with hormones. It is a major component of hyaluronic acid, heparin, mucoitin sulphate and chondriotine which are found in blood, skin and cartilage.

(ii) *Reduction.* Free aldehyde and ketone groups of monosaccharides are reduced to alcoholic hydroxy groups by sodium—mercury amalgam and water. D-glucose after reduction yields a mixture of sorbitol and mannitol.

(iii) *Condensation.* Aldehyde groups of monosacharides condense with primary amines to form Schiff's base, when treated with hydroxylamine, glucose forms glucose oxime.

(iv) *Esterification.* The hydroxyl groups of alcohols in the carbohydrate may be converted to esters by treating with the appropriate acetylating agents. When D-glucose is treated with acetic anhydrine in the presence of pyridine, penta-acetyl glucose is formed.

(v) *Methylation.* Methylating agents such as Ag_2O, CH_3OH react with monosaccharide to yield glycoside.

(vi) *Fermentation.* Monosaccharides such as glucose and fructose undergo alcholic fermentation by micro-organisms such as yeasts and produce ethanol and carbon dioxide.

Oligosaccharides

The oligosaccharides are those carbohydrates which on hydrolysis give two to five simple monosaccharide molecules. The oligosaccharides are composed of two to five monosaccharide units. During union of monosaccharde units water molecule is eleminated and the units are linked through an oxygen bridge. It is a glycosidic linkage. Their general formula is $(CH_2O)_{n-1}$.

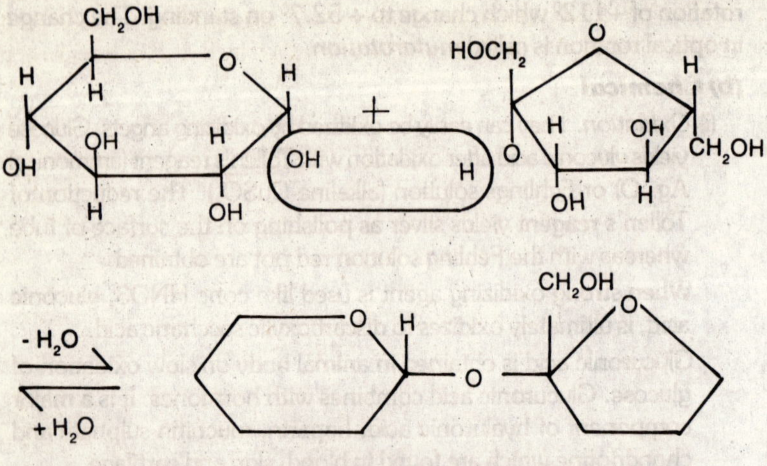

Sucrose (disaccharide)

The disaccharides have been classified into two groups namely, reducing and nonreducing. The reducing disaccharides are maltose, lactose, cellobiose, gentibiose and the non-reducing are sucrose and trehalose.

Maltose. It is composed of two units of D-glucose joined together through their 1 and 4 carbon atoms. It is obtained as a hydrolytic product by the action of amylase on strach

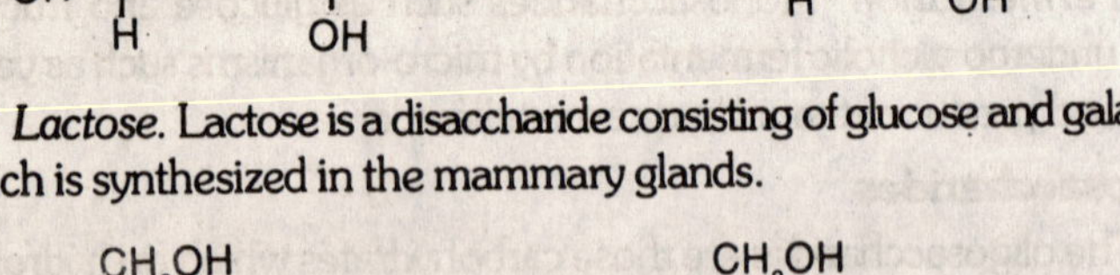

Lactose. Lactose is a disaccharide consisting of glucose and galactose which is synthesized in the mammary glands.

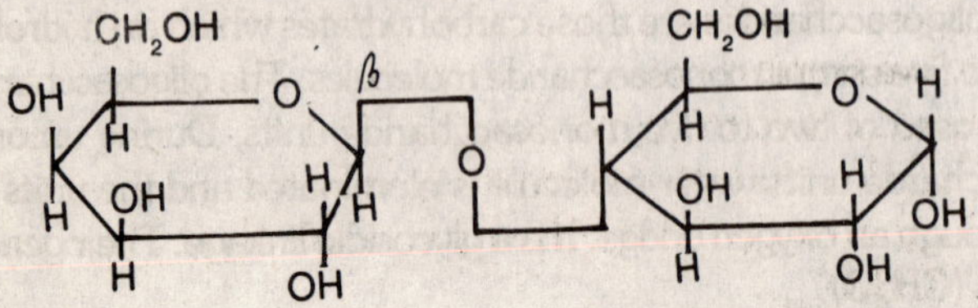

Cellobiose. Incomplete hydrolysis of cellulose given cellobiose. It is composed of two molecules of glucose linked by β-1, 4 glucosidic bond.

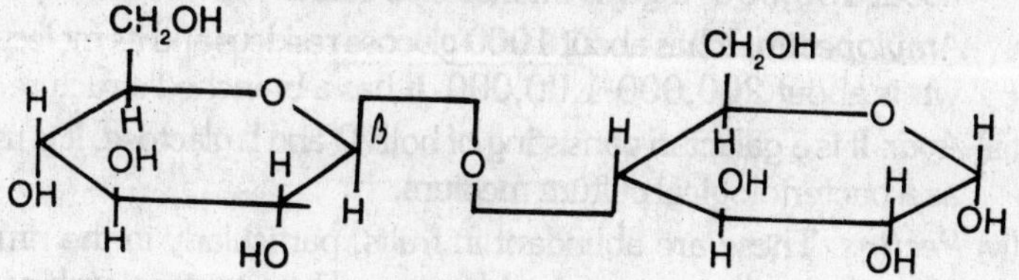

Raffinose. It is a trisaccharide consisting of fructose—glucose—galactose. On hydrolysis it yields *Melibiose* (glucose—galactose) and fructose.

Sucrose. It is formed by the union of one α-D-glucose and one β-D-fructose units with the elemination of one H_2O molecule. Hydrolysis of sucrose by dilute acid or sucrase produces a molecule of glucose and a molecule of fructose.

Polysaccharides

Polysaccharides may be regarded as carbohydrates formed by linking of a number of monosaccharides by glucosidic linkages. A carbohydrate having minimally 6 or more monosaccharide units may be regarded as polysaccharides. Their general formula is $(C_6H_{10}O_5)n$. They are tasteless, colourless amorphous powders which are insoluble in water. because of insolubility and large size, they form colloidal solutions and will not pass across natural animal membranes. They are chemically inert and do not ionize and for this reason very much suited as reserve food material such as glycogen in animals and starch in plants.

Polysaccharides consisting of only one type of monosaccharide units are called *homopolysaccharides* while those with different types of monosaccharde derivatives are *heteropolysaccharides*.

I. Homopolysacchardes

(i) *Glycogen.* It forms the carbohydrate reserve of the animal tissues and it is mainly stored in the muscles and liver. Fungi and yeast do also contain glycogen. It is a branched chain polymer having 6000—30,000 glucose units.

(ii) *Starch.* It is the most important food source of carbohydrate and is found in potatoes, rice, cereals etc. It is hydrolysed in gut yielding dextrins and maltose and eventually glucose.

Treatment of starch with hot water dissolves amylose, while amylopectine remains as such:

Amylose. Contains about 200-500 glucose units which are arranged in the form of a straight chain. The molecular wt. is about 150,000. It gives intense blue colour with iodine.

Amylopectin. It has about 1000 glucose resideus and its molecular wt. is about 200,000-1,00,000. It has a branched structure.

(iii) *Agar.* It is a galactan consisting of both D and L glactose. It is used as a bacteriological culture medium.

(iv) *Pectins.* These are abundant in fruits, particularly in the rim of citrus fruits like orange and lemons. They contain arabinose, glactose and galacturonic acid.

(v) *Xylan.* In addition to cellulose all plants contain xylan. It is a hemicellulose and consists of D-xylose.

(vi) *Inulin.* It is a starch found in the tubers and roots of *Dahlia*, *Dandelions* etc. On hydrolysis it yields fructose. M. Wt. is about 5,000 and about 30-35 fructose units per mole are present.

(vii) *Dextrins.* They are the intermediate products formed during hydrolysis of starch to the glucose.

(viii) *Cellulose.* The cell wall of plants is formed by cellulose. The cellulose consists of a lenier chain of glucose units ranging 900-2000. It does not occur in the animal tissue and it cannot be utilized by man for energy production purposes. Herbivorous animals can, however, utilize cellulose with the help of microorganisms as they can digest cellulose.

(ix) *Chitin.* It is an important structural polysaccharide in invertebrates particularly in arthropods. Structurally, it consists of N-acetyl-D glucosamine units.

II. Heteropolysaccharides

A. *Glycoproteins.* These are protein-polysaccharide compounds occuring in tissue, particularly in mucus secretion. Examples are ovalbumin, fibrinogen, γ-globulin of serum, human chorionic gonadotropins, luteinizing hormones and the blood group substances of RBC (antigen).

B. *Mucopolysaccharides.* The most important mucopolysaccharides from biological point of view are:

(i) *Heparin.* It is present in the liver, lungs and spleen having a molecular wt. about 20,000. The *exect* structure of heparin in unknown. It is found to be constituted by glucuronic acid, glucosamine and sulfuric acid. Sulfuric acid is linked to hydroxyl group of sugar derivatives as well as to amino group of glucosamine. Heparin prevents clotting of blood in the vessels.

(ii) *Hyaluronic acid.* It is a mucopolysaccharide present in connective tissue and acts as an intercellular connecting material. It is abundently present in the umbilical cord, vitreous fluid of eyes and in the synovial fluid present at the joints. This polysaccharide consists of N-acetyl glucosamine and glucuronic acid.

(iii) *Chondrotin sulfates.* They are found in the cartilage, adult bones, skin, cornea, tendons and heart valves. It conists of N-acetylgalactosamine and glucuronic acid.

Functions

The carbohydrates are very important biologically as a source of energy for the cell. The main source of energy is glucose. It the main form of crbohydrates whch is transported form cell to cell by blood in animal and by sap in plants. Carbohydrates also help in the formation of cell wall in plants. Some carbohydrates may serve as the prosthetic group of certain proteins. Glycogen and starch are major storage materials in animals and plants respectively.

Lipids

Term lipid includes fats and fat-like substances. Strucrurally, the different substances of this group may not be similar to fats, but all are soluble in fat-solvents like alcohols, ether, chloroform, carbon tetrachloride, acetone etc. They are insoluble in water. Lipids, like carbohydrates also, contain carbon, oxygen and hydrogen. The carbon and hydrogen are present in larger quantity than oxygen. Some lipids have phosphorous and nitrogen. The lipids in the cell may serve as condensed reserve of energy as well as form the membranous structures.

Classification of Lipids

Lipids are classified into:

I. Simple Lipids

II. Compound Lipids

III. Derived Lipids

I. Simple Lipids

These are esters of fatty acids with various alcohols.

1. Natural Fats

These are the esters of fatty acids with glycerol. The chemical term for a natural fat is *triglycerides.* The triglycerides are formed by the combination of three molecules of fatty acids joined to one molecule of glycerol. As already stated triglycerides are neurtral fats, neutral because the acid ions are neutralized during their unification with glycerol; fats because the bulk of the molecule is devoid of electro-negative elements which can unite with hydrogen to form water molecule.

CH_2OH		$HOOC—R_1$		CH_2OCOR_1
|				|
CHOH	+	$HOOC—R_2$	→	$CHOCOR_2$
|				|
CH_2OH		$HOOC—R_3$		CH_2OCOR_3
Glycerol molecule		3 fatty acids molecules		Triglyceride + $3H_2O$

Triglycerides can exist in the solid or liquid form. It is based on the kind of fatty acid residues in its structure. The triglycerides which are blow 20°C called *oils* containing a large proportion of short-chained unsaturated fatty acids like *oleic, linoleic acid* etc. The triglycerides which are solid above 20°C are called *fats*. They contain long chained saturated fatty acids like *palmitic acid* and *stearic acid*.

2. Fatty acids

These are obtained by the hydrolysis of fats. These are monocarboxylic acids. The molecules of a fatty acid has a polar carboxyl group soluble in water and a non-polar hydrocarbon chain soluble only in fat solvents.

The fatty acids are classifed under three groups based on their degree of saturation and unsaturation.

(i) *Saturated fatty acids.* A saturated fatty acid contains as many hydrogen atoms as its carbon chain can hold. General formula for saturated fatty acid is R—COOH, where R is CH_3 $(CH_2)n$. n is varying from zero in acetic acid to 86 in mycolic acid. The most abunant saturated fatty acids in nature *palmitic* (C_{18}) and *Stearic* (C_{16}) *acids*.

(ii) *Unsaturated fatty acids.* They have one or more double bonds in their carbon chain. When there is a single double bond, results inthe loss of 2 hydrogen atoms, such fatty acids are called *monous saturated fatty acids*. Their genral formula is CnH_{2n}—1—COOH, examples are *palmitoletic acid* and *oleic acid*. When there ae 2,3,4 or more double bonds in the carbon chain with the consequent absence of 2,6,8 or more hydrogen atoms, such fatty acids are *polyunsaturated fatty acids.* The general formula is CnH_{2n}—2 COOH or CnH_{2n}—3 COOH. Examples are *linoleic acid* and *archidonic acids.*

Linoleic, linolenic and archidonic acid are often termed as *essential fatty acids.*

Table 2.3. Showing a few common fatty acids found in lipids.

Fatty acid		*Formula*
Saturated fatty acid	→	
Butyric acid	→	$CH_3(CH_2)_2$ COOH
Caproic acid	→	$CH_3(CH_2)_4$ COOH
Palmitic acid	→	CH_3 $(CH_2)_{14}$ COOH
Steraric acid	→	CH_3 $(CH_2)_{16}$ COOH
Unsaturated fatty acids		
Palmitoleic acid		CH_3 $(CH_2)_5$ (H=CH $(CH_2)_7$ COOH
Oleic acid		CH_3 $(CH_2)_7$ CH=CH $(CH_2)_7$ COOH
Linoleic acid		CH_3 $(CH_2)_4$ CH=CH CH_2 CH=CH $(CH_2)_7$ COOH
Linolenic acid		CH_3 CH_2 CH=CH CH_2 CH=CH CH_2 CH=CH $(CH_2)_7$ COOH

3. Waxes

Wexes are another class of simple lipis containing one molecule of fatty acid and one high molecular weight alcohol. The constituent acids and alcohols have usually 24-36 carbon atoms. The waxes has high melting points. They are chemically inert as they do not have double bonds in their hydrocarbon chains and are highly insoluble in water. They serve as protective converings on leaf surface (plants). *Bees wax* is an ester of palmitic acid with myricyl alcohol ($C_{30}H_{61}OH$) and *spermaceti* from the sperm whale is an ester of palmitic acid with cetyl alcohol ($C_{16}H_{33}OH$).

II. Compound lipids

These are esters of fatty acids and alcohol with additional compounds, such as phosphoric acid, sugars, proteins etc. These are classified as follows:

(i) Phospholipids

Lipids containing phosphorous are phospholipids. They also have nitrogen containing bases and other substituents. The phospholipids are abundant in brain and nervous tissues. The different types of phospholipids are:

(a) *Phosphatidic acids.* These are compounds consisting of glycerol, two fatty acids and a phosphate group.

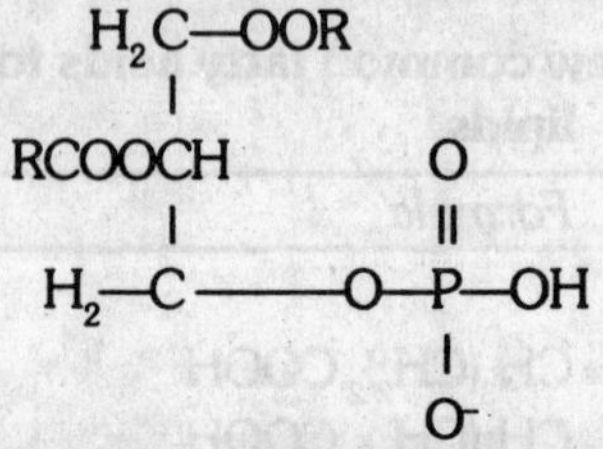

Here R_1 and R_2 represent the residues of the molecules of fatty acids. The phosphatidic acids do not found in any great quantity in tissues but these are important as in intermediate biosynthesis of triglycerides and other phospholipids.

(b) *Lecithins.* Lecithins are choline esters of phosphatidic acid. On hydrolysis they yield one molecule of glycerol, 2 molecules of fatty acids and one molecule of phosphoric acid to which a nitrogenous base choline is attached.

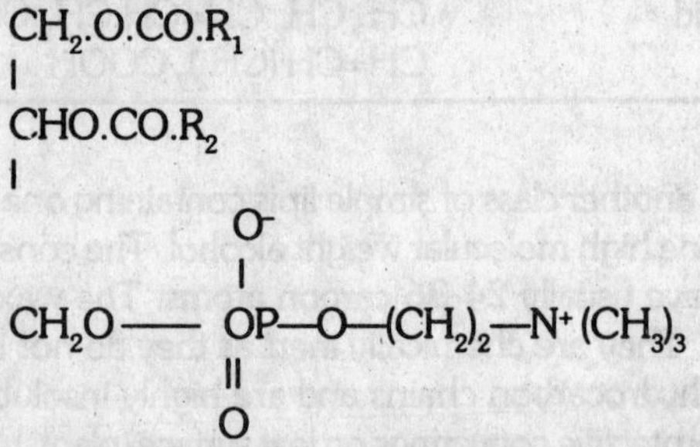

The fatty acids are palmitic, stearic, oleic, linolic and arachidonic acids. Lecithins are yellowish grey solids, soluble in ether and alcohols but insoluble in acetone. Lecithins get broken down by the enzyme lecithins to lysolecithin. This product (lysolecithin) has the ability to hemolyze the red blood corpuscles. Enzyme lecithinase occurs in the venom of snake cobra. and bee.

Lecithin is an important constituent of lipoproteins. Egg-yolk is a rich source of lecithin which also has an important role in fat metabolism in the liver.

(c) *Cephalins.* They resemble lecithins in most properties. The fundamental difference between the lecithins and cephalins is the nature of nitrogenous base. Cephalins contain ethanolamine, colamine and sometimes serine in place of choline.

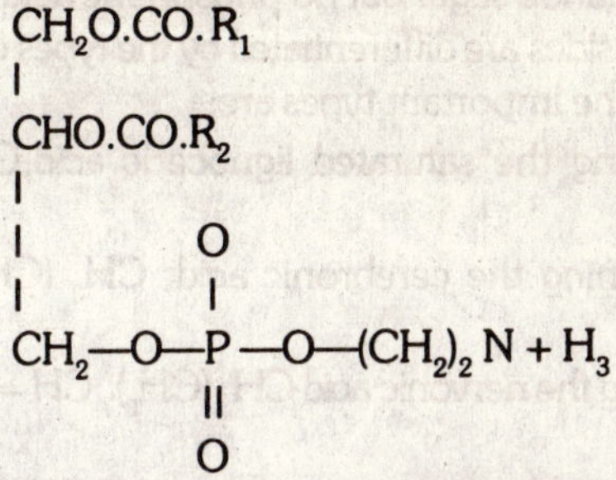

(d) *Plasmalogens.* These phospholipids are abundant in brian and muscle. They resemble lecithins and cephalins in structure but possess an aldehyde group in place of one of the fatty acids in typical phospholipid molecule.

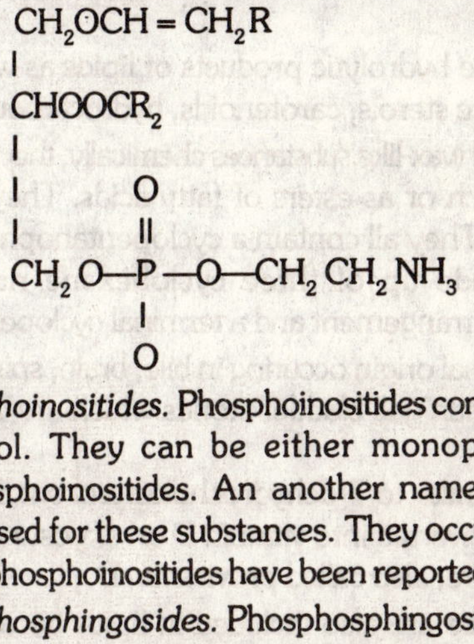

(e) *Phosphoinositides.* Phosphoinositides contain hexahydric alcohol inositol. They can be either monophosphoinositides or diphosphoinositides. An another name lipoinositol was also proposed for these substances. They occur in animal and plants but diphosphoinositides have been reported from brain tissue only.

(f) *Phosphosphingosides.* Phosphosphingosides have been found to occur in high concentration in brain, nerves, lungs and spleen tissue. These contain a nitrogenous base sphingosine or its derivative which remains attached to long-chain fatty acid by its amino group. On hydrolysis they yield fatty acids, phosphoric acid, choline and sphingosine.

(ii) Glycolipids

The glycolipids occurs in brain tissue and in the myelin sheath of nerves, lungs, kidneys, spleen, liver, retina, egg-yolk etc. The glycolipids include two major catagories, cerebrosides and gangliosides.

(a) *Cerebrosides.* They occur in the brain and myelin sheath of nerves. they are based on sphinge mine and have in addition a fatty acid and a monosaccharide sugar but no phosphoric acid or glycerol. Individual cerebrosides are differentiated by the types of fatty acids in the molecule. the important types are:

Kerasin containing the saturated lignoceric acid, $CH_3\ (CH_2)_{22}$ COOH.

Cerebron containing the cerebronic acid, $CH_3\ (CH_2)_{21}CHOH$ COOH.

Nervon containing the nervonic acid $CH_3\ (CH_2)_7\ CH = CH\ (CH_2)_{13}$ COOH.

Oxynervone containing hydroxy derivatives of nervonic acid CH_3 $(CH_2)_7\ CH{=}CH\ (CH_2)_{12}$ CHOH COOH.

(b) *Gangliosides.* These contain N-acetylneuraminic acid (sialic acid), fatty acids, sphingosine and 3 molecules of hexose.

III. Derived Lipids

These lipids include hydrolytic products of lipids as well as other lipid like compounds like sterols, carotenoids, hydrocarbous etc.

The *sterols* are solid wax-like substances chemically, they are alcohols and occur either as such or as esters of fatty acids. They are highly soluble in fat solvents. They all contain a cyclopentanophenanthrene (sterane) nucleus (made up of three cyclohexane rings, in the phenanthrane type of arrangement and a terminal cyclopentane ring).

Cholesterol is animal origin occuring in bile, brain, spinal cord etc. It is obtained form human gall bladder stones which are deposited in the bile duct.

Ergosterol. It is similar to 7-dehydrocholesterol but differs in the side chain. It can be converted into vitamin D on exposure to the ultra violet light and, therefore, also called provitamin D.

Carotenoids. These are also included in lipids because of their solubility in fat solvent. Carontenes consists of carbon and hydrogen only where as xanthophyll contain oxygen in addition. Vitamin A is derived from carotene. *Estrogens, progesterone, testosterone* and *anderosterone* are steroidal hormoens.

Prostaglandins

Prostaglandins are derivatives of fatty acids. They were first discovered in human seminal fluid secreted by prostate gland, hence the name. It has been shown now that prostaglandins are synthesized and released by many other tissues such as kidneys, testis, placenta, lungs, liver, uterus, gastrointestinal tract, brain and heart.

The prostaglandins are 20 carbon fatty acids including a five membered ring in their molecular structure. Different prostaglandins differ with one another in the number and position of double bonds and hydroxyl group substituents.

Table 2.4. Showing Elements that Occur in Protoplasm

S.No.	*Name of elements*	*Approximate percentage*	*Function*
Major Elements			
1.	Carbon	18	Forms backbone of all organic molecules.
2.	Hydrogen	10	Present in most organic compounds and major components of water.
3.	Oxygen	65	Cellular respiration; in organic compound and component of water.
4.	Nitrogen	03	Major components of all the amino acids, proteins and nucleic acids.
Tracer Elements			
5.	Sodium	0.2	For water balance, conduction of nerve impulse.
6.	Potssium	0.5	Conduction of nerve impulse, muscular contraction.
7.	Phosphorus	1.0	Nucleic acids formation; bones formation and in energy transfer.
8.	Magnesium	0.1	Constituent of certain enzymes (ATPase).
9.	Calcium	1.5	Blood clotting, muscle contraction, bones and teeth formation.
10.	Sulfur	0.3	Constituent of most proteins.
11.	Chloride	0.1	Negative ion of interstitial fluid.
12.	Iron	0.01	Component of haemoglobin and ceriain enzymes.

Prostaglandins PGE, PGE_1, PGE_3, $PGF_{1\alpha}$, $PGF_{2\alpha}$ and $PGF_{3\alpha}$ are considered as *primary prostaglandins*. Most of their action appear to be binding of hormones to membrane.

Recently, new type of prostaglandin has been isolated from human seminal fluid which has been designated as PGX[2].

Amino Acids

There is a common plan of construction for the thousands kinds of proteins in living systems. The 20 kinds of naturally occurring amino acid monomers are strung together in unbranched, linear polymer chains of proteins. These are the 20 amino acids specified in the genetic code that is universal to all organisms. Some other kins of amino acids are also found in cells, but they are either degradation products or residues that have been modified form oen of the 20 commonly occurring amino acids after this latter has been inserted into the polymer chain. Hydroxyproline is a major amino acid constituent of collagen in connective tissue, but proline residues are initially included in the protein and are converted to hydroxyproline after polymerization. Hydroxyproline is not one of the encoded acid, but proline is. Many proteins contain fewer than 20 kinds of amino acids. The relative proportions and the absolute number of the amino acid repertory vary from one protein to another, as a reflection of the specific in formation in genes, which are the blue prints for protein construction.

Nucleotides and Nucleic Acids

Nucleotides are involved in at least two major cellular functions: (1) they are monomeric units from which DNA and RNA polymers are constructed, and (2) they act as agents in certain energy-transferring reactions during metabolism. A mononucleotide is made up of one nitrogen-containing organic base, one pentose sugar, and one phosphate residue derivative from phosphoric acid. When there is no phosphate group, the sugar-base combination is called a *nucleoside* (Table 7.5). For this reason, *nucleotides* (phosphate-sugar-base) are also called nucleoside phosphates. Nucleoside mono-, di-, and tri- phosphates contain one, two or three phosphate groups respectively.

The nitrogenous bases commonly found in nucleic acids and their nucleotide building blocks are derivatives of purine and pyrimidine. The commonly occuring purines *adenine* and *guanine* are found in both DNA and RNA, as in the pyrimidine compound *cytosine*, the second kind of pyrimidine in DNA is *thymine*, while its demethylated form, *uracil*, occurs in RNA.

Table 2.5. Showing nomenclature of nucleic acids and their constituent units.

Base	*Nucleoside*	*Nucleotide*	*Nucleic Acid*
Purines			
Adenine (A)	Adenosine	Adenylic acid	RNA
	Deoxyadenosine	Deoxyadenylic acid	DNA
Guamine (G)	Guanosine	Guanylic acid	RNA
	Deoxyguanosine	Deoxyadenylic acid	DNA
Pyrimidines			
Cytosine (C)	Cytidine	Cytidylic acid	RNA
	Deoxycytidine	Deoxyadenylic acid	DNA
Thymine (T)	Thymidine	Thymidylic acid	DNA
Uracil (V)	Uridine	Uridylic acid	RNA

Since each kind of nucleic acid contains one unique pyrimidine, it is convenient to study synthesis and activity of DNA or RNA using isotopically-labelled precursors containing one or the other of these bases. Usually the nucleosides uridine or thymidine, or other nucleotide forms, are added to the biological system under study.

The only difference in the pentose sugars of nucleotides is the presence of a hydroxyl group at carbon atom 2 of D-ribose in RNA, but a hydrogen at carbon-2 of 2-deoxy-D-ribose in DNA monomers and polymers. This seemingly simple difference is partly responsible for profound differences in stabilities, pairing potential, and functions of DNA and RNA.

Polynucleotides of both DNA and RNA varieties are built from mononucleotides that are linked convalently via phosphodiester bridges between the 3' position of one unit and the 5' position of hte next. Since there is no restriction on the ventrical sequence of adjacent mononucleotides in either DNA or RNA, a considerable variety of molecules is possible even though only 4 kinds of nucleotide monomers (one for each of the four kinds of bases in the combination with sugar and phosphate) are used in polymer construction. The theoretical variety is calcualted as 4", where 4 is the number of different kinds of nucleotides, and *n* is the number of monomers in the polymer. For a molecule made of only 75 monomeric units, as in some of the smallest RNAs, there may be 4^{75} different arrangements of the constituent units. Each arrangement theoretically constitutes a moleule of different specificity. Where the average gene may include about 500 nucleotides in a DNA sequence, 4^{500} different sequences are theoretically possible

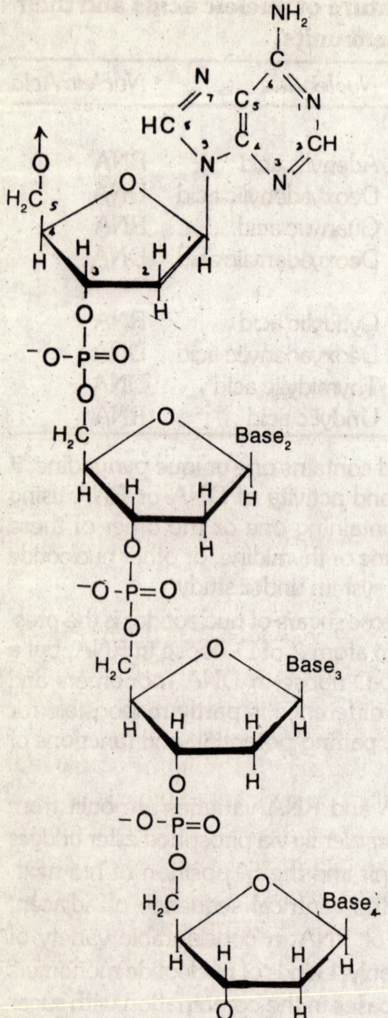

Fig. 2.7. The molecular organization of a backbone of DNA showing phosphodiester and glycosidic bond.

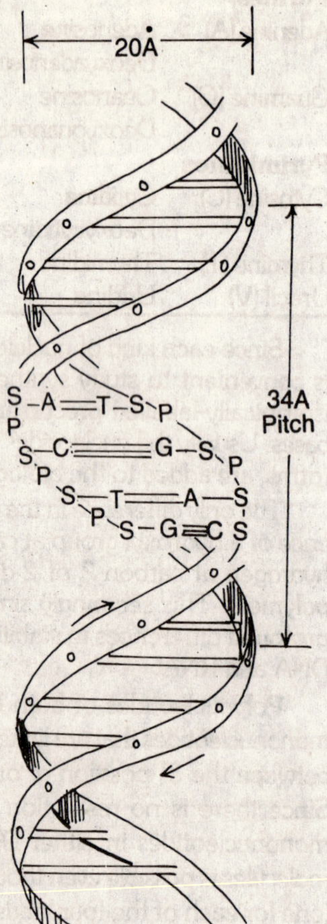

Fig. 2.8. DNA and its paired bases.

and, therefore, that many different and specific genes. Despite the apparently meager number of monomer types an astronomically high number of possible genes can be constructed.

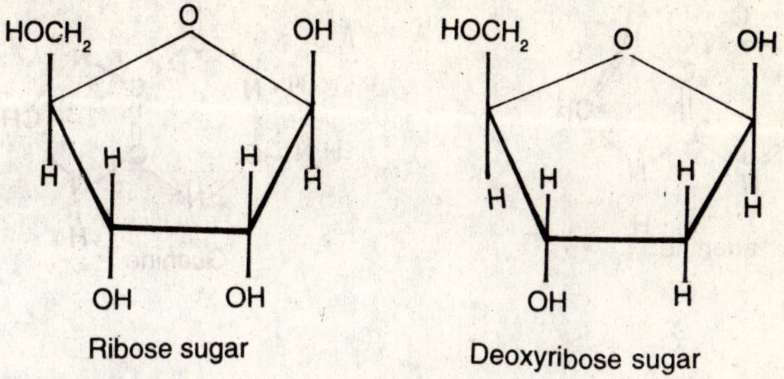

Fig. 2.9. Pentose Sugar.

Such variety can easily account for all past and present life forms.

The DNA Double Helix

DNA molecules usually have regular helical configurations because most DNA molecules consists of two *complementary polynucleotide strands*. The two strands are held together by *hydrogen bonds* between complementary pairs of purines and pyrimidines. Adenine always bind with thymine while guanine always binds with cytosine. This repeated hydrogen bonding within the double helix structure and bonding between virtually all the surface atoms in the sugar and phosphate groups with water molecules serve to stabilize the structure.

Since the purine-pyrimidine pairs are found in the center of the molecule, their flat surface can stack on top of each other and thereby limit their contact with water. In double-helical molecules, a regular structure is possible because the complementary base pairs are exactly the same size. Single polynucleotide chains could not have a regular backbone structure because pyrimidines are smaller than purines, which would cause the angle of helical rotation to vary with the sequence of bases.

DNA double helix molecules are very stable at physiological temperatures because: (1) disruption of the double helix breaks hydrogen bounds and brings hydrophobic purines and pyrimidines into contact with water, which is energetically unsatisfactory; and (2) there are many weak bonds within the DNA molecule, arranged so that most of them cannot break without many others breaking at the same time. Even though some hydrogen bonds may be broken by thermal motion, hydrogen bonds in the rest of the molecule remain intact and the molecule does not fall apart. In fact, when held together by more than ten nucleotide

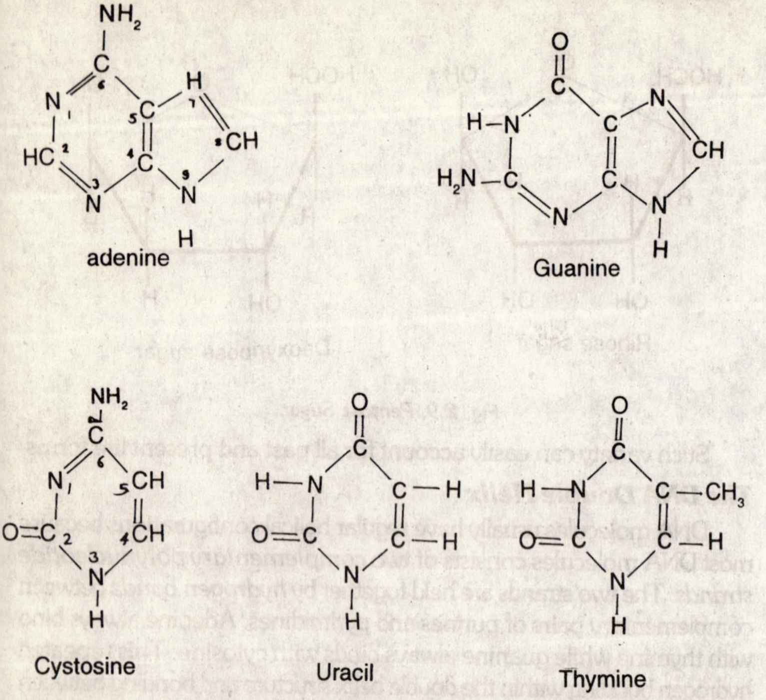

Fig. 2.10. Various heterogenous bases.

pairs the double helices are quite stable. At room temperature weak bonds is the stability of molecular shape, in proteins as well as in nucleic acids. At abnormally high temperatures there is more frequent breakage of weak bond, which become less stable as temperatures rise above physiological levels. Once a significant number of weak bonds have been broken, a protein or nucleic acid molecule usually loses its original form and changes to an inactive or denatured form.

3

Enzymes

The cell is a highly sophisticated factory. It derives its raw materials either from some circulatory system or directly from its external environment. It produces a great variety of end production from these raw materials. A cell contains many thousands of different compounds present, ranging from simple inorganic ions to extremely large and complex molecules of DNA. Besides this production capacity, the cell possesses the ability to divide and reproduce itself, to replace and repair its basic structure continuously and ability to respond extremely rapidly to a variety of internal and external stimuli.

All the informations for the performance of these activities are inherited by the cell in the form of deoxyribonucleic acid and each cell makes its own machinery to carry them out. This machinery is the enzymes of the cell and the aim of this chapter is to show how enzymes are able to carry out the function alloted to them.

Term enzyme was first introduced by *W. Khune* in 1878, derived from Greek means 'in yeast'. *Berzelius,* was first who recognized and defined the nature of enzymes. In 1827, he proposed that forments were the catalysts produced by living cells. *J.B. Sumner* (1926) for the first time isolated the enzymes *urease* in crystalline form. He was awarded Noble Prize in 1946 for this work. Since that time about 2000 enzymes have been isolated in highly purified form and more than 150 have been isolated in crystelline forms. The enzymes can be defined as "soluble, collidal, organic, catalysts, which are produced by living cells but are capable of acting independently of the cells." All the enzymes are protein in nature without any exception and exhibit all the properties of proteins.

Classification of Enzymes

Enzymes are generally named by adding *ase* to the root indicating the substrate on which the enzyme acts. According to IUB (International Union of Biochemists) the enzymes are classified on the basis of their reaction specificity. There are six classes of the enzyme according to IUB.

1. *Oxidoreductase.* This class includes enzymes which are concerned with biological oxidation and reduction and therefore, concerned with respiration and fermentation processes. About 221 enzymes have been listed in this category.
2. *Transferases.* These enzymes catalyze the transfer of a group from one substance to another. The groups may be alkyl, glycosyl, aldehydes, sulphur or phosphorus.
3. *Hydrolases.* Enzymes, which catalyze the hydrolysis of their substrates by adding constituents of water across the bond they split, are called hydrolases. The substrates may be esters, peptides, glycosides etc.
4. *Lyases.* These enzymes catalyze removal of groups from substrates by mechanism other than hydrolysis leaving double bond in the substrate. The enzymes act on the following bonds: C-C, C-N, C-O, C-S, etc.
5. *Isomerases.* These enzymes catalyze reactions which brings about intramolecular rearrangement of atoms in the substrates.
6. *Ligases.* These are also called *Synthetases.* These enzymes catalyze the linking together of compounds utilizing energy made available by simultaneous breakdown of pyrophosphate bond of ATP or similar high energy bond compounds. These enzymes form C-O, C-S, C-C and C-N bonds.

Nature of Enzymes

All the enzymes are proteins with one or more polypeptide chains or their derivatives. The molecular weight vary from 10,000 upto 50,000 daltons. Each enzyme is made up of amino acids. There are 20 amino acids which participate. These are linked together by peptide bonds. The sequence of amino acids is specific for a particular enzyme and determines the properties of the enzymes.

The enzymes are of two types:

(i) Simple Proteins

Some enzymes such as amylase, pepsin, trypsin and Chymotrypsin and called simple proteins. These on hydrolysis yield only amino acids.

(ii) Conjugated Proteins

Other enzymes are conjugated or derived proteins formed by the combination with other groups (non-proteins) called *cofactors*. Usually the enzyme complex is called *holoenzyme*. Thus the protein part characterized as heat labile and colloidal is called *apoenzyme* and the heat stable and dialysable part as *co-factor*. Three types of co-factors may be identified, *prosthetic group, coenzyme* and *metal ions*.

(a) Prosthetic group

The prosthetic group is organic compound. These are permanently bonded to the enzymes. If the prosthetic group is removed, the enzyme activity is lost e.g. catalase and peroxidase.

(b) Coenzymes

These are also organic compounds but their association with the apoenzyme is transient, usually occuring only during the course of catalysis. The coenzyme molecule may serve as the cofactor in a number of different enzymes catalyzed reactions. The essential chemical component of many coenzymes are vitamins. The B group vitamins (except biotin and lipoic acid) function as part of coenzyme such as Co-A is a derivative of panthothenic acid (B_3), flavin adenine dinucleotide (FAD) contains riboflavin (B_2), Thiamine pyrophosphate contains riboflavin (B_2), Thiamine pyrophosphate contains thiamine (B_1), Nicotinamide adenine dinucleotide (NAD) and Nicotinamide adenine dinucleotide phosphate (NADP) contains the vitamin niacin (B_5).

(c) Metalic ions

A number of enzymes require metal ions for their activity. These ions form co-ordination bonds with specific side chains at the active site and at the same time form one or more co-ordinate bonds with the substrate. The common ions are Zn^{++}, Mg^{++}, Cu^{++}, Fe^{++}, Co^{++} etc.

General Properties of Enzymes

1. ***Protein Nature.*** All the enzymes are protein in nature. The protein structure provides for weak associations due to side groups of amino acids within it. The enzyme-substrate complex involves the association of the enzyme and substrate by weak associations rather than covalent bondings.
2. ***Colloidal Nautre.*** They are colloidal in nature due to which they present a large surface area for reactions to take place.
3. ***Required in small amount.*** A small amount of enzymes is required by biological systems because the enzyme is reused. It

has been estimated that a single enzyme can act upon 5,00,000 substrate molecules per minute. This value is called *turn over number.*

4. ***Remain unaltered at the end.*** At the end of a specific reaction, the enzymes remain unchanged.
5. ***Accelerate the rate of reaction.*** An enzyme changes the rate of reaction only, not the direction of the equilibrium. A reaction will proceed at a faster rate in the presence of an enyzmes, but the equilibrium between substrate and product will remain constant with or without the intervention of the enzyme.
6. ***Reversibility of reactions.*** Like a true catalyst, enzymes have been found to accelerate the chemical reaction in either direction i.e., onwards and the backwards depending upon the availability of suitable energy source.
7. ***Denaturation.*** If the natural conditions are changed. The enzymes undergo denaturation which is accompanied by a parallel loss of enzymic activity.
8. ***Enzyme specificity.*** Enzymes are highly specific in nature i.e., a particular enzymes can catalyze only a particular type of reaction. Thus specificity is an important criterion of enzyme action. The specificity may be of following types:

(a) *Reaction specificity.* Different enzymes bring out different reactions on same substrate.

(b) *Substrate specificity.* This is of following two types

(i) *Absolute specificity.* A particular enzyme acts on a particular substrate.

(ii) *Group specificity.* Several enzyme hydrolyse polypeptide, but each has a preference for peptide linkages involving specific sites of polypeptide chains or bonds involving specific amino acids.

(iii) *Stereospecificity or optical specificity.* An enzyme may have optical specificity for a D or L optical isomers. Thus *L-amino acid oxidase* attacks only L-amino acid and not the D-amino acid.

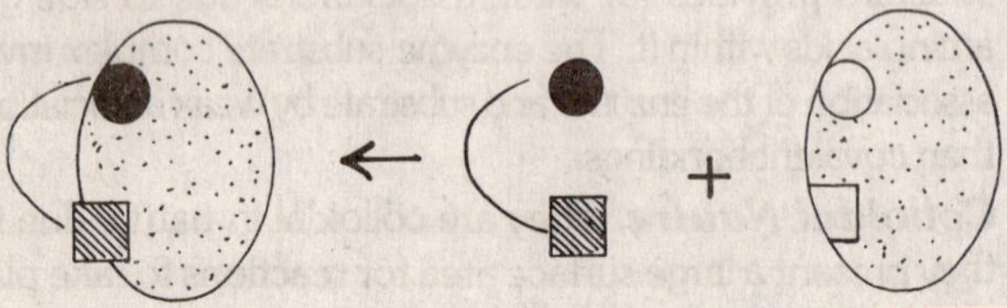

Fig. 3.1. Combination of Enzyme with the substrate in form ES Complex according to Fisher's hypothesis.

Mechanism of Enzyme Action

The enzyme speeds up a given reaction, but remains unchanged at the end of the reaction. The most recent view regarding mechanism of enzyme action is that the enzyme combines with the substrate (on which enzyme acts) to form an *enzyme-substrate complex* (*Michaelis* and *Menten,* 1931). For this complex the energy is liberated and the substrate is broken down into the products of reaction.

$$\text{E enzyme (E) + Substrate (S)} \xrightarrow{\text{binding}} \underset{\text{Catalysis}}{\underset{\text{(ES)}}{\text{Enzyme – Substrate}}}$$

$$\text{Products (P) + Enzyme (E)} \xrightarrow{\text{release}} \underset{\text{(EP)}}{\text{Enzyme – Products}}$$

The reaction takes place in three steps i.e., binding, catalysis and release. The ES is also known as *Michaelis Complex.*

Most of the catalyzed reactions are reversible in nature i.e., the same enzyme can catalyse a particular reaction in the forward as well as in backward direction. The equilibrium of the reactions may be shifted towards either direction at different times depending upon the concentration of the substrate or products in the reaction medium.

Enzymes Reduce the Energy of Activation

Energy of activation is that minimal amount of energy, which is required by a molecule to take part in a reaction. It is designated by $\Delta E\neq$ or $\Delta G\neq$. For example, decomposition of H_2O_2 (hydrogen peroxide) without a catalyst has an energy of activation of some 18,000 cal/mole. When the enzyme catalyse is added, it is less than 2,000 Cal/mole.

Enzymes accelerate reactions by reducing free-energy of activation, this may be interpretated in term of a *transition state theory* illustrated in the figure. The chemical compound A is in a metastable state and requires energy of activation E to convert the compound in a new activated state of B. The reaction then becomes spontaneous, converting B into C and energy E is recovered. A further amount of energy is released according to the difference of enthalpy (*H) of A and C. Thus enzymes increase the rate of conversion of A into C with a lower energy activation (e).

The Catalytic Site

Biochemists are of the opinion that some restricted regions the large sized enzyme is concerned with the process of catalysis. This region

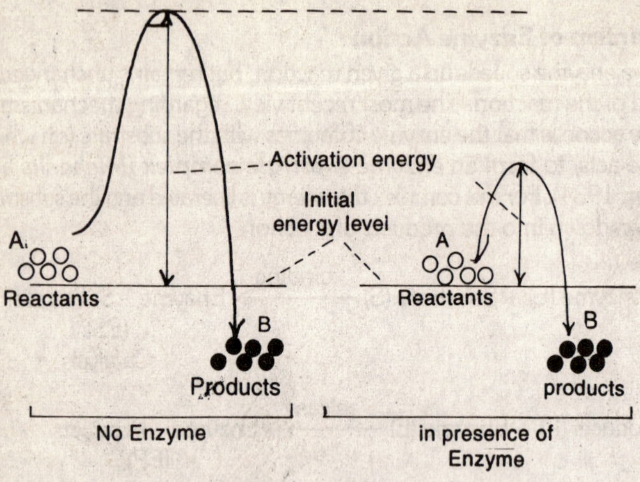

Fig. 3.2. Graphic representation showing that activation energy of an enzyme catalysed reaction is lower than that of an uncatalysed reaction.

was termed as the *active site.* Nowadays, scientists refer to the *catalytic* site, since it is now known that other sites of enzymes are also 'active'.

The catalytic sites exert a binding force on the substrate molecule both by hydrophilic and hydrophobic groups. In most of the cases the enzyme substrate formation occurs by the multiple bonding with the substrate. This may involve formation of covalent bonds, hydrogen bonds or electrostatic bonds.

The sites of catalysis and binding of substrate have been reported to lie at different places in the enzyme molecules which remain quite apart from each other in the absence of substrate. However, in the presence of substrate both the sites come closer to each other in such a way that the substrate fits into enzyme in a most convenient manner. If the tertiary and quaternary structure of enzyme protein is destroyed, the enzyme property of protein is lost. The enzyme besides possessing site for the actual substrate, does also possess certain sites for attachment of coenzyme or cofactor (in the enzymes which require these additional factors for their activity) During the course of activity of such enzymes, *enzyme-substrate-cofactor* complexes are formed instead of enzyme-substrate complexes.

Interaction of Enzyme-substrate

The mechanism of enzyme - action has been explained by two theories:

1. Lock and key theory.
2. Induced-fit theory.

1. Lock and Key theory

This theory was proposed by *Emil Fisher* in 1890. According to this theory the enzyme and the substrate have a complementary shape. It was concluded that a very tight fit between enzyme-substrate complex would be required. The theory can be explained easily by the fact that a particular lock can be opened by a particular key. Similarly enzymes have specific sites whose a particular substrate can only by attached.

2. Induced -fit theory

In recent years the lock and key theory has been given up and a new theory has been proposed. It is called induced-fit theory proposed by *Daniel Koshland* (1960). According to this theory the catalytic site of the enzyme does not initially exist in a shape that is complementary to the substrate but is induced to assume the complementary shapes as the substrate bound. Thus according to this theory the enzyme is flexible.

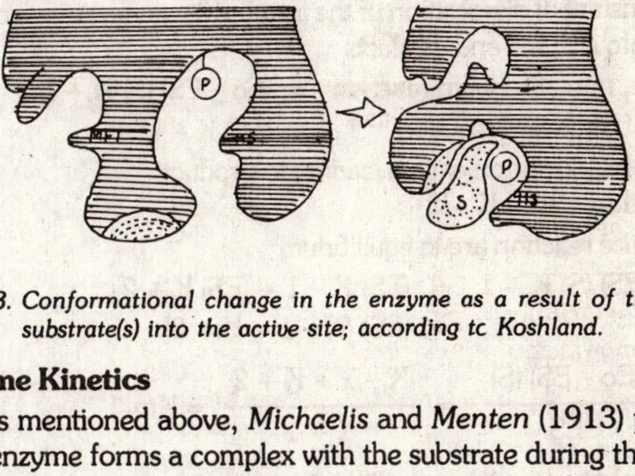

Fig. 3.3. Conformational change in the enzyme as a result of the entry of substrate(s) into the active site; according tc Koshland.

Enzyme Kinetics

As mentioned above, *Michcelis* and *Menten* (1913) postulated that, enzyme forms a complex with the substrate during the reaction called enzyme-substrate complex. The complex then undergoes dissociation to form product (P) and the enzyme (E). The rate of these reactions are governed by constant $K + 1$ (For the complex formation) and $K - 1$ (for complex dissociation and $K + 2$ (for the product formation) according to the following equation:

$$E + S \underset{K-1}{\overset{K+1}{\rightleftharpoons}} ES \underset{K-2}{\overset{K+2}{\rightleftharpoons}} E + P \quad (1)$$

Here $K - 1$ and $K - 2$ represent the rate constants of the reverse reactions. For simplicity the conversion of substrate to product and the release of product from the enzyme have been combined into one step. For many reactions *K* – 2 is negligable.

The dissociation constant *Km* of the enzyme-substrate complex is equal to:

$$Km = \frac{(E)\,(S)}{(ES)} \qquad (2)$$

Figures in brackets indicate their molar concentration. If *Eo* is the initial total concentration of the enzyme and *E* is the concentration of the free enzyme in the reaction and *ES* is the concentration of the bound enzyme the above equation may be represented as follows:

$$Km = \frac{(Eo\text{–}ES)\,(S)}{(ES)} \qquad (3)$$

In equation (1), *K* – 1 is the velocity constant of forward reaction, *K* – 1 is the velocity constant of backward reaction and *K* + 2 is the velocity constant of dissociation of the intermediary enzyme – substrate complex into enzyme and products

Hence, rate of forward reaction = (Eo – ES) (S) K + 1

Rate of backward reaction = (ES) K – 1

Role of dissociation of (ES) leading to product formation + [ES] K + 2

when the reaction are in equilibrium

$$(Eo - ES)\,(S)\,K + 1 = (ES)\,K - 1 + (ES)\,K + 2$$
$$= (ES)\,(K - 1 + K + 2)$$

$$\text{or} \quad \frac{(Eo - ES)\,(S)}{ES} = \frac{K - 1 + K + 2}{K + 1} = Km \qquad (4)$$

In actual practice, it is very difficult to get a pure enzyme, hence its molecular weight determination is quite difficult. It is, therefore, better to eleminate (*Eo*) and (*ES*) from the above equation.

The reaction velocity for dissociation (*ES*) 'v' will be proportional to its concentration in the system.

$$\upsilon \,\alpha\, (ES)$$
$$= K\,(ES)$$

$$\text{or } (ES) = \frac{v}{K} \qquad (5)$$

By placing (V/K) for (ES) in equation no, 4 we get.

$$Km = \frac{(Eo - ES)(S)}{(ES)} = \frac{(Eo - v/K)(s)}{v/K} \qquad (6)$$

If there is large excess of substrate, (S) will be very much greater than (Eo) and whole of the enzyme will be converted into ES and (ES) will be equal to (Eo). Under such conditions, reaction velocity will be maximum and equal to *V*

$$V \alpha (Eo)$$
$$= K (Eo)$$

$$\text{or } (Eo) = V/K \qquad (7)$$

Dividing (5) by (7) we get

$$(ES)(ES) = \frac{\upsilon/K}{V/K} = \upsilon/V \qquad (8)$$

Now equation no. 5 can be further simplified as

$$Km = \frac{(Eo - ES)(S)}{(ES)}$$

$$\text{or } Km (ES) = (Eo - ES)(S) = (Eo)(S) - (ES)(S) \qquad (9)$$
$$\text{or } Km (ES) + (ES)(S) = (Eo)(S)$$
$$\text{or } (ES)(Km + S) = (S)(Eo)$$

$$\text{or } \frac{(ES)(Km + S)}{(Eo)} = (S) \qquad (10)$$

By substituting the value of ES/Eo in equation 10

$$\frac{\upsilon}{V} (Km + S) = (S)$$

$$\text{Therefore, } v \quad (Km + S) = V(S)$$
$$\text{or} \quad Km + (S) = (V/v)(S)$$
$$\text{or } Km = (V/v)(S) - (S)$$
$$= (S)(V/v - 1) \qquad (11)$$

In equation no. 11, *V* is the limiting velocity, *υ* is reaction velocity and *S* is concentration of the substrate used.

In equation no, 11, when the ratio of limiting and observed velocities (V/v) equals to 2, then this equation may be written as

$$Km = (S)(2 - 1)$$
$$= S \qquad (12)$$

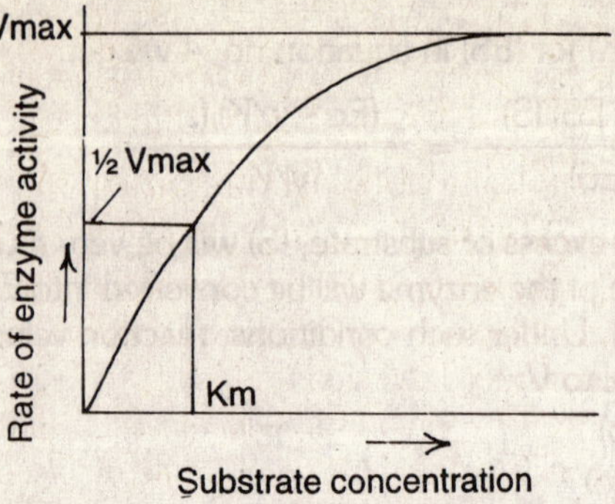

Fig. 3.4. Graph for experimental determination of Michaelis constant (Km).

Thus, the substrate concentration, which is expressed in moles per litre, required for half attainment of the maximum activity of an enzyme is a characteristic constant of that enzyme and is known as its *Km*. The *Km* is known as *Michaelis constant.* Km Value is very much helpful in evaluating affinity of the enzyme for a substrate. If Km value for a substrate of an enzyme is low, this indicates high affinity or *vice versa*. If also gives an idea regarding the type of inhibition caused by an inhibitor.

Michaelis equation no. 11 may be rewritten in the following form

$$v = \frac{V(S)}{Km = (S)} \qquad (13)$$

$$\frac{1}{v} = \frac{Km + (S)}{V(S)}$$

Separating the components of numerator on the right side the equation gives

$$= \frac{1}{V} + \frac{Km}{V} \quad \frac{1}{(S)} + \frac{1}{V(S)}$$

which simplifies to

$$= \frac{1}{V} = \frac{Km}{V} \quad \frac{1}{(S)} + \frac{1}{V(S)} \qquad (14)$$

Equation (14) is a transformation of Michaelis — Mention equation called the *Lineweaver Burke equation*. For enzymes obeying the Michaelis—Mention relationship exactly, a plot of 1/v against 1/(E) yields a straight line. This line will have a slope of Km/v an entercept of 1/V on the 1/v axis and an intercep—1/Km on the 1/(S) axis.

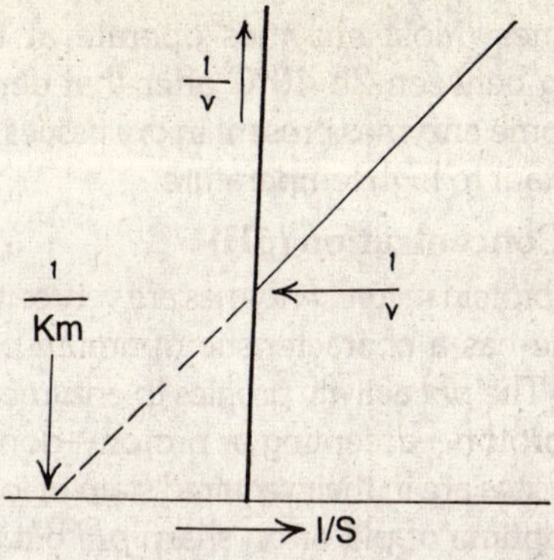

Fig. 3.5. Graphic representation of Lineweaver-Burk plot.

The Michaelis-Mention equation is basic to all aspects of kinetics of enzyme action. If we know Km and V we can calculate the reaction rate of an enzyme at any given concentration of its substrate.

FACTOR AFFECTING ENZYME ACTIVITY

1. Temperature

Between a range of 5°C to 40°C the activity of an enzyme is increased. When the temperature is raised by 10°C, the reaction speed

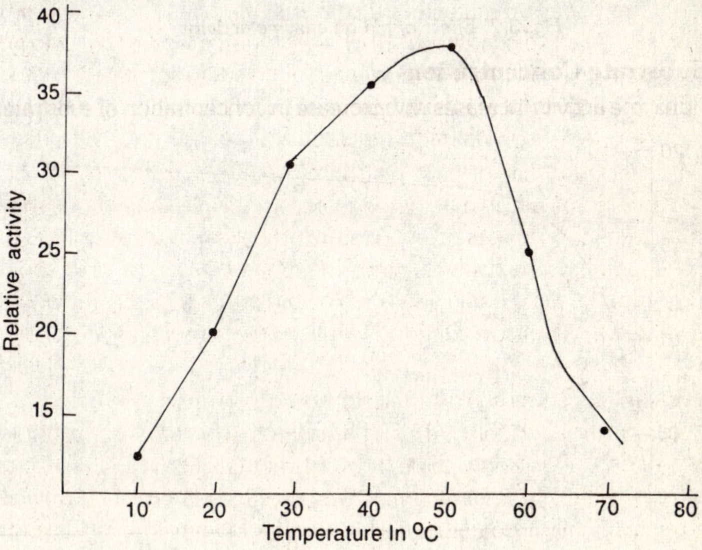

Fig. 3.6. Effect of Temperature on the rate of Enzyme Activity.

increase 2 or 3 times. Most enzymes operate at their best in a temperature ranging between 25-40^{0}C after that denaturation take place, as however, some enzymes present in dry tissues such seeds and spores are very resistant to high temperature.

2. Hydrogen Ion Concentration (pH)

Because of their protein nature, enzymes are very sensitive to change in *pH*. Each enzyme has a characteristic optimum *pH* at which its activity is maximum. The *pH* activity profiles to enzymes reflect the pH at which important proton—accepting or proton—donating groups in the enzyme catalytic sites are in their required state of ionisation. Some enzymes have two optima of pH or on sharp *pH* but a range of *pH*. Generally, the optimum pH lies between 4 and 9.

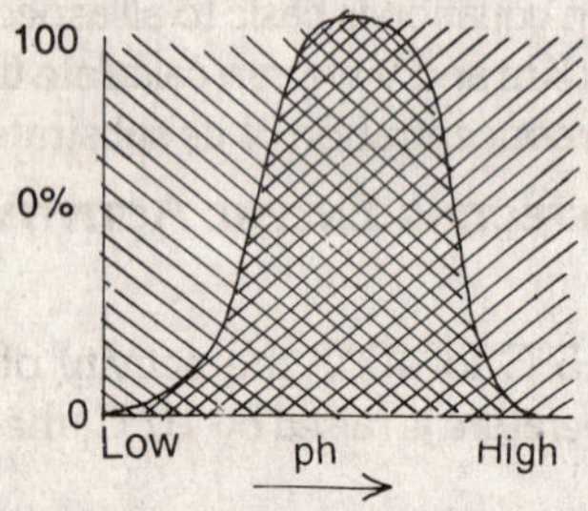

Fig. 3.7. Effect of pH on enzyme activity.

3. Substrate Concentration

Enzyme activity increases with increase in concentration of substrate

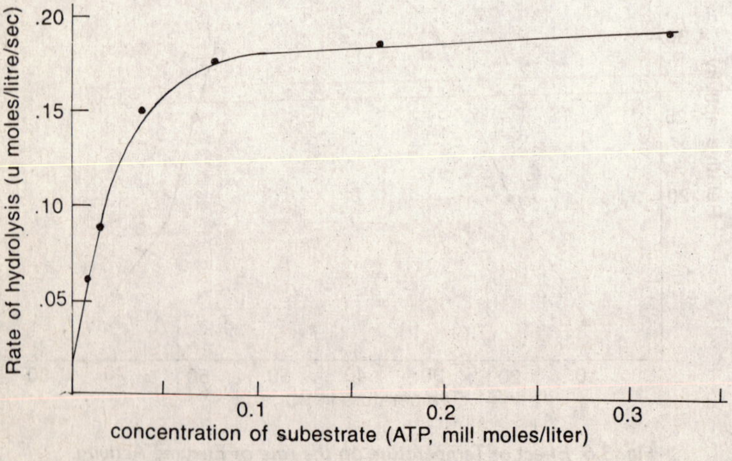

Fig. 3.8. Effect of substrate concentration on the rate of reaction.

to a maximum and than it levels off. Some enzymes are inhibited by high concentrations of substrate.

Two types of graphs are obtained with substrate concentration and enzyme activity plot. Michaelis type of curve of straight line curve is obtained with the enzymes which are made up of one unit. Sigmoid type of curve is obtained when enzyme which is made up of the several subunits and show interaction amongst subunits.

4. Enzyme Concentration

If an excess of substrate is present, doubling the enzyme concentration usually doubles the rate of reaction. This phenomenon usually applies only at the start of the reaction for the end product of the reaction often have an inhibitory effect on the enzyme and decreases its efficiency.

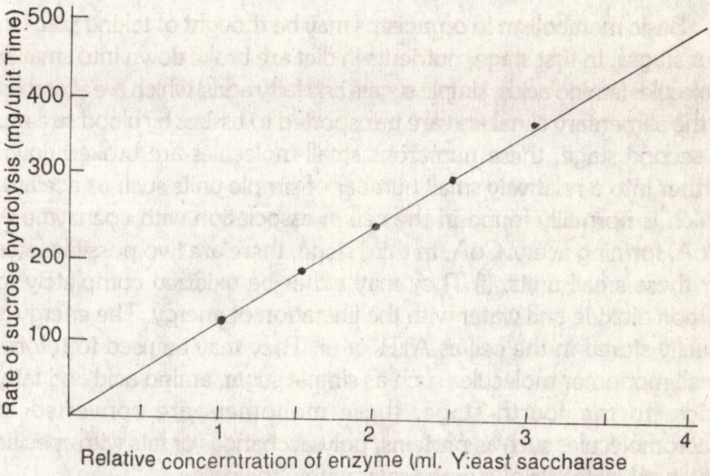

Fig. 3.9. Effect of Enzyme Concentration on the rate of reaction when an excess of substrate is present.

5. Effect of Ions

Many enzymes remain inactive they only become active in the presence of cations such as Mg^{++}, Ca^{++}, Mn^{+}, Zn^{+}, Ma^{+} or K^{+}. In some cases, the cations are loosely bound to the enzyme, while in other cases, they bind to the substrate. Anions are also effective to increase the activity of enzyme. For example, Cl– (chloride) enhances the activity of salivary amylase.

6. Redox Potential

The enzymes are sensitive to redox potential of the cell. Enzymes such as oxidising and reducing are involved in changing the redox components of a cell and hence they are influenced by redox potential. It is due to the presence of readily oxidisable –SH group in their molecule.

7. Effect of Ultra-violet Length

The U.V. light has inactivating effect upon the enzyme activity.

8. Effect of End Products

If the end products of the reactions are allowed to accumulate then these will inhibit the activity of enzyme (negative feed back). It has also been suggested that the end product may combine with the enzyme at a site other than catalytic site and cause a conformational change in the enzyme protein. It has been referred to as *allosteric effect.*

Control of Enzyme Action

Basic metabolism in organisms may be thought of taking place in four stages. In first stage, nutrients in diet are broke down into smaller molecules (amino acids, simple sugars and fatty acids) which are absorbed in the alimentary canal and are transported to tissues by blood stream. In second stage, these numerous small molecules are broken down further into a relatively small number of simple units such as acetate, which is normally found in the cell in association with coenzyme A (CoA) forming acetyl CoA. In third stage, there are two possible fates for these small units. (i) They may either be oxidized completely to carbon dioxide and water with the liberation of energy. The energy is usually stored in the cell as ATP. or (ii) They may be used to reform small monomer molecules such as simple sugar, amino acid and fatty acids. In the fourth stage, these monomers are converted to macromolecules such as proteins, polysaccharides or fats with specific structural, catalytic or storage roles within the cells.

It is obvious that these various metabolic reactions must be strictly controlled. Every enzyme within the cell is not acting at its maximal at all times. Some enzymes may be functioning at greatly reduced levels, while other may be completely inactive. The control mechanisms of the cell are highly complex and, as yet, in completely understood, but this section will attempt to show some of the basic principles underlying such control.

1. Control by Substrate

This is the simplest type of control. Since the rate of an enzyme-catalyzed reaction is proportional at low substrate concentrations to the concentration of substrate, as the concentration of any particular substrate builts up, more of it will be converted to product in an attempt to reduces its concentration.

2. Control by Inhibitors

Some of the elements or compounds can reduce the rate of enzymatic activity. These substances may be of physiological or unphysiological in nature. This phenomenon is known as *enzyme inhibition* and such substances are known as *enzyme inhibitors.*

Enzyme inhibition may be divided into three main types:

(i) Competitive inhibition

(ii) Non-competitive inhibition

(iii) Allosteric inhibition.

(i) Competitive inhibition

Competitive inhibitors act by combining with the catalytic site of enzyme preventing access by substrate. Because of their competitive nature, their structure is often very similar to that of the substrate and the activity of the enzyme will depend upon the relative concentrations of substrate and inhibitor and the relative binding affinities of these molecules for its catalytic site. Enzyme diphosphoglycerate mutase is an enzyme in the metabolism of glucose in RBCs and catalyses the inter conversion of 1, 3-diphosphoglycerate and 2, 3-diphosphoglycerate

$$\begin{array}{c} COOP \\ HO-C-H \\ CH_2OP \end{array} \rightleftharpoons \begin{array}{c} COO^- \\ PO-C-H \\ CH_2OP \end{array}$$

It is competitively inhibited by low concentration of 2, 3-diphosphoglycerate and hence if this metabolite builds up, the interconversion is slowed.

(ii) Non-Competitive inhibition

The non-competitive inhibitors are equally important in metabolism. Here substrate and inhibitor do not exhibit mutual competition. Most probably, the sites of attachment of substrate and non-competitive inhibitor are different on the enzyme molecules. The inhibitor are most structurally dissimilar to the substrate. If the inhibitor can be removed from its site of attachment preserving the activity of the enzyme, it is known as *reversible non-competitive inhibition*. If it cannot be removed at all, or only removed after loss of enzymic activity, it is known as *irreversible non-competitive inhibition*. Both reversible and irreversible inhibitions have been found to exhibit similar reaction kinetics.

(iii) Allosteric inhibitions

In mixed inhibitions, the inhibitor molecule attached to the enzyme prevents the breakdown of enzyme-substrate complex, and also at the same time, it interfers with the binding of the substrate of the enzyme. Thus, such inhibitors exhibit partially competitive and partially non-competitive kinetics. The sites other than the enzymatic sites, lying on the different region in the enzyme molecule where an inhibitor molecule may get attached is known as *allosteric site.* Presence of inhibitor on allosteric site changes the conformation around the catalytic site. Thus results into reduced enzyme-substrate complex formation. If the substrate molecules have preoccupied the catalytic site, it may affect conformation around the allosteric site thereby reducing enzyme-inhibitor complex formation and thus reduces the extent of inhibition.

3. Control by Covalent Modification Cascades

Some enzymes exist in the cell as inactive precursors. To activate these enzyme, it is necessary to remove a small peptide from the precursor form called a *zymogen*. For example, trypsinogen in duodenum is released as zymogen and a small amount is activated by another enzyme, enterokinase. Once a small quantity of active trypsin is produced, this in turn activates more trypsinogen and the whole process is autocatalytic. Chymotrypsinogen is another zymogen which is converted to chymotrypsin by trypsin. The net effect in which one enzyme acts to convert a zymogen to its active enzyme is termed as cascade.

Allosteric Enzymes

These are also called as *regulatory enzymes.* The kinetic properties of these enzymes cannot be accounted for by the Michaelis—Menten model. When a product is formed after a series of enzymatic reactions and it influences the activity of first enzyme by which the rate of product formation itself is controlled, under such circumstances. The first enzyme is called an *allosteric enzyme.*

$$a \xrightarrow{E_1} b \xrightarrow{E_2} c \xrightarrow{E_3} d \xrightarrow{E_4} e \xrightarrow{E_5} P$$

(Product)

If accumulation of product P influences the activity of enzyme E_1 then E_1 is known as allosteric enzyme.

Isoenzymes

It is now known that some of the enzymes exist in many physiologically distinct forms and all the forms of a single enzyme can catalyse the same reaction. The only differences noted in the different forms of a single enzyme are that of electrophoretic mobility and possible of the quaternary structure of the enzyme protein. Such forms of a single enzyme, having same catalytic activity are known as *isoenzymes.* According to International Union of Biochemists (IUB) the term isoenzyme has been described as the multiple forms of an enzyme occuring within a single species. About 100 enzymes at present are known to exist in such multiple forms. For example *lactic dehydrogenase* of human and animal tissues have been shown to exist in five isoenzymatic forms, namely LD_1, LD_2, LD_3, LD_4, and LD_5 and their relative distribution varies considerably from tissue to tissue and from species to species.

Isoenzymes may be *homologous* or *analogous*. Homologous isoenzymes have essentially similar molecular structure and catalytic activity. Cytochrome –C from different animals is the example of homologous isoenzymes. Analogous isoenzymes have similar reactions but different molecular structure and catalytic properties.

ANTIENZYMES

The wall of alimentary canal is not digested by the proteolytic enzymes secreted by itself. This has led to the discovery of *antienzymes.* It has been reported that such tissues possess certain proteins having antagonistic property to the proteolytic enzyme.

4

Vitamins

Although a diet may be adequate from the point of view of protein, carbohydrate, and fat, it has now been definitely established that certain accessory food substances are necessary in the diet to maintain good health, especially of growing animals. Experience gained in expeditions, in the old sailing-ship days, and of the health of prisoners, had long indicated that mere calories and proteins were insufficient to sustain life, but its was not, however, until Hopkins of Cambridge, in 1912, from his investigation of restricted diets in relation to the essential amino-acids, placed the subject on an experimental basis that the importance of these accessory food factors was fully appreciated, although admittedly much uncorrelated and not generally recognised clinical evidence, notably that of Eijkmann, in 1890, was available.

The factors are, for the most part, products of the plant world, and from this source they are acquired by animals. Many are relatively simple chemical substances and have now been obtained in a crystalline form. They may be distinguished by differences in their source, in their action, and in the maladies (deficiency diseases) which result from their absence. In addition to those described below there are, however, many other substances which are necessary for the health of lower animals and the growth of bacteria. Only those important to man are discussed.

A considerable number of vitamins have now been identified, largely in the first instance by their solubility in fat and in water respectively. It is now possible to feed suitable animals on restricted diets and to demonstrate the effect of the individual deficiencies.

In 1912, Hopkins fed young rats on pure food mixtures containing approximately the constituents of milk, *viz.* caseinogen, fat, starch, cane-sugar, and inorganic salts. The energy-value was adequate but growth ceased. His famous growth curves are shown in Fig. The addition of a minute amunt of fresh milk caused the recommencement of growth. Now it is known that milk contains several vitamins- A, D, B and C all concerned in growth. In 1881 Bunge had come to a similar conclusion from work on mice. In 1913 McCollum and Davis showed that this essential substance for growth was present in the fats of butter and egg-yolk. They also differentiated the fat-soluble vitamins A and D and crude lactose.

The fat-soluble Vitamin A

Source

The *source* of this vitamin is the pro-vitamin A of the green plant, whence it is acquired by animals, and becomes dissolved in their fat. Thus it is present in milk and butter, provided the cows are fed on green food. Animals have the power of storing the vitamin in their livers from which it may be obtained in the oil. The oils of fishes which feed on green plant organisms in the sea, either directly or by eating smaller fishes which do so, are particularly rich in the vitamin. Hence their virtue in therapeutics. Butter substitutes, such as margarines made from vegetable fats, have usually no vitamin A and little D. Both A and D are present in egg-yolk.

Properties

It is important to observe that vitamin A is very liable to destruction on heating, but it is now clear that this depends on oxidation processes.

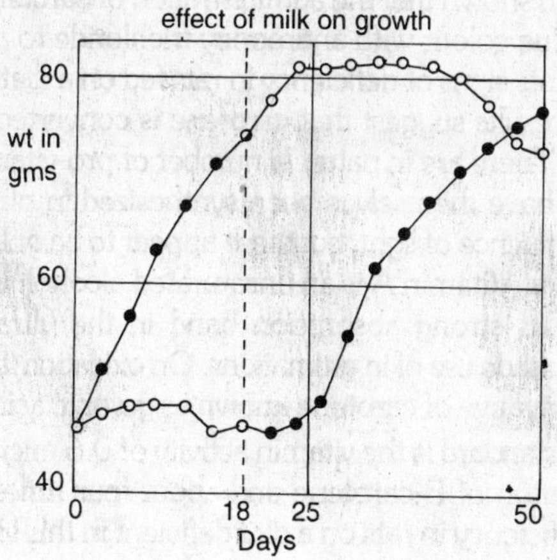

Fig. 4.1. Curves of growth of two sets of rats.

Milk can, therefore, be sterilized at quite high temperatures in an autoclave (air free) without losing vitamin A, which, however, is destroyed by ordinary boiling or even free aeration at all temperatures. Speaking generally, condensed milk still has enough vitamin A, and dried milk also, provided the latter has been made by a process such as passing it over a heated drum, which dries without permitting too long exposure.

Deficiency in Vitamin A has been studied especially by E. Mellanby in puppies. Its absence leads to an overgrowth, especially of the spinal column, and to compression of the spinal cord, which may degenerate and lead to diverse upset-notably difficulty in walking.

The vitamin is also concerned with the replacement of the epithelial cells of the mucous membranes, which are liable to become dry, keratinised, and liable to infection, especially the conjunctiva of the eyes. It cannot, however, be called a general anti-infection vitamin as has sometimes been said. Human volunteers, controlled by the Medical Research Council, showed no serious impairment or special liability to infection when deprived for well over a year. Only occasionally night blindness occurred, due to lack of visual purple. The absence of impairment appears to be due to the fact that body normally has considerable stores, especially in the liver but also in the ovary, testis, and adrenal glands.

Chemistry

The parallelism between the yellow colour and potency of vitamin A in foods led to the finding that the vitamin is related to the pigment carotene. It has been shown that the administration of carotene causes the characteristic blue colour with antimoney trichloride to appear in the livers and prevents signs of deficiency in rats fed on a diet deficient in vitamin A. The results suggest that carotene is converted into the vitamin in the liver. There are in nature a number of pro-vitamins A or carotenoids which have themselves been synthesized in plants from precursors by the influence of light, but these appear to be only of value to man as precursors. Vitamin A is an unsaturated alcohol having the formula $C_{20}H_{30}O$. Its strong absorption band in the ultra-violet is characteristic and is made use of in estimations. On oxidation it becomes converted into a derivative of carotene known as geronic acid.

The vitamin A standard is the vitamin activity of 0.6 microgram of a standard preparation of B-carotene and about four times this will prevent signs of deficiency in rats on a diet deficient in this vitamin.

A vitamin A which differs in its absorption spectrum has been differ—entiated, but its action appears to be similar.

The Antirachitic or Calcifying Vitamin D and Sunlight

In 1921, *Mellanby* recognized the need for a factor in food (cod-liver oil) that could prevent rickets. *McCollum* demonstrated that this vitamin was not the same as vitamin A, and in 1924, *Steenbock* and *Hess* were able to show that many foods acquire vitamin D activity after exposure to ultraviolet irradiation.

Chemistry

The structure of vitamin D_3 is the same as that of D_2 except that the side chain on position 17 is that of cholesterol.

Sources

The vitamins D are actually a group of compounds. All are sterols which occur in nature, chiefly in the animal organism. For nutritional purposes, there are 2 important D vitamins:

(i) D_2 *(Activated Ergosterol; Ergocalciferol or Viosterol)*. It occurs in the plant kingdom (e.g., in ergot and in yeast).

(ii) D_3 *(Activated 7-Dehydrocholesterol, Cholecalciferol)* 7-Dehydrocholesterol is the form in which provitamin D_3 is found in natural foods. Man and other mammals can synthesize provitamin D_3 in the body. The vitamin is then activated in the skin by exposure to ultraviolet rays and carried to various organs in the body for utilization or storage (in liver). In its active form, vitamin D is not well distributed in nature, the only rich sources being the liver and viscera of fish and the liver of animals which fed on fish. Eggs and butter also contain vitamin D. Milk is a poor source of the vitamin. Vitamin D is stable in foods; storage, processing, and cooking do not affect its activity.

Functions

The principal action of vitamin D is to regulate the absorption and utilization of calcium and phosphorus. The vitamin also has a direct effect on the calcification process.

This vitamin has the same source as vitamin A, but may be distinguished from the latter by the fact that it is less easily oxidised; for example, cod-liver oil if exposed to the action of ultra-violet light or of oxygen at 100°C. for twenty-eight hours loses its power to cure xerophthalmia in rats while the action of vitamin D remains (McCollum). It therefore withstands ordinary cooking and preserving better than A. Butter, on the other hand, is less active than cod-liver oil in curing rickets, although more effective in curing xerophthalmia. Vitamin A may also be distinguished by it giving a blue colour with antimony trichloride. Vitamin D is present in small quantities in vegetables oils which contain no vitamin A. Tunny fish-liver oil contains most. This vitamin is present in yeast and a wide range of lower plants when fresh, but they rapidly lose their vitamin when cut. As in the case of A, ordinary winter milk, therefore, contains very little of the vitamin unless the cows have been specially fed on fresh green blood.

Deficiency

The absence of vitamin D or of sunlight leads to imperfect calcification of the bones (rickets) and of the teeth, and in this connection it may be noted that, in 1906, Hopkins suggested this possibility, which was subsequently confirmed experimentally on dogs by E. Mellanby, to be due to the absence of a fat soluble vitamin, although at that stage vitamin D had not been separated from A. In rickets the bones are so soft that the legs bend under the weight of the body. Bow-legs, knock-knees, and many other deformites especially of the spine and pelvis may occur. The child is described as having "the head of a philosopher, the legs of a grand piano, and the belly of a poisoned pup." There is lowered phosphorus (1.5 mg. per 100 c.c. instead of 5 mg.) and calcium in the

Fig. 4.2. Rickets in man.

blood. The bones of rachitic animals will calcify if incubated in normal serum, but not in rachitic serum, showing that the bones has the power of calcify if the salts are available.

That lack of vitamin D produces dental caries has now been abundantly proved by Lady Mellanby, although, apparently, some persons require more vitamin D than others. There seems to be little doubt too that many children in receipt of a full quantity of vitamin D have caries because of their difficulty in digesting fat or other digestive disturbances. Attention to the diet and the giving of vitamin without the fat may result in striking improvement. For complete dental development carotene appears to be also an advantage. Dental caries and rickets do not, however, always occur in the same individuals, so presumably other factors are concerned in the former—e.g. fluorine deficiency or bacteria.

The function of the vitamin is, apparently, to correct any improper balance between the calcium and phosphorus intake, and the greater the disproportion of these two elements the more important the vitamin, but even if these substances are present in proper amounts the vitamin is necessary. The vitamin appears to increase the retention in the body of calcium which may be available in the food, probably by increasing the amount absorbed from the intestine.

Fig. 4.3. Rickets in a dog.

There were at first opposing views on rickets and its treatment. One group of workers insisted on the importance of diet (E. Mellanby and others), and another insisted on exercise and sunlight (L. Findlay, Paton). The discovery by Huldschinsky in 1919 that ultra—violet light would cure rickets and that vitamin D may be produced in foods by similar treatment has shown that both were right. This production of the vitamin D_2 by irradiation depends on the presence of ergosterol (a sterol first obtained from ergot), but it has been found that 7-dehydro-cholesterol also becomes active when irradiated (D_3). In man, it is assumed that the sunlight activates ergosterol in the skin, for the feeding of excised human skin if irradiated will protect rats against rickets but not if not irradiated. How this occurs is not yet clear, since the penetrating power of ultra-violet light through skin is very poor, being only a millimeter. Excess of the vitamin leads to excessive precipitation of

VITAMINS

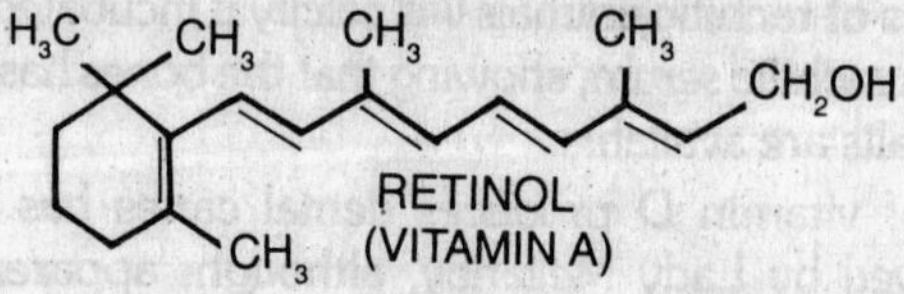

RETINOL
(VITAMIN A)

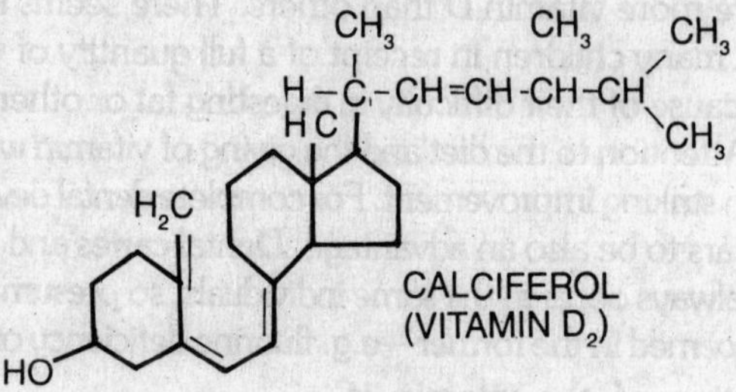

CALCIFEROL
(VITAMIN D_2)

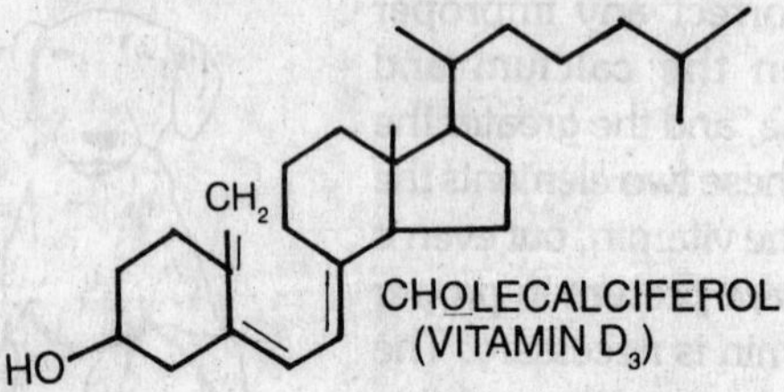

CHOLECALCIFEROL
(VITAMIN D_3)

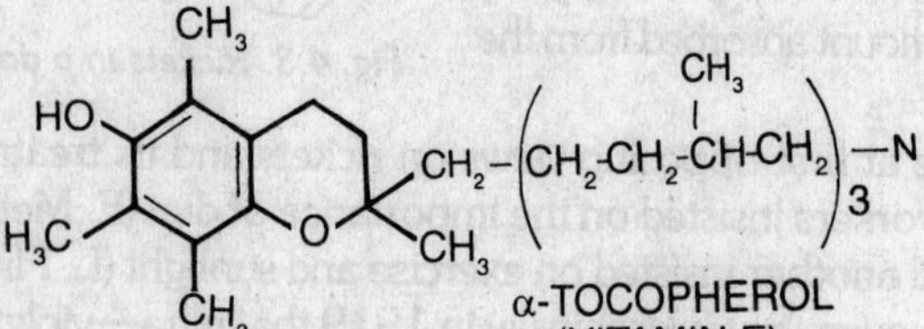

α-TOCOPHEROL
(VITAMIN E)

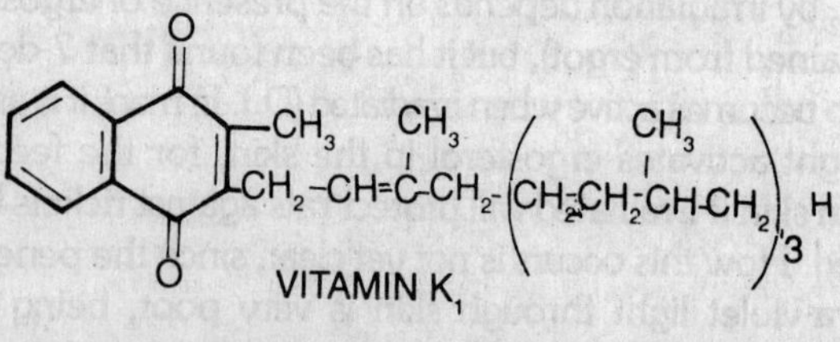

VITAMIN K_1

Fig. 4.4. Structure of some fat-soluble vitamins.

calcium throughout the body especially in the arteries and kidneys (Kreitmar and Moll).

Chemistry

A pure crystalline substance "calciferol," which is a distillate of irradiated ergosterol which is even more active than vitamin D, has not been prepared (Bourdillon) and found to be the isomer of ergosterol. The activity of 0.05 microgram of this has been adopted as the international unit.

There is increasing agreement that children should be supplied with 600-700 units daily and adults with about 200 units (Ministry of Health) associated with an adequate supply of calcium and phosphorus.

Vitamin D is *antagonized* by feeding animals on certain cereals, especially maize and oatmeal, which *contain physic acid*. If, however, the oatmeal is cooked slowly the physic acid is destroyed by a phytase normally present. Wholemeal bread is therefore rachitic, but not porridge properly made and especially when taken with whole milk.

In conclusion it may be added that the making of a few dogs rickety has resulted in almost banishing from Britain a disease which, until comparatively recently, was responsible for the death or crippling of thousands of children. This is a fact worthy of contemplation by those who condemn animal experiments.

Less than fifty years ago persons with bow-legs or knock-knees were a common sight in all our streets and were often seen on football fields, while individuals with more severe deformities crowded our hospitals. The present generation of young readers can scarcely by expected to appreciate that vast improvements which our knowledge of vitamins has brought about.

Toxicity of Vitamin D

Excessive quantities of vitamin D (ten times the normally required daily level) given over prolonged either to children or to adults produce demineralization of bone, and multiple fractures may occur after only minimal trauma. Serum calcium and phosphorus are both markedly elevated, resulting in metastatic calcification of many soft tissues and formation of renal calculi. The latter may block the renal tubules sufficiently to cause a secondary hydronephrosis.

The Vitamins E

In 1936, *Evans* and *Emerson* isolated vitamin E from wheat-germ oil. It was given the name tocopherol (Gk. *tokos,* childbirth, *phero,* to bear; *ol,* an alcohol).

Chemistry

The tocopherols are unsaturated alcohols. There are 4 varieties of tocopherols: alpha, beta, gamma and delta tocopherol; of these, alpha tocopherol ($C_{29}H_{50}O_2$) has the widest distribution and greatest biological activity. It has a "chromane ring system" containing four methyl (—CH_3) groups, one phenolic group and a phytyl side chain. The beta and gamma tocopherols have one methyl group less in the ring than α-tocopherol.

Sources

The richest sources of tocopherol are plant oils, e.g. wheat germ, rice, cotton seed, as well as the lipids of green leaves. Fish-liver oils, rich in vitamins A and D, are devoid of tocopherol. Good soruces of vitamin E include eggs, muscle meats, liver, fish and chicken.

Functions

1. The most striking chemical characteristic of the vitamin E is their antioxidant property. Polyunsaturated fatty acids are easily attacked by molecular oxygen, resulting in the formation of peroxides. The tocopherols prevent this.
2. It is required for the normal reproductive function in rats, hence also called "antisterility" vitamin. In the absence of vitamin E, the germinal epithelium of the testes of rats is destroyed. In the female rate, ovulation and fertilization takes place, but there is death and resorption of the foetus. This situation can be repaired by including vitamin E in the diet.
3. In some animal e.g., rabbit a lack of vitamin E produces muscular dystrophy.

Deficiency

The wide distribution of vitamin E in vegetable oils, cereal grains, and animal fats makes it unlikely that a deficiency of this nutrient occurs in normal humans.

Vitamin K

Vitamin K (*Koagulations* vitamin) was identified in 1935 by *Dam* as a factor present in green leaves which prevented haemorrhage.

Chemistry

The vitamin isolated from alfalfa is called vitamin K_1 (*Phylloquinone*). It is a derivative of 2-methyl-1, 4-naphthoquinone with a phytyl radical on position 3. Another compound with antihaemophilic properties was isolated from bacteria and called vitamin K_2 (*Farnoquinone*).

Sources

Vitamin K is present in alfalfa and in such dark green vegetables as spinach, cabbage leaves as well as in cauliflower and peas. Tomatoes, cheese, egg-yolk,a nd liver are good sources. Fruits are poor sources, as are moulds, yeast and fungi.

Functions

The best known function of vitamin K is to catalyze the synthesis of prothrombin by the liver. It also regulates the synthesis of other plasma clotting factors dependent on vitamin K (factor VII, IX and X).

Deficiency

In the absence of vitamin K a hypoprothrombinemia occurs in which blood clotting time may be greatly prolonged. A dietary deficiency of vitamin K is not likely to occur since the vitamin is fairly well distributed in foods and the intestinal micro-organisms synthesize considerable vitamin K in the intention. However, a deficiency may occur as a result of prolonged oral therapy with drugs capable of vitamin K-producing bacteria (Sulfaguanidine). Furthermore, the absorption of vitamin K from the intestine depends on the presence of bile. A deficiency state will therefore result as a consequence of biliary tracts obstruction or if there is a defect in fat absorption, such as in sprue and celiac disease.

WATER-SOLUBLE VITAMINS

The water-soluble vitamins make up a large group. All but one belong to the B complex. The remaining water-soluble vitamin is ascorbic acid or vitamin C. For the most part, water-soluble vitamins are not stored in the body. Excesses are largely excreted, thus eliminating the possibilities for toxicity that exist with overdosage of fat-soluble vitamins.

At least 11 vitamins compose the B complex. Seven of them are essential in human nutrition and must be provided in daily meals; 3 of these are included on the allowance table of the Food and Nutrition Board. These 3 are thiamine (vitamin B_1), riboflavin (vitamin B_2), and niacin. They were once though to be one substance. Gradually, their separate identities were established. Certain overlapping characteristics and functions delayed the process of separation. For example, the fact that each is soluble in water led to difficulties in recognition. And all are concerned together in the task of providing the body with energy. Their function basically is converting sources of energy in the body, especially carbohydrate, into "living" energy. In effect, they make it possible for carbohydrate to "burn" at body temperature. In addition, each has its own characteristic contributions to make in nourishing the body. All 3 occur in very small amounts in every body cell.

Thiamine (B_1) : (Antiberiberi substance, Antineuritic factor)

The structure of thiamine, which was first isolated in crystalline form by *Jansen* and *Windaus* in 1926, was established by *Williams* and his coworkers in 1936.

Chemistry

Thiamine ($C_{12}H_{17}N_4OS$) is 2, 5-dimethyl-6-aminopyrimidine bonded through a methylene linkage to 4-methyl-5-hydroxyethylthiazole.

Sources

The outer layers of the seeds of plants are especially rich in thiamine. Whole-wheat bread, therefore, is an excellent source, whereas ordinary white bread is a poor source of the vitamin, since most of the thiamine is removed in the milling process. Among the more abundant sources are unrefined cereal grains, legumes, brewer's yeast, liver, kidney and lean cuts of pork. Excessive cooking leashes the water-soluble thiamine from many food-stuffs.

Fig. 4.5. Pigeon suffering from vitamin B_1 deficiency. *Fig. 4.6. Same pigeon after the administration of the vitamin.*

Functions

It is a constituent of the coenzyme *thiamine pyrophosphate* (TPP) and is essential for carbohydrate metabolism (oxidative decarboxylation of pyruvate or of α-ketoglutarate).

Deficiency

Thiamine deficiency causes *Beriberi* which was first described by Admiral *Takaki* in 1887 among Japanese sailors. The disease is still an important public health problem in Philippines, Vietnam, Thailand and Burma where polished milled rice is a staple food. Beriberi affects predominantly the peripheal nervous system, the gastrointestinal tract and the cardiovascular system. There are two forms of the disease:

"dry" and "wet" beriberi. In the dry form, there is rapid loss of weight and muscle wasting. Marked peripheral neuritis and muscular weakness result in the patient's becoming almost helpless. Deep reflexes are lost, sensory changes may occur, and anxiety states and mental confusion are evident. The heart becomes enlarged. In wet beriberi there is a generalized oedema that may obscure the weakness and muscular wasting. Acute cardiac symptoms may develop rapidly.

Riboflavin (B_2)

Warburg and *Christian* isolated this *yellow enzyme* from yeast in 1932.

Chemistry

Riboflavin corresponds to 6, 7-dimethyl-9-oribityl-isoalloxazine and contains the alcohol (ribitol) derived from the sugar ribose attached at position 9.

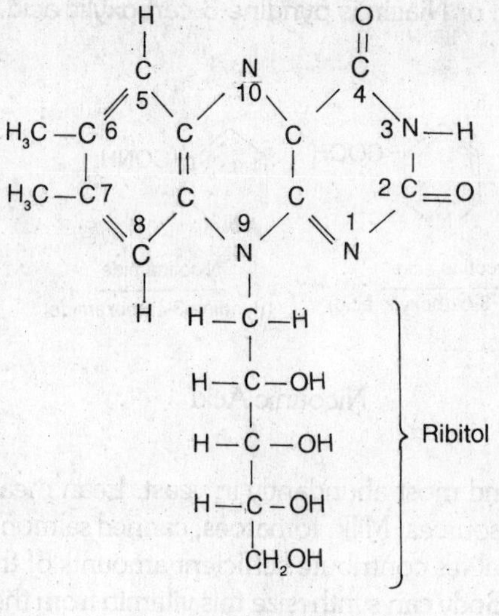

Riboflavin

Source

Riboflavin is widely distributed throughout the plant and animal kingdom, with very rich sources in anaerobic fermenting bacteria. Milk, liver, kidney, and heart are excellent sources. Many vegetables are also good sources, but the cereals are rather low in riboflavin content.

Functions

It is a constituent of two coenzymes—flavin mono-nucleotide (FMN), and flavin adenine dinucleotide (FAD), which are involved in intermediary metabolism. All known biological functions of riboflavin involve these two coenzymes.

Deficiency

Riboflavin deficiency in man is characterised by a particular type of glossitis (magenta-coloured tongue), fissures at the corners of the mouth and lips (cheilosis), localized seborrheic dermatitis of the face and corneal vascularization.

Nicotinic Acid or Niacin : (Pellagra-Preventing, PP, Factor of Goldberger)

Pellagra was recognised as early as 1735 by *Gasper Casal,* physician to Philip V of Spain. The identification of nicotinic acid as the responsible factor was made by *Elvehjem, Woolley,* and their associations in 1937.

Chemistry

Nicotinic acid or Niacin is pyridine-3-carboxylic acid. Its amide is nicotinamide.

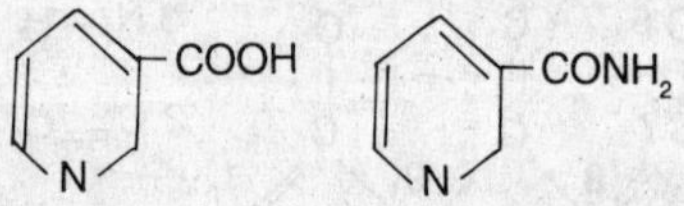

Nicotine acid (Pyridine-3-carboxylic acid) — Nicotinamide (Pyridine-3-carboxamide)

Nicotinic Acid

Sources

Niacin is found most abundantly in yeast. Lean meats, liver, and poultry are good sources. Milk, tomatoes, canned salmon, and several leafy green vegetables contribute sufficient amounts of the vitamin to prevent disease. Body can synthesize this vitamin from the amino-acid tryptophan.

Functions

It is convertible into nicotinamide and thus functions metabolically as a component of the 2 coenzymes: nicotinamide adenine dinucleotide (NAD) and nicotinamide adenine dinucleotide phosphate (NADP). These coenzymes, which operate as hydrogen and electron transfer agents by virtue of reversible oxidation and reduction, play a vital role in metabolism, leading to the formation of ATP in the electron transport system (ETS).

Deficiency

Pellagra was established as deficiency disease by *Goldberger.* The disease is symptomatized by three D's, namely diarrhoea, dermatitis of those areas that are exposed to sunlight and dementia, due to disturbances of the central nervous system. Stomatitis and glossitis are also associated with the other symptoms.

Alcoholism causes deficiency of the vitamin and is an important precipitating factor of the disease.

Pyridoxine (B_6) : (Rat Antidermatitis Factor)

Pyridoxine was first discovered in 1938 and synthesized by *Stiller* in 1939. Natural soruces contain two other forms of this vitamin, *pyridoxal* and *pyridoxamine.* The three substances as a group are designated vitamin B_6; no single one is considered the vitamin, since all three are equally effective in animal nutrition.

Chemistry

Pyridoxine is 2-methyl-3-hydroxy-4, 5-di (hydroxymethyl) pyridine. The active form of the vitamin is pyridoxal phosphate.

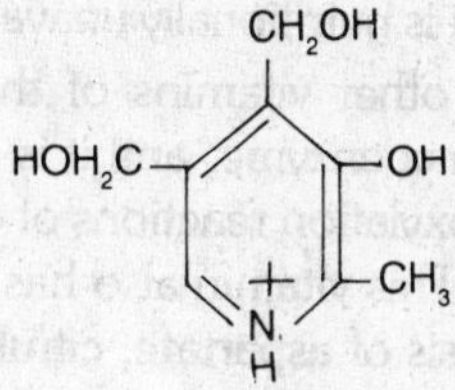

Pyridoxine

Sources

The B_6 group is widely distributed in nature, and those foods rich in other members of the B complex are excellent sources of these materials, e.g., the germs of various grains and seeds, egg yolk, yeast and meat, particularly liver and kidney.

Functions

1. It is a constituent of coenzymes *pyridoxal phosphate,* which is involved in the non-oxidative degradation of amino-acids, viz., transmination, deamination, decarboxylation etc.
2. It is essential in the synthesis of heam for haemoglobin and cytochromes.

Deficiency

There is little evidence that diets containing a reasonable balance of naturally occuring foodstuffs are ever seriously defficient in vitamin B_6. However, occasional cases of B_6 deficiency do arises as a result of malabsorption, alcoholism, antagonism to drugs etc. Pyridoxine deficiency in humans may be associated with a reversible hypochromic microcytic anaemia with a high serum iron similar to that observed in pyridoxine deficient animals.

Function

The most important function of vitamin D is that it increases calcium and phosphate absorption from the gut. It also promotes calcification by a direct action on bone. It prevents and cure rickets.

Biotin

Biotin is slightly soluble in cold water, but quite soluble in hot water, and is heat stable although readily oxidized. Biotin combines with a glycoprotein found in raw (but not cooked) egg white called avidin to form a stable complex which is not hydrolyzed by the digestive enzymes. Hence in this special circumstance biotin cannot be absorbed from the intestine so that it is nutritionally unavailable.

In similarly with the other vitamins of the B complex, biotin presumably functions as a coenzyme, and it is believed to act in the carboxylation and decarboxylation reactions of oxaloacetate, malate, aspartate, and succinate. This vitamin also has been implicated as a cofactor in the bio-synthesis of aspartate, citrulline, and unsaturated fatty acids as well in pyruvate oxidation.

There is no distinct disease known in man caused by an underlying bitoin deficiency, although experimentally induced lack of this vitamin

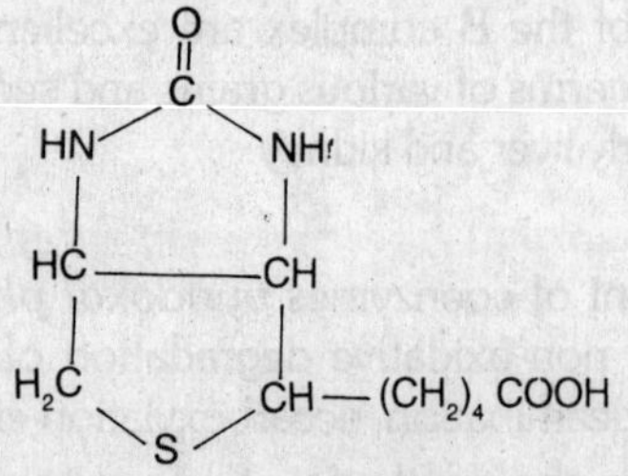

Biotin

Fig. 4.7. Structural formula of biotin.

produces in humans a dermatitis, lassitude, anorexia, insomnia, muscular pain, precordial distress, and mild anemia. All of these symptoms are ameliorated promptly when biotin is included in the diet once again.

Studies performed on human subjects have indicated that biotin is synthesized in sufficient quantities by the intestinal flora to supply the nutritional requirements of humans. An exogenous source of this vitamin is unnecessary.

Biotin is found in high concentrations, in meats, egg yolks, peanuts, mushrooms, and chocolate.

Pantothenic Acid

Pantothenic acid is a water soluble compound which is produced by the chemical combination of pantoic acid and β-alanine and has a molecular weight of 219. Pantothenic acid is stable in neutral solution but is unstable in acid or alkali. The hydroxy analog of pantothenic acid, panthenol, is a viscous oil, and it exhibits the same biologic activity as the vitamin itself. The calcium salt of pantothenic acid is a crystalline solid, and this compound is the most extensively used form of the vitamin.

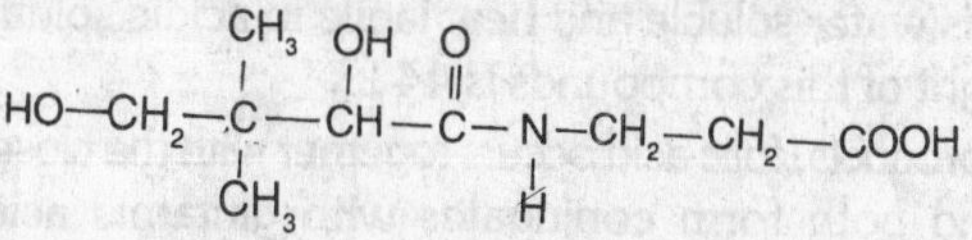

Pantothenic Acid

Fig. 4.8. Structural formula of pantothenic acid.

Coenzyme A requires pantothenic acid in its structure, and this coenzyme in turn is essential for many reversible acetylation reactions in carbohydrate, amino acid, and fat metabolism. Thus a specific protein (apoenzyme) in combination with coenzyme A (abbreviated Co-A) comprises the enzyme system.

Acetyl coenzyme A (abbreviated acetyl Co-A), which is the activated molecule, may act either as an acetyl receptor or an acetyl donor, and this compound enhances condensation reactions such as that involved in the formation of citrate from oxaloacetate in the Krebs cycle. Acetyl Co-A also functions as the recipient of acetyl radicals generated during the metabolic oxidation of pyruvate, citrate, and fatty acids; these radicals then are transferred to other compounds.

Acetyl Co-A also conjugates with actyl (R-COO-) groups in addition to acetyl radicals; hence it is active in the synthesis of fatty acids, cholesterol, and other sterols.

Pentothenic acid is a vital constituent of coenzyme A; the importance of this vitamin in cellular metabolism is evident.

Since there is an intimate relationship between the tissue level of pantothenic acid and adrenocortical function, pantothenic acid (functioning as a constituent of coenzyme A) presumably is involved in the activity of this endocrine gland.

No clinical syndrome which is caused by a deficiency of pantothenic acid in the diet is known, despite the importance of this vitamin to many facets of intermediary metabolism. The daily requirement for this substance is satisfied even by poor diets. Pantothenic acid (the name of this compound means "derived from everywhere") is ubiquitous in its occurrence in both plant and animal tissues. Organ meats, egg yolk, peanuts, whole grains, cereal brans, and vegetables all are good sources of this vitamin, whereas meats, milk, and fruits also contain reasonable although somewhat lower quantities. It appears that ordinary cooking procedures do not lead to an appreciable loss of pantothenic acid.

Folic Acid (Petroylglutamic Acid, Folacin)

Folic acid is water soluble and heat labile in acidic solutions. The molecular weight of this compounds is 441.

In natural products, folic acid occurs together with the closely related folinic acid and both form conjugates with glutamic acid. These conjugates are hydrolyzed during digestion; therefore free folic and folinic acids are released. Although both of these compounds are biologically active, folinic acid is the more potent of the two.

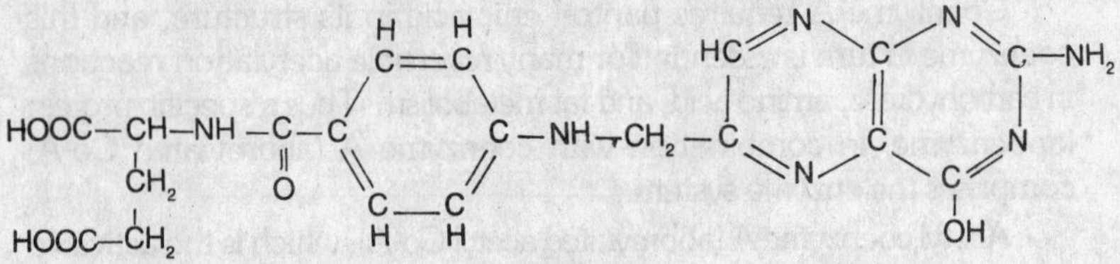

Folic Acid

Fig. 4.9. Structural formula of folic (pteroyglutamic) acid.

Within the body, folic acid is converted into folinic acid by reduction and the addition of a formyl group, a process that is enhanced by the presence of ascorbic acid (vitamin C). Presumably other metabolic reactions then convert folinic (and folic) acid into an active coenzyme, although the metabolic activities of folic and folinic acids in vivo have not been established precisely. Experimental evidence, favours the view that a folinic acid coenzyme functions as a catalyst in the transfer of 1-carbon units, even as pantothenic acid functions in the metabolism of 2-carbon units. When folic or folinic acid is present as a coenzyme linked to various substance specific proteins (apoenzymes), the resulting enzymes are essential to the catalysis of such reactions as the conversion of glycine to serine; the methylation of homocysteine to form methionine; the production of creatine from guanidoacetic acid; and the synthesis of ribose nucleic acids, thymine desoxyriboside, as well as the synthesis of prophyrin compounds, including hemoglobin.

Clinically, the most significant role of folic acid resides in its ability to facilitate the synthesis of nucleoproteins, as these compounds are critical to the maturation of the erythrocytes. Therefore, when an insufficient quantity of nucleoproteins is present, hematopoiesis is arrested at the megaloblast stage of red cell maturation in the bone marrow. The result is a characteristic and abnormal cytologic pattern in the blood which includes macrocytic anemia, leukopenia, thrombocytopenia, and multilobed neutrophils. Thus in humans a deficiency of folic acid in the diet results in a macrocytic anemia which is similar to pernicious anemia, except that the deranged central nervous functions which accompany pernicious anemia are absent. Additional nonspecific symptoms of folic acid deficiency may include gastrointestinal mal-absorption, diarrhoea, gastrointestinal lesions, and glossitis. Folic acid therapy induces reversion of the anemia to a normal blood picture, as well as amelioration of the other symptoms. Folic acid also is useful in the therapy of the malabsorption syndrome (sprue) as well as in certain other types of anemia, eg. the megaloblastid anemias of infancy and pregnancy.

Considerable amounts of folic acid are present in liver, kidney, leafy vegetables, lima beans, nuts, and whole grainier cereals. However cooking and storage of the food may cause up to half of this vitamin to be lost. The intestinal flora also synthesize folic acid, although the quantity derived from this source in unknown.

The daily human requirement for folic acid is known precisely; however, it appears that a dietary intake of between 0.5 and 1.0 mg per day should be adequate to meet the normal physiologic requirements for this vitamin. This statement is based upon the fact that anemias that have been caused by a deficiency of folic acid respond to therapeutic doses of the vitamin administered at levels ranging between 5 and 15 mg per day. Furthermore, a dose of only 200 μg per day of folic acid was found to be the minimum quantity of the vitamin which was effective therapeutically in cases of infantile megaloblastic anemia.

Cyanocobalamin (Vitamin B_{12})

Cyanocobalamin is unique among the vitamins in two respects. It is the only vitamin that contains an essential mineral element (cobalt), and in addition it is the only compound known that contains cobalt which is also essential to life. Vitamin B_{12} is water soluble, and the compound is unstable in acids, alkalies, and light. The molecular weight of cyanocobalamin is approximately 1,450.

Cyanocobalamin is believed to be involved in the metabolic transfer of single carbon intermediates, principally methyl groups. For example, such transmethylation reactions are involved in the synthesis of choline from methionine, of glycine from serine, and of methionine from homocysteine. Cyanocobalamin is also involved in purine metabolism, as well as in the synthesis of pyrimidine bases. This vitamin affects in some manner the metabolism of folic acid, and it may be necessary for production of the folinic acid coenzyme moiety. By virtue of its role in the metabolism of purines and pyrimidines, cyanocobalamin is essential to the synthesis of nucleic acids as well as nucleoproteins, and thus it is essential to the normal production of erythrocytes by the bone marrow. Lastly, vitamin B_{12} is essential to the normal metabolism of neural tissue, although the exact manner in which it functions in this regard is unknown.

The absorption of cyanocobalamin from the gastrointestinal tract, and the role of intrinsic factor in this process, were discussed earlier. It has been postulated that intrinsic factor either liberates cyanocobalamin from a strong protein vitamin complex (in which state it is present in foods) or else it participates in the transport of the vitamin across the intestinal mucosa during absorption. Regardless of the mechanism, however, intrinsic factor is known to be essential for the normal absorption of cyanocobalamin, and in patients suffering from pernicious anemia, the fundamental defect is in the gastric mucosa (atrophy) so that no intrinsic factor is secreted, cyanocobalamin no longer is absorbed from foods, and pernicious anemia develops. In addition to the classic Addisonian blood picture found in this disease, a progressive neurologic degeneration occur, with symptoms, ranging from functional impairment of the peripheral sensory receptors and peripheral neuritis to sclerosis of the posterior and lateral columns of the spinal cord. This disease ultimately terminates in death unless parenteral cyanocobalamin therapy is instituted and sustained, as the fundamental gastric malfunction remains uncorrected. The abnormal blood picture in pernicious anemia

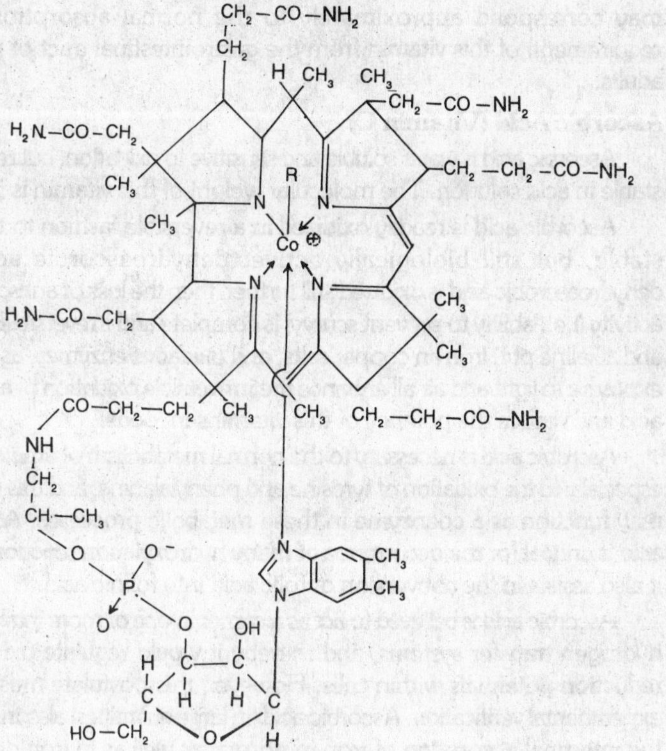

Cobalamin

Fig. 4.10. Structural formula of cabalamin.

is reverted to normal by folic acid therapy. The deranged neural function is not ameliorated by therapy with folic acid. Treatment of pernicious anemia with folic acid alone is hazardous because of the continuing neurologic deterioration which takes place in the absence of vitamin B_{12}.

Cyanocobalamin is found primarily in foods of animal origin; consequently, liver and kidney are excellent foodstuffs insofar as their content of this vitamin are concerned. Muscle meats and fish contain vitamin B_{12} in moderate quantities, whereas milk, vegetables, and cereals are relatively poor sources.

The daily human requirement for cyanocobalamin has not been established definitely; however, as 1 µg of crystalline vitamin per day is sufficient for maintenance of a patient with pernicious anemia, this level may correspond approximately to the normal absorption (and requirement) of this vitamin from the gastrointestinal tract of normal adults.

Ascorbic Acid (Vitamin C)

Ascorbic acid is water soluble and sensitive to oxidation, but relatively stable in acid solution. The molecular weight of this vitamin is 176.

Ascorbic acid is readily oxidized in a reversible fashion to the less stable, but still biologically active, dehydroascorbic acid. (If dehydroascrobic acid is oxidized still further, then the loss of antiscorbutic activity (i.e., ability to prevent scurvy) is complete and irreversible. Heat, and alkaline pH, iron an copper salts, and oxidative enzymes, as well as exposure to light and air all enhance the irreversible oxidation of ascrobic acid and reduce the potency of this vitamins in foods.

Ascrobic acid is necessary to the normal metabolism of amino acids, especially to the oxidation of tyrosine and phenylalanine, and this vitamin may function as a coenzyme in these metabolic processes. Ascrobic acid is critical for the occurrence of many hydroxylation reactions, and it also assists in the conversion of folic acid into folinic acid.

Ascorbic acid is believed to act as a carrier in one or more intracellular hydrogen transfer systems, and thereby it would regulate oxidation-reduction potentials within cells. However, this postulate must await experimental verification. Ascorbic acid (in large quantities) also increases the intestinal absorption of iron in normal as well as in iron-deficient persons, an effect that is due, presumably, to the chemical reducing property of the vitamin.

Vitamin C is essential to the production of the intercellular matrices in collagenous and fibrous connective tissue as well as for the normal synthesis of mucoproteins and collagen in such structures as cartilage, bone, teeth, and skin. Consequently, vitamin C is essential to the normal formation of bones and teeth through its effects upon the synthesis of chondroid and osteoid. Ascorbic acid is also necessary for the production of callus during the physiologic repair of fractures, as well as in the healing of burns and wounds in which the formation of a collagenous intercellular matrix is the first stage in the regenerative process of the damaged tissues. Ascrobic acid is also necessary for the maintenance of a normal physiologic permeability of the capillary walls.

Ascrobic acid is absorbed from the gut by simple diffusion, and as the rate of this process depends upon the concentation gradient across the mucos, intestinal absorption of vitamin C is regulated directly by

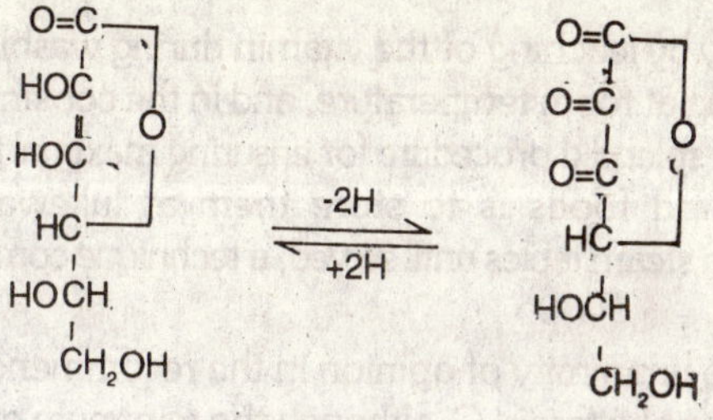

L-Ascorbic Acid L-Dehydroascorbic acid

Fig. 4.11. Structural formula of ascorbic acid (vitamin C) and dehydroascrobic acid.

the quantity of the vitamin ingested. High concentrations of ascrobic acid normally are found in the pituitary gland, adrenals, corpus luteum, and thymus, whereas this vitamin is present at lower levels in such tissues as the brain and viscera. Such a saturation of certain bodily tissues and organs with ascrobic acid results in stores of the vitamin which are sufficient to prevent deficiency symptoms for several months on an inadequate intake of this nutrient.

Man is one of the few animal species which must derive his vitamin C from exogenous sources. Severe and prolonged deficiency of ascrobic acid in the diet results in the classic symptoms of scurvy, in which state the symptoms related to a weakened intercellular substance are most obvious. The patient with scurvy exhibits swollen, tender joints; prolonged would healing; friable and spongy gums; loose teeth; and an overall weakness. There are abundant diffuse hemorrhages, and these hemorrhages appear most frequently under the skin and mucous membranes as well as near the bones and joints. Infantile scurvy is the form of this disease seen most frequently in the United States. The response of the patient to vitamin C therapy is quite rapid.

Among the foods in which vitamin C is found in quantity are citrus fruits (and their juices), raw or slightly cooked vegetables including peppers, cauliflowers, kale, turnip greens, potatoes, tomatoes, and cabbage.

It is important to realise that the ultimate concentration of ascrobic acid in fruits and vegetables that reach the consumer is dependent upon the conditions under which they were grown, stored, and prepared for consumption. For example, the total quantity of sunlight available during the ripening of tomatoes governs their final vitamin C concentration to a large extent. The ascorbic acid content of fruits and vegetables is reduced sharply by leaching of the vitamin during washing, oxidation through standing at room temperature, and in the cooking (or canning) process itself. A splendid procedure for ensuring maximal loss of vitamin C from prepared foods is to store them at lukewarm to warm temperatures on steam tables until served, a technique commonly utilized in cafeterias.

There is no unanimity of opinion in the recommendations for an adult daily intake of vitamin C, although the minimum quantity of this vitamin that is necessary to prevent scurvy in adults is on the order of 10 mg per day. Currently the Food and Nutrition Board suggests a daily allowance of 30 mg per day for infants, 75 mg per day for adult, and still larger quantities for pregnant and lactating women. Actually, a daily allowance of vitamin C ranging between 50 and 100 mg does not actually saturate the tissues, although some tissue retention and excretion of the vitamin in the urine do take place at these levels.

5

FEEDING AND DIGESTION

Animals need food (1) for energy to keep alive and carry on their functions and (2) for building and maintaining their cellular and metabolic machinery.

Most plants use the energy of sunlight and carbon dioxide from the atmosphere to synthesize sugars and, indirectly, all the complicated compounds that constitute a plant. All animals use chemical compounds to supply energy as well as building materials; they must obain these from outside sources, either directly from plants by eating them or by eating other organic material. Ultimately the chemical energy and organic compounds animals need are derived from plants and thus indirectly from the energy of sunlight.

The acquisition and ingestion of food are referred to as *feeding*. Virtually all food, whether of plant or animal origin, consists of highly complex compounds that cannot be incorporated by the organism or used for fuel without being broken down to simpler compounds. In fact, these complex substances usually cannot even be taken into or absorbed by the organism without such breakdown, called *digestion*.

A variety of chemical compounds can be used to provide energy, but in addition animals have specific needs for compounds they cannot synthesize; amino acids, vitamins, certain salts, and so on. Both the need for food to provide energy and the need for specific food components belong to the subject of *nutrition*.

In this chapter we shall discuss food intake, beginning with the mechanism of feeding, proceeding to the processes of digestion that make the food compounds available to the organism, and then briefly specifying some needs with regard to the quality of the food and specific nutritional requirements. Finally, we shall discuss how plants may defend themselves by producing substances that make them *inedible* to many animals.

Feeding

Food is obtained by a diversity of mechanical means, and these determine the nature of the food a given animal can obtain and utilize. Major feeding mechanisms and examples of animals that use the different means to obtain food are listed in table 5.1.

Table 5.1. Methods used by various animals to obtain food, classified by character of the food

Type of food	*Feeding method*	*Animals using method*
Small particles	Formation of digestive vacuoles	Amebas, radiolarians
	Use of cilia	Ciliates, sponges, bivalves
	Formation of mucus traps	tadpoles Gastropods, tunicates
	Use of tentacles	Sea cucumbers
	Use of setae, filtering	Small crustaceans (e.g., daphnia), herring, baleen whales, flamingos, petrels
Large particles or masses	Ingestion of inactive masses	Detritus feeders, earthworms
	Scraping, chewing, boring	Sea urchins, snails, insects, vertebrates, Coelenterates, fish,
	Capture and swallowing of prey	snakes, birds, bats
Fluids or soft tissues	Sucking plant sap, nectar	Aphids, bees, humming birds
	Ingestion of blood	Leeches, ticks, insects, vampire
	Sucking of milk or milklike secretion External digestion	bats Young mammals, young birds Spiders
	Uptake through body surface	Parasites, tapeworms
Dissolved organic material	Uptake from dilute solutions	Aquatic invertebrates (probably of secondary importance)
Symbiotic supply of nutrients	Action of intracellular symbiotic algae	Paramecia, sponges, corals, hydras, flatworms, clams.

Small Particles

Microscopic algae and bacteria can be taken directly into a cell (e.g., into the digestive vacuole of an amoeba). A variety of organisms, even some quite large ones, catch microscopic organisms with the aid of cilia. A few animals, notably tunicates and some gastropods, use a sheet of mucus that traps fine suspended particles; the muscus sheet is then ingested and the food organisms digested.

Tentacular feeding is used by sea cucumbers, which live burrowed in the mud with their tentacles extended above the mud surface. The tentacles entangle fine particles and at intervals are retracted into the animal's mouth, where digestible material is removed.

Various filtering devices are common in a wide variety of small planktonic crustaceans such as copepods and in amphipods, sponges, and bivalves. Quite a few vertebrates, ranging from fish to mammals, also use filtering to obtain food, but their filters usually strain off organisms larger than the microscopic food commonly utilized by invertebrates.

Many pelagic fish are plankton eaters. Herring and mackerel have gill rakers structured so that they function as a sieve that catches plankton, predominantly small crustaceans. Some of the largest sharks, the basking sharks and whale sharks, feed exclusively on plankton, which is strained from the water that enters the mouth and flows over the gills. It has been estimated that a large basking shark, in 1 hour, strains the plankton from 2000 tons of water (Matthews and Parker 1950).

A few marine birds also feed on plankton. One of the petrels, the whale bird or prion, has a series of lamellae extending along the edge of the upper break. It feeds by staining crustaceans from the surface water in a way similar to what baleen whales do on a much larger scale. The flamingo, also, is a plankton eater, using its beak to strain small organisms from the water.

The baleen whale is entirely specialized for feeding on plankton. Its filtering apparatus consists of a series of horny plates attached to the upper jaw and hanging down on both sides. As the whale swims, water flows over and between the plates, and plankton is caught in the hairlike edges of the plates. The plankton-eating whales include the largest living animal, the blue whale, which may reach a weight of over 100 tons. It is interesting that both the largest whales and the largest sharks are plankton feeders rather than carnivores and thus avoid any extra links in the food chain.

Large Food Masses

Methods for dealing with large particles or masses of food include a tremendous variety of mechanisms and structures. Some animals ingest the medium in which they live and digest the organic material it contains, whether small living organisms or dead organic matter. Such animals frequently pass large amounts of relatively inert material through their digestive tracts and utilize whatever organic material it contains.

Many animals use mechanical methods such as chewing and scraping to obtain their food, which is frequently of plant origin. The range of these animals includes such a variety that to list them would be pointless. We need only think of the large numbers of insects and other invertebrates, as well as the many vertebrates, that are herbivores. Some carnivores are also included in this group. They capture prey and tear, shred, or chew it before swallowing.

Some carnivores capture and swallow their prey whole. Relatively few invertebrates feed this way, but many vertebrates do, including representatives of all classes: fish, amphibians, reptiles, birds, and mammals.

The mouth and oral cavity may have various kinds of jaws or teeth that serve in the mechanical breakdown of food into smaller particles (*mastication*), making it more accessible to the digestive enzymes.

In some animals the mechanical breakdown of food occurs in a special compartment with thick muscular walls, the gizzard, where a grinding action takes place. An earthworm, for example, has a gizzard for such grinding, called *trituration*. A bird, which has no teeth, also has a gizzard and often swallows small stones that help in the trituration of food. This is particularly important in a seed-eating bird, for its food is hard and would, in the absence of grinding, take a long time to soften in the digestive tract. The gizzard may be preceded by a simple storage chamber, a *crop,* in whch the moisturing of softening begins, but without any important grinding action.

At the beginning of the digestive tract there is usually a special larger compartment, a *stomach,* which serves as a storage organ and in which digestive action also takes place. In all vertebrates, with a few exceptions, the stomach has a major role in the digestion of proteins, and the most important enzyme in the stomach is the protein-digesting pepsin, which acts best in highly acidic solution. The stomach of a vertebrate also has special cells that secrete acid (hydrochloric acid).

The stomach is followed by a hollow tube, the *intestine,* in which further digestion takes place and which is particularly important for the absorption of the products of digestion. These products are mainly simply sugars, glycerol and fatty acids, and amino acids. The length of the intestine varies a great deal, and it can usually be subdivided into several parts. The posterior portion is of little importance in digestion and serves mainly in absorption. Periodically the remaining material is expelled as feces through the anus.

Liquids

Many liquids are of value as foods. Animals feeding on liquids are often highly specialized and adapted to their food sources, and in some cases there is a mutual benefit to the feeding animal and the provider. For example, insects obtain nectar from flowers and in turn provide pollination by moving from flower to flower. In many cases, fluid feeders are parasitic; this is true of aphids, which suck plant sap, and of animals such as leeches, mosquitoes, and ticks that suck the blood of other animals.

All mammals begin their lives as fluid feeders when for a period they live exclusively on the milk produced by their mothers. It is less well known that some birds feed their young on a milklike secretion. In the pigeon this secretion is formed in the crop, it is known as *crop milk*, and is regurgitated to feed the nestlings. Curiously, the formation of crop milk is stimulated by the same hormone, prolactin, that in mammals stimulates the mammary glands to produce milk. The biological advantage of feeding the young this way is that it allows the parents to be opportunistic in their own feeding and frees them from the need to find special kinds of food (e.g., insects) for the young. It also buffers the young against fluctuations and shortages in the food supply.

Table 5.2. Composition of "milk" from Pigeons and Emperor Penguins, as percentage of total dry matter. The composition of Rabbit Milk is given for comparision (Prevost and Vilter 1962).

Constituent	*Pigeon*	*Penguin*	*Rabbit*
Protein	57.4	59.3	50.6
Lipid	34.2	28.3	34.3
Carbohydrate	0	7.8	6.4
Ash	6.5	4.6	8.4
Total	98.1	100.0	99.7

This advantage is particularly striking in the emperor penguin, which can feed its young on "milk" secreted by the esophagus (Prevost and Vilter 1962). The emperor penguin reproduces in mid-winter. When the Antarctic winter approaches in March, the birds leave the sea and walk over the ice to the rookeries, where the female leaves a single egg with her mate to be incubated for over 2 months while he stands on the ice in the cold and dark of winter. In the meantime, the female walks back to the sea to feed and returns later to the fating male to feed the young. If she is delayed, however, the still-fating male begins to feed "milk" to the young, which is kept alive and even may gain in weight. In protein and fat content pigeon and penguin milk (table 5.2) is similar to mammalian milk (although the milk of many mammals has a much higher carbohydrate content).

Spider provide a special case of fluid feeding. A spider's prey is often as large or larger than its own body, usually covered with hard chitin, and not easily torn apart. Spiders have solved the problem of obtaining food from their insect prey by piercing it with their hollow jaws and pumping digestive juices into its body. These liquefy the tissues and the spider then sucks the prey empty.

Some parasites (e.g., tepeworms) live in highly nutritive media and have completely lost their digetive tracts. Because their hosts provide the digestive system, the parasites lack both digestive tract and the digestive enzymes that are otherwise necessary before food can be absorbed, and all their nutrients are taken up directly through the body surface.

Dissolved Organic Material

Whether organic matter dissolved in the water can be utilized by aquatic animals is a question that remains controversial. It was suggested early in this century by the German biologist Putter that dissolved organic matter in the sea, although present in a highly dilute state, could be used directly by aquatic animals. Putter even suggested that such dissolved organic matter was more important than plankton, for plankton organisms are scattered so far apart in the water that often their capture is not feasible.

Whether the uptake of dissolved organic material is important has been difficult to study experimentally because of the extremely dilute solutions to be examined. This changed with the introduction of radioisotopes to label organic compounds, and it become possible to demonstrate clearly that many aquatic invertebrates indeed can absorb both glucose and amino acids from exceedingly dilute solutions (Stephens 1962, 1964).

In such experiments it is not difficult to show that some uptake has taken place. This is beyond doubt; the difficulty is to show that the amount of a compound taken up exceeds the amount given off in the same period of time. The uptake must exceed the loss to be of net value to the animal. A comparison of simultaneous uptake and release of free amino acids in sea water has been made by Johanes et. al. (1969). In experiments with a single species, a turbellarian, they showed that the release of free amino acids to the water exceeded by about fourfold the simultaneous uptake of 14C-labelled amino acid. Another question that remains open is whether bacteria play a role in the uptake. The bacteria could absorb organic solutes and later be ingested by the animal. Bacteria can effectively remove glucose or acetate from solutions containing only 1 to 10 μg per liter; for common fresh-water algae this is impossible (Wright and Hobbie 1966).

Although present evidence is uncertain, there is no strong indication that any aquatic invertebrate depends primarily on the uptake of organic materials from the highly dilute solution in which it lives. The elaborate mechanical feeding mechanisms animals possess also speak against any major role of dilute dissolved organic substances.

Symbiotic Supply of Nutrients

Some invertebrate animals contain within their cells symbiotic algae. These algae are of importance in supplying nutrients to the host, providing a shorter pathway between photosynthesis and animal than ispossible in any other way. The algae, in turn, utilize ammonia given off by the animal host for synthesis of protein. Such symbiotic relationship between animal and plant occur among protozoans, sponges, corals, hydras, flatworms, and clams.

In this relationship the animal obtains organic compounds (primarily carbohydrates) from the living cells of the plant, and not from the destruction and digestion of the plant cells. In this respect the symbiotic supply of nutrients differs in principle from all the feeding methods discussed above, which always involve destruction of part or the whole of the organism that provides the food.

The importance of the symbiosis has been demonstrated in hydras by comparing animals that carry large (green hydras) with animals that do not. The presence of symbiotic aglae greatly prolongs the survival of hydras under conditions of starvation, and if a limited amount of food is given, the green hydras grow faster. The oxygen consumption of the green hydras is lower in light than in the dark and lower than in animals without symbiotic algae (Muscatine and Lenhoff 1965; Pardy and Dieckmann 1975).

Two major types of algae occur in the host animals; green algae (Zoochlorellae) or brown algae (Zooxanthellae). If we include in the term carbohydrate not only sugars, but also derivatives such as sugar alcohols and glycerol, we can say that the bulk of material transferred is of carbohydrate nature. The soluble carbohydrate give off to the host animal is invariably different from the major intracellular carbohydrate of the algae. Zoochlorellae, for example, synthesize and retain in their cells sucrose, but they excrete maltose or glucose to the host animal.

The amount of organic material obtained in a symbiotic relationship is substantial. The algae in a green hydras release as much as 45 to 50% of the carbon fixed in photosynthesis to their host. Isolated algae release carbohydrate, mainly maltose, to the medium, and the amount depends on the pH of the solution. At pH 4.5 these algae release as much as 85% of their photosyntheiszed carbon, but in alkaline medium the release declines to a few percent. This opens the possibility that the host animal may be able to control the amount of carbohydrate released by the algae by adjusting its intracellular pH.

When the Zooxanthellae are isolated from their host and incubated in sea water, they release some carbohydrate, but the amount increases if homogenized host tissue is added. In the case of Zooxanthellae from *Tridacna* (the giant clam) a 16-fold increase in carbohydrate release was obtained in the presence of homogenate of the host.

Feeding Mechanisms

The intake of food is called *feeding*. The method of feeding appears to be determined largely by the habitat and the food available to a particular organisms. In most cases the form and function of the digestive tract also depends upon the types of food and feeding mechanism. Accroding to *C.M. Yonge* (1928), the feeding mechanisms of animals belong to the following three categories:

(a) mechanisms for dealing with small particles.
(b) mechanisms for dealing with large particles (or masses).
(c) mechanisms for taking in fluids (or liquid food).

(a) Mechanisms for Dealing with Small Particles

Minute particles in the form of detritus, diatoms, flagellates, and bacteria, etc. are used as food by a large number of sessile or sluggish aquatic animals. They collect the food particles by any of the following methods:

1. Pseudopodial feeding

Feeding by the use of pseudopodia is a characteristic of rhizopods. In Amoeba, the food is captured by the pseudopodia which extend around the food particle and engulf it together with some amount of water. Particles are caught by the reticulate pseudopodia of radiolarians and foraminiferans and then brought to the main part of the body for digestion.

2. Mucoid feeding

Many animals secrete mucous sheets which filter out particles, which are then eaten by the animals. The tube-living polychaete, *Chaetopterus* and the echiuroid worm, *Urechis*, secrete a mucous net, pump water through it by the activity of posterior appendages and swallow the trapped food. An interesting example of mucoid feeding is provided by a sessile gastropod, *Vermetus*. Some protochordates also have a mucous-ciliary type of feeding, in which food particles entangled in mucous are ingested by the animal.

3. Ciliary and flagellar feeding

This mechanism is commonly used by a large number of sedentary animals like ciliate protozoans, sponges, many tubiculous annelids, lamellibranches, and many tunicates and cephalochordates, etc. In most of such animals the beating of cilia or flagella continuously produces the respiratory-cum-feeding currents of water and helps in directing the food particles towards the mouth.

4. Tentacular and setose feeding

Some animals depend largely upon their freely movable ciliated tentacles for collecting food particles. In certain coelenterates the nematocytes on the tentacle help to paralyse the prey which is often attached to the tentacles by a sticky secretion. In some crustaceans the setae (hair like chitinous structures) on various appendages are used to create the currents of water. Copepods produce currents by antennal vibrations.

(b) Mechanisms for Dealing with Large Particles

Some burrowing animals swallow *inactive food* such as mud. Certain holothurians use their tentacles to force mud into their digestive tracts. Annelids such *Lumbricus* and *Arenicola* use their reversible pharynx for this purpose. The organic content of the mud is utilized and the residue ejected as faeces.

Some animals have structures *for scraping and boring* food stuffs. Examples are the Aristotle's lantern of echinoid echinoderms, radular apparatus of some molluscs, boring valves of the shipworm *Teredo*, and heavy mouth parts of termites.

In certain animals specialized organs serve to seize the prey. They may be for seizing only as the protruded proboscis of *Didinium*, the jawed pharynx of many polychaetes, and the teeth of a shark etc. The other category of such organs are the *structures for seizing and masticating*. Examples are the jaws and radulas of molluscs, the jaws and other mouth parts of crustaceans, insects, arachnids and myriapods, and the toothed jaws of vertebrates. In the jawed vertebrates the anterior teeth often bite or gnaw, posterior teeth masticate. However, there is much variation ranging from amphibians and billed birds which chew very litle to the ungulates with heavy chewing molars.

A third category of the prey-feeding mechanism employs the *structures for seizing followed by external digestion.* An interesting example of this is furnished by the sarcodinian, *Vampyrella,* which digests a hole through the walls of the algae on which it feeds. Some carnivorous gastropods, cephalopods, spiders and echinoderms seize their prey also carry out some external digestion.

(c) Mechanisms for Taking Fluid Food

Some parasitic animals feed on fluids and soft tissues. The structures for *piercing and sucking* are found in hookworms, leeches, ticks, mites and cyclostomes, etc. The structures for *sucking* only are met with in suctorian protozoans, nematodes, adult butterflies and suckling mammals. The endoparasites such as porozoans, cestodes and *Sacculina* etc., depend on the absorption of the nutrient materials from the internal body surface of their hosts.

DIGESTION

The Digestive System as a Wrecking Company

The ultimate role of the digestive system is the chemical breakdown of food macromolecules. Digestion yields molecular components that (1) are small enough to be soluble in water and to be absorbed from the divestive tract and (2) are in the form most useful to the tissues as a source of energy and raw materials and thus for growth and repair. All chemical phases of digestion are achieved by hydrolysis of the bonds that join the component molecules of foodstuffs. The digestive system is therefore not unlike a wrecking company that deals in recycled materials suitable for construction, repair, and fuel. What is "wrecked" are the macromolecules of the diet. The "bricks" and "boards" are the amino acids, sugars, and other chemical components of the diet. The "customers", of course, are the body's cells and tissues.

Protozoa

Although protozoans (Greek for "primitive animal") are protists, not animals, they are closely allied to animals and were included with them in order classifications. Their digestion is usually intracellular, within food vacuoles, the food usually being ingested by phagocytosis. Lysosomes secrete digestive enzymes into the vacuoles. As in most animals, digestion in the vacuole begins with an acid phase, followed by an alkaline phase. The acid, which may reach a high concentration, kills many bacteria and other live food, dissolves calcium salts, and provides

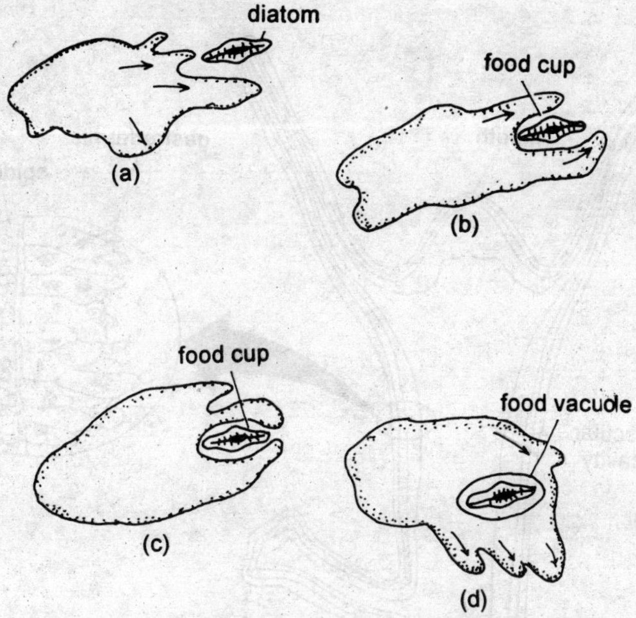

Fig. 5.1. Phagocytosis in an amoeba. (a-d) Stages in the ingestion of a small protist (a) by an amoeba, showing food cup and food vacuole formation.

an optimum pH for the functioning of some enzymes. In many protozoans, the final phase of digestion is completed close to neutrality. As the food is reduced to molecular size, it is absorbed into the cytoplasm. Finally, indigestible materials are expelled by exocytosis as the vacuolar membrane fuses with the plasma membrane in a process essentially the reverse of food vacuole formation.

Lower Invertebrates

These organisms have no specialized digestive organs. In many, digestion takes place within individual cells.

Sponges and Coelenterates

Sponges have no digestive cavity. Instead, specialized cells engulf food particles, enclosing them in food vacuoles, much as in protozoans. But before digestion is complete, the material is passed along to other cells called *amoebocytes,* which complete the process and distribute the products of digestion to other cells.

Coelenterates, or cnidarians, possess a *gastrovascular cavity,* which serves for both digestion and circulation. It has but one opening, the

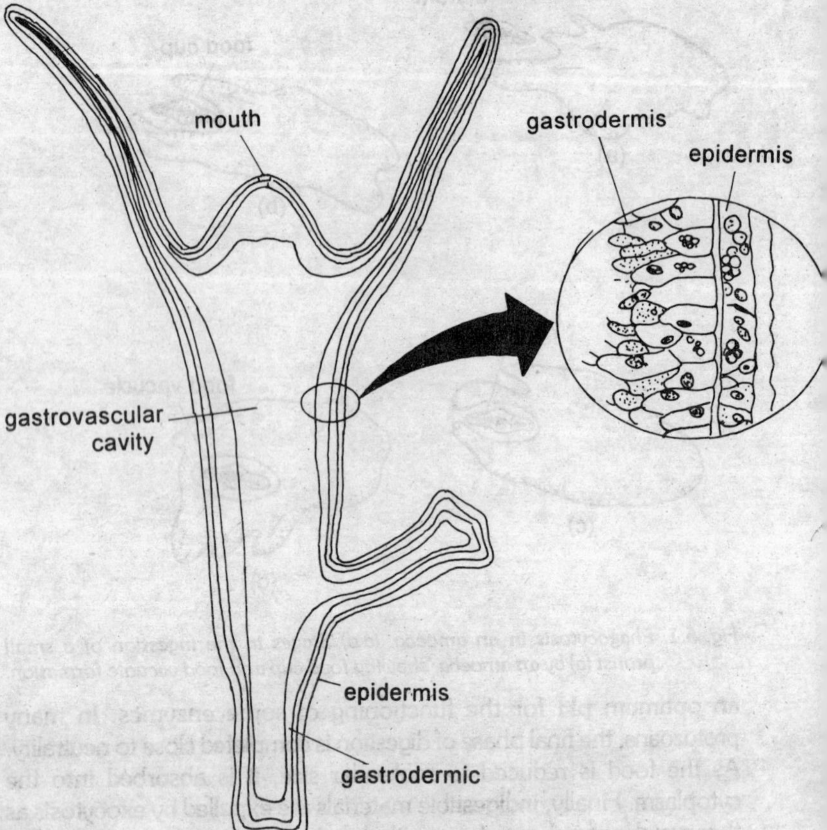

Fig. 5.2. The gastrovascular cavity of a cnidarian (coelenterate), Hydra, shown in longitudinal section.

mouth, into which all food enters and out of which all residue is expelled. The vascular inefficiency of this cavity is unimportant because, as in sponges, most digestion in intracellular. However, before the intracellular phase, the gastrodermis, the cell layer lining the cavity, secretes digestive enzymes into the gastrovascullar cavity where gross degradation occurs. The digestion of the resulting small particles is then completed intracellularly in food vacuoles. In an evolutionary sense, coelentrates are very little advanced over sponges in their mode of digestion.

Flatworms

Platyhelminthes, or flatworms, exhibit an interesting array of approaches to nutrition. Among free-living species, or nonparasites,

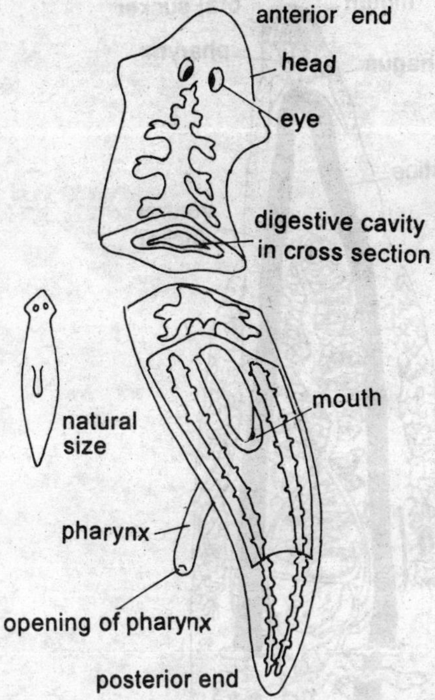

Fig. 5.3. A planarian. Sketches show the gross structure of the digestive system.

some have a mouth but no digestive cavity, the food particles being digested by the cells of a loosely organized tissue near the mouth. Other free-living forms have a simple digestive cavity, essentially a coelenterate like gastrovascular cavity, which usually has multiple branches. Among parasitic flatworms, the flukes have a gut with only two branches. Its simplicity is probably related to the fact that these parasites have an unvarying diet, the nature of which depends on which organs of their host they parasitize. Finally, tapeworms have no digestive organs at all, depending entirely on products of digestion furnished by their host—which is why adult tapeworms never lie anywhere but in the digestive tracts of other animals. The outer surfaces of tapeworms, through which they absorb the products of digestion, possess fine projections called microvilli similar to those lining the absorptive surface of the mammalian small intestine.

Digestion in Higher Animals

Beginning with the nemerteans and nematodes, the digestive tracts of all higher animals are one-way traffic routes, an important evolutionary

Fig. 5.4. An adult trematode (the Chinese liver fluke).

breakthrough that greatly increased nutritional efficiency. In primitive systems, the mouth is used to discharge undigested matter. But with the evolution of a mouth and an anus, one specialized for receiving food and the other for discharging residue, new possibilities for efficiency emerged. Various regions of the digestive tract became specialized in a kind of reverse assembly line that performed food-processing tasks in a far more orderly line that performed food-processing tasks in a far more orderly and efficient way than was otherwise possible.

The Saga of a Sandwich

Merely serving a sandwich to a hungry person sets in operation a number of preparatory actions within the digestive system, the most obvious of which is that the mouth "waters," or produces an unusual amount of saliva. Taking a bite of the sandwich immediately involves other senses and prompts some decisions by the diner regarding the food. The mouth is well suited to monitor what enters the body because of its rich supply of touch, temperature, and chemical receptors that permit decisions over whether the sandwich is fit for consumption. Is it too hot? Is it toasted? Does it taste good? The sense of smell also plays an important role in the chọice of food.

Once a bite of the sandwich is accepted into the mouth, it may be moistened if it is too dry or chewed into small pieces if it too large to be easily swallowed. Chewing, or mastication, not only makes food small enough for swallowing but actually constitutes the first stage of the digestive process, mechanically "opening the gates" to a large array of enzymes. Without chewing, digestion of some meals would take many hours longer.

Teeth accomplish the mechanical breakdown of foods and help mix food with saliva, preparing it for the chemical stages of digestion to follow. In the human, two sets of teeth develop, the 20 deciduous, or "milk," teeth of childhood and the 32 permanent teeth that replace them. The permanent teeth of mammals are of four types, eight type adapted to a different function. Humans have eight incisors, used for biting; four canines, used for tearing, and eight premolars, used for

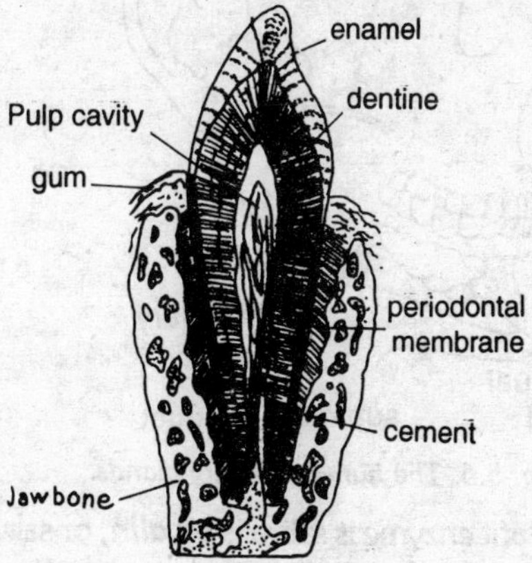

Fig. 5.5. Diagrammatic section through a typical canine tooth.

bitting; four canines, used for tearing, and eight premolars, or bicuspids, and 12 molars, or tricuspids, used for grinding. The last four tricuspids to erupt in humans are called the "wisdom teeth."

In teeth, as in other anatomical parts, form is related to function. For example, rodent incisors, which are specialized for gnawing, are continually worn away and never cease growing. Teeth of herbivores such as the elephant are broad and flat—highly suited to grinding plant tissue with thick cell walls. In several mammals, some teeth are elongated as tusks, highly efficient in offense and defense. Elephant tusks are incisors, whereas the tusks of boars and the now extinct sabertoothed tiger are canines.

Each tooth is composed of a *crown* and *root*. Most of the tooth is composed of bonelike *dentin*, but the crown is covered with *enamel*, the hardest material in the body. A good supply of dietary calcium is needed in early youth for optimum tooth development.

Saliva

The mouth is served by three pairs of salivary glands. One pair is at the angle of the jaw, another pair just below the lower jaw at the base of the tongue, and a third pair below the tongue. The pair at the angle of the jaw produce a watery secretion rich in enzymes. The others produce a mucus as well. Different foods elicit different types of salivary secretion.

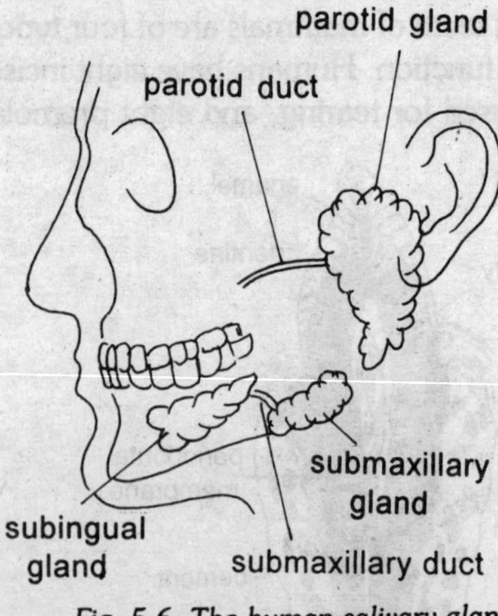

Fig. 5.6. The human salivary glands.

The most important enzyme is saliva is *ptyalin*, or salivary amylase, which partially digests starches before they are swallowed. A more important function of saliva is to moisten and lubricate food materials.

This not only makes them easier to swallow but dissolves the soluble ones and suspends others, thereby readying them for action by the digestive juices of the stomach and intestine.

The Oesophagus

Once the bite taken out of a sandwich has been moistened and chewed, it is moved by the tongue to the pharynx at the rear of the oral cavity. Here its presence induces a swallowing reflex. By peristalsis, the food—now referred to as a food bolus—is passed down the oesophagus to the stomach.

Peristalsis is characterized by a relaxation-contraction sequence of the muscles in an organ's wall that passes in waves down the length of the organ in a "milking" action that moves the contents along. Both the circular and longitudinal muscles are normally in a state of partial contraction called tonus (1). Relaxation of the circular musculature results in a widening of the lumen, the cavity within (2). This is followed immediately by a contraction of adjacent circular fibers, pushing the contents into the relaxed portion (3). The entire process is repeated down the length of the organ (4, 5). When properly stimulated, organs capable of peristalsis can usually conduct it in the reverse direction as well, as in vomiting.

The Stomach

The food next enters the stomach, a curved, muscular, saclike organ, the internal surface of which is thrown into coarse folds called *rugae,* which allow the stomach to expand and thereby accommodate food. The inner surface is marked by numerous, small gastric pits at the bases of which are millions of microscopic *gastric glands.* The mammalian stomach serves to some degree as a storage organ, an important function, considering the time-consuming nature of digestion. Very few nutrients are absorbed here. It also functions to dissolve or suspend food particles. The warmth and churning action of the stomach serve to reduce to fine droplets, or emulsify, lipids and lipid-containing foods such as butter, cheese, oils, or bits of fat, all of which might be found in a meat sandwich. The most important digestion that occurs in the stomach is that which is accomplished through the action of *pepsin,* the enzyme that initiates the breakdown of the proteins found in food. The sole role of pepsin is to hydrolyze peptide bonds between certain amino acids. This converts large proteins bonds between certain amino acids. This converts large proteins to somewhat smaller molecules called peptones. One early insight into the nature of this process was provided by William Beaumont.

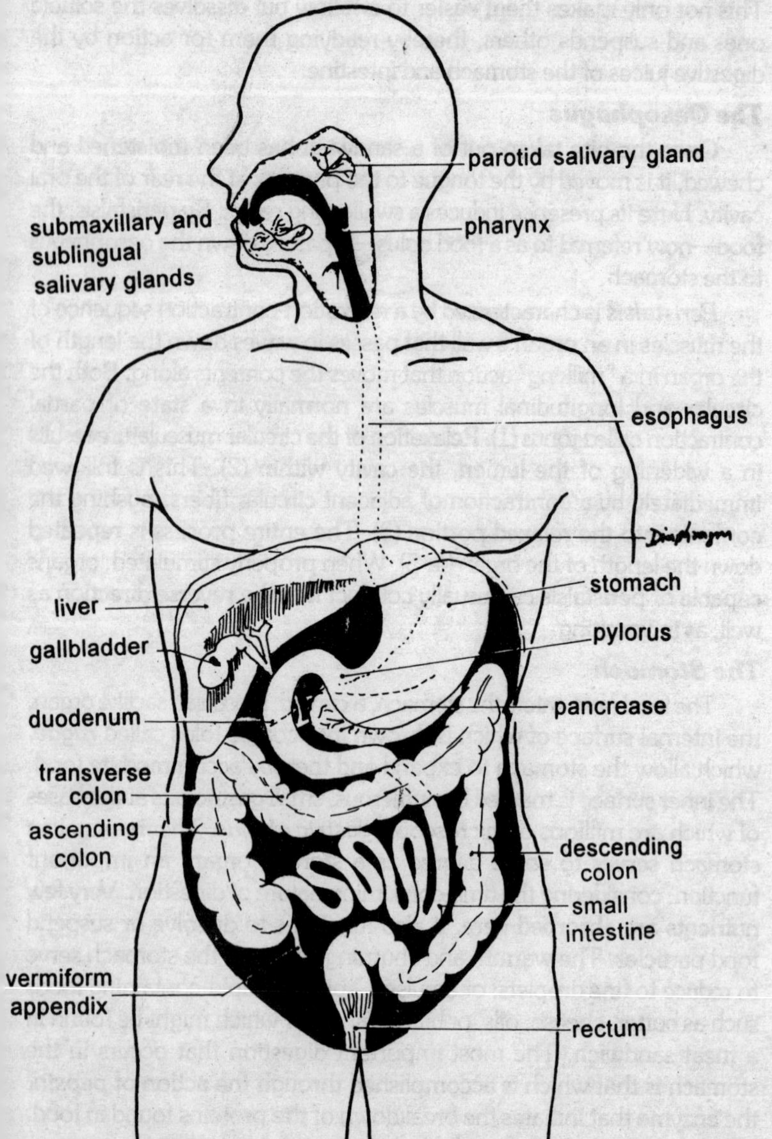

Fig. 5.7. Diagram of the human body showing the parts of the digestive system.

The secretory epithelium lining the stomach includes three cell types: *chief cells,* which secrete *pepsinogen,* an inactive precursor of the enzyme pepsin; *parietal cells,* which secrete hydrochloric acid (HCl); and *clear columnar cells,* which secrete *mucus,* a ropy proteinaceous material that in normal amounts protects the gastric epitehlium from the ulcers or open sores that can result from its being digested by its own secretions.

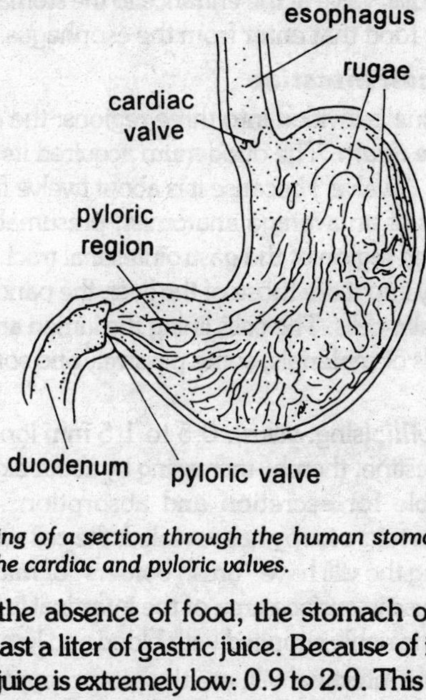

Fig. 5.8. Drawing of a section through the human stomach to show the rugae and the cardiac and pyloric valves.

Even in the absence of food, the stomach of the average adult contains at least a liter of gastric juice. Because of its HCl content, the pH of gastric juice is extremely low: 0.9 to 2.0. This creates a formidable barrier to most bacteria and other live cells, quickly killing them as they enter. The acid has other roles as well. It hydrolyzes some foods. It coagulates proteins, thereby preparing them chemically for hydrolysis by various enzymes. It also provides the optimum pH for the action of the powerful protease, pepsin. The inactive form of pepsin, pepsinogen, is activated by HCl.

Just as the sight and smell of food promotes of flow of saliva, the taste of the first bite of a sandwich prompts increased flow of gastric juice into the stomach. The arrival in the stomach of a good bolus resulting from that first bite also promotes an increase in gastric secretion. This phase of gastric secretion has been shown to be due in part to a direct, triggering effect of dissolved food molecules on the gastric epithelium. A similar secretory response is elicited in the small intestine by the arrival of food. Gastric secretion involves a combination of nervous and hormonal controls.

The result of all this gastric activity in that the food is reduced to a thick broth called *chyme.* At this stage, the *pylorus*—the narrow, muscular, funnel-shaped posterior end of the stomach—squirts the chyme through the pyloric valve into the small intestine. The pyloric valve normally prevents material that has entered the intestine from reentering the stomach. The cardiac valve at the entrance to the stomach performs a similar function for food that enter from the esophagus.

Tissues of the Small Intestine

The small intestine is divided into three regions: the *duodenum,* the *jejunum,* and the *ileum.* The duodenum acquired its name (from the Latin *duodecim,* "twelve") because it is about twelve finger widths in length (the fingers of an average anatomist, presumably). It is the most highly specialized region of the gastrointestinal tract. Among the secretions that empty into it are those of the liver, the pancreas, and its own epithelial glandular cells. The cells lining the lumen are capable of absorbing many kinds of materials, some passively and some by active transport.

Fine, fingerlike *villi,* (sing. villus), 0.5 to 1.5 mm long, cover the lining of the small intestine, thereby increasing by about eight times the surface area available for secretion and absorption. Even more important, as electron microscopy has revealed, the cell surfaces of the epithelial cells forming the villi have "brush borders" of *microvilli.* This brings the total increase in surface area of the intestinal lining to about 150-fold. This tissue is richly supplied with blood and lymph vessels.

At the base of the intestinal villi are the pits or crypts into which empty the thick mucus secretion of Brunner's glands, providing a coating to the cell surfaces that is regarded as critical in the protection against duodenal ulcer. The epithelial lining of the intestine is continually renewed, being completely replaced every three to six days. It is estimated that 250 g (8.8 ounces) of tissue are shed into the intestine of the average adult each day.

Small Intestine Secretions

As soon as the chyme enters the small intestine, it mixes with the secretions found there, and the pace of digestion quickens. The secretions of the small intestine are of three types.

1. *Bile* is produced by the liver, stored in the gallbladder, and released when fatty foods enter the duodenum. It contains *bile salts,*

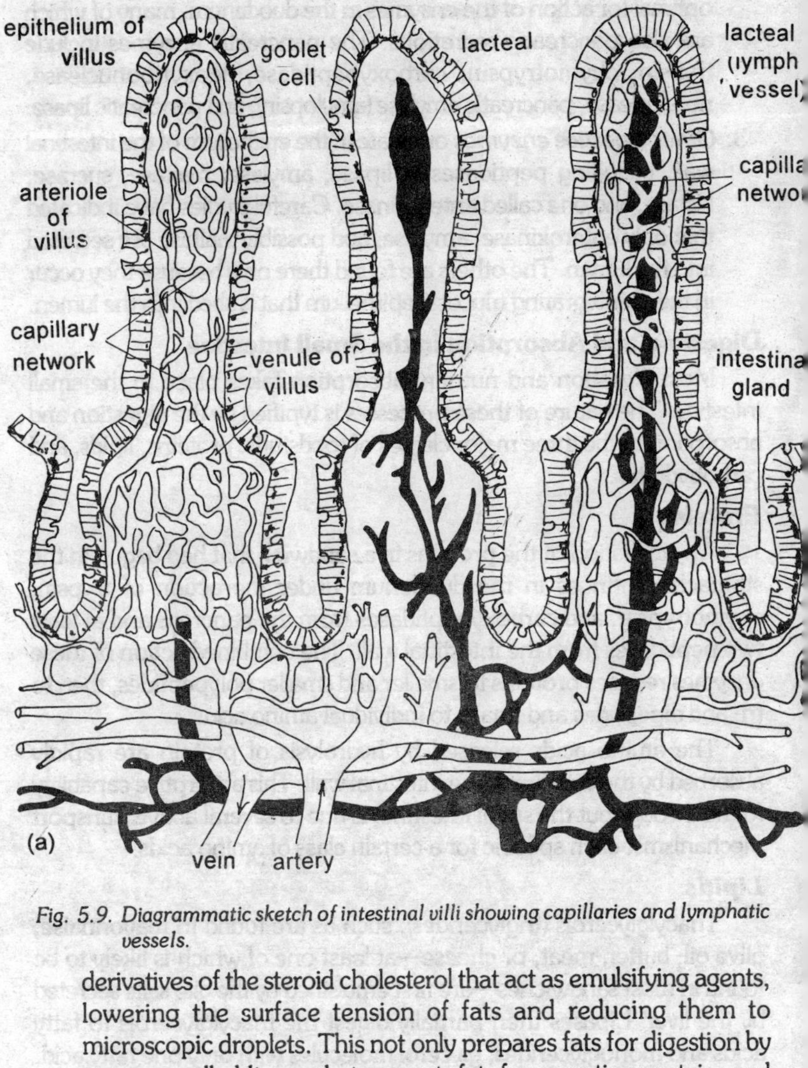

Fig. 5.9. Diagrammatic sketch of intestinal villi showing capillaries and lymphatic vessels.

derivatives of the steroid cholesterol that act as emulsifying agents, lowering the surface tension of fats and reducing them to microscopic droplets. This not only prepares fats for digestion by enzymes called *lipases* but prevents fats from coating proteins and other foods, thereby shielding them from the action of their own specific enzymes. Later, bile salts combine chemically with the products of fat digestion, thus facilitating their absorption.

2. *Pancreatic juice* is alkaline (pH 7.2–8.2) and neutralizes the chyme as it leaves the acid stomach, abruptly halting the action of pepsin, which can only function in an acid pH. The alkaline pH is also

optimal for action of the enzymes in the duodenum, many of which are also pancreatic secretions. The pancreatic enzymes include trypsin, chymotrypsin, carboxypeptidase, deoxyribonuclease, ribonuclease, pancreatic amylase (amylopsin), and pancreatic lipase.

3. *Other digestive enzymes* originate in the epithelium of the intestinal wall, including peptidases, a lipase, amylase, maltase, sucrase, lactase, and one called enterokinase. Careful studies have indicated that only enterokinase, amylase, and possibly maltase are secreted into the lumen. The others are found there only because they occur in the disintegrating mucosal epithelium that is shed into the lumen.

Digestion and Absorption in the Small Intestine

Most digestion and nutrient absorption takes place in the small intestine. The nature of these processes is typified by the digestion and absorption of the three major classes of foodstuffs: proteins, lipids, and carbohydrates.

Proteins

Any digestion of the proteins in a sandwich that had begun in the stomach continues in the duodenum under the action of trypsin, chymotrypsin, and carboxypeptidases from the pancreas and at least two peptidases from the intestinal wall. The combined action of these enzymes reduces proteins to smaller and smaller polypeptides, then to tri- and dipeptides and finally to individual amino acids.

The amino acids released by hydrolysis of protein are rapidly absorbed by the cells lining the intestinal wall. This absorptive capability found throughout the small intestine, is due to several active transport mechanisms, each specific for a certain class of amino acids.

Lipids

Triacylglycerols (triglycerides), such as are found in mayonnaise, olive oil, butter, meat, or cheese—at least one of which is likely to be found in most sandwiches—are first emulsified by the bile salts secreted by the liver. Lipases then partially digest the triacylglycerols to fatty acids and monoglycerides, glycerol molecules with only one fatty acid. These two products and the bile salts form a complex that becomes highly emulsified. The fatty acids and monoglycerides then enter the cells, but the bile salts remain in the intestinal lumen. Lipid absorption occurs prinicipally in the duodenum and upper jejunum. The mechanism of absorption is apparently passive, a matter of diffusion rather than of active transport.

Within the epithelial cells, the newly absorbed products of lipid digestion are immediately resynthesized into triacylglycerols and phosphoglycerides. This part of the process requires energy and expends ATP. The newly synthesized lipid is coated with albumin, a low molecular weight protein, and then secreted, as very fine droplets called chylomicrons, into the lymphatic vessels, giving the lymph a milky appearance after a meal. Hence, the small lymphatic vessels are called lacteals (from Latin *lacteus,* "milk"). Short-chain fatty acids, however, make their way from the intestinal lumen, through the epithelial cells, through the lymph, and into the bloodstream without being resynthesized into lipid. A brief account of the forms in which lipids are transported in the blood is found in Panel.

Carbohydrates

Any sandwich contains an abundance of starch, the main polysaccharide in bread. If the sandwich contains meat, then another polysaccharide, glycogen, may also be present. If the bite of the sandwich is followed by a swallow of milk, then lactose, a disaccharide, is included. If the drink is sweetened, another disaccharide, sucrose, may be ingested. Appropriate enzymes digest these carbohydrates to monosaccharides such as glucose, fructose, and galactose. Salivary amylase (ptyalin), pancreatic amylase (amylopsin), and an amylase secreted into the lumen by the intestinal epithelium break down starch, at first into smaller polysaccharides called dextrins and then into the disaccharide maltose. Because ptyalin in inactivated by gastric HCl, most starch is digested by amylopsin and intestinal amylase. The maltose produced in reduced to glucose by a maltase secreted by the intestinal mucosa. Sucrose (table sugar) and lactose (milk sugar) are hydrolyzed by sucrase and lactase, respectively, which are also secreted by the intestinal mucosa. The products are glucose and fructose from sucrose and glucose and galactose from lactose.

It has long been known that glucose is actively absorbed from the lumen of the small intestine. Although the process has been studied intensively, it was only recently elucidated. It is complex and investigators do not yet fully agree on its details. Glucose dissolved in the intestinal juice enters the "brush border" of the mucosal epithelium, where it binds to a symport carrier protein that also binds Na^+. The glucose and Na^+ are released into the cytoplasm whereupon the Na^+ is pumped out again. The process, which involves a carrier in the epithelial cell plasma membrane, is affected by differences in concentration of sodium and potassium ions between the inside and outside of the cells. This active transport mechanism is capable of absorbing all glucose present in the intestinal lumen. As the absorbed glucose accumulates in the epithelial cells, it passes by diffusion into the blood-stream. Although galactose appears to be absorbed by the same mechanism as glucose, fructose is exceptional in being absorbed by diffusion alone. Despite this ability to absorbed by diffusion alone. Despite this ability to absorb monosaccharides, all three hexoses usually cross the plasma membranes in disaccharide form and are only then digested to monosaccharides within the cells of the intestinal epithelium.

What Happens in the Colon

The colon is separated from the ileum by the ileocecal valve. The human *cecum* is a large, blind pouch below the entrance of the ileum. The *vermiform appendix,* which appears to be a vestige of a more extensive cecum, joins the cecum proper. The large intestine, or *colon,* extends upward from the cecum, crosses the abdomen from right to left, and then courses downward as the descending colon, sigmoid (S-shaped) colon, and rectum, opening to the exterior at the anus. The anus is normally kept closed by external and internal sphincter valves.

The colon has three principal functions: (1) promotion of bacterial decay of undigested residues; (2) recalamation of water (about three-fourths is resorbed); and (3) temporary storage of feces prior to defecation. Contrary to a common belief, rather than being composed principally of undigested food, human feces are almost entirely a product of the digestive system.

Feces consists principally of water, bacteria, and epithelial cells from the intestinal mucosa. Also included are inorganic salts and whatever undigested food has not been metabolized by the bacteria. Muscus is also secreted into the lumen and acts as a lubricant for the feces. The composition of feces is almost entirely unaffected by diet, and, indeed normal feces continue to be produced during early stages of fasting. In humans, eating a large amount of vegetable or cereal grains, however, results in the appearance in the feces of undigested fiber, small amounts of which are normally digested by bacteria. Although the role of bacteria is not nearly as important in humans as in herbivores, these microorganisms do contribute to our nurtition, most notably by producing vitamin K, a blood-clotting factor.

As the feces pass through the colon, most of the water is reabsorbed. If allowed to remain too long in the colon, too much water is reabsorbed and constipation results. When irritants such a raw vegetable are included in the diet, they stimulate peristalsis, preventing constipation.

Overirritation, however, results in such rapid movement and insufficient absorption of water that diarrhea results.

By weight, normal feces are usually 40 percent water but can be as much as 75 percent. Of the dry weight, about 5 percent is nitrogen, up to 20 percent salts, up to 18 percent lipids, and up to 50 percent bacteria. The color of feces is due to bile pigments derived from the heme of hemoglobin.

Control of Gastrointestinal Activity and Secretion

Like most internal organs, the organs involved in digestion are innervated by the autonomic nervous system, a part of our nervous system that controls our automatic functions. Reflex peristaltic responses to the presence of food are an example. Hormones supplement the action of the nervous system in controlling digestive activity. Of the two, hormones are more important in this regard. The control of pancreatic secretion is one of the best examples.

When a substantial amount of hydrochloric acid enters the duodenum from the stomach, the duodenal mucosa is stimulated to secrete the hormone secretion into the bloodstream. When the blood-borne hormone reaches the pancreas, it stimulates the release of pancreatic secretions into the pancreatic duct. The bicarbonates in the pancreatic juice neutralize the HCl from the stomach. The mucosa detects the concentration of acid to be neutralized and responds accordingly.

Special Cases: Ruminant and Nonruminant Herbivores

Perhaps the most unusual modifications of the stomach are seen in ruminant mammals, animals that "chew their cud". This method of nutrition has evolved in animals that need to eat large amount of food relatively quickly, chewing being done later at a more comfortable or safer location away from the teeding grounds. But more important, by providing an opportunity for hordes of microorganisms to digest the thick cellulose cell walls of grass and other vegetation, ruminants can utilize the vast amounts of forage available to herbivores. Animals generally lack the ability to produce a *cellulase,* an enzyme needed to digest cellulose. The intestinal bacteria and protozoans that have such enzymes have made the herbivorous lifestyle possible. In ruminants, the upper portion of the stomach expands to form a large pouch, the *rumen,* and a smaller, honeycomblike *reticulum*. The lower portion of the stomach consists of a small antechamber, the *omasum,* with the "true" stomach, or *abomasum,* behind it. Food first enters the rumen, where it encounters a remarkable community of specialized anaerobic bacteria

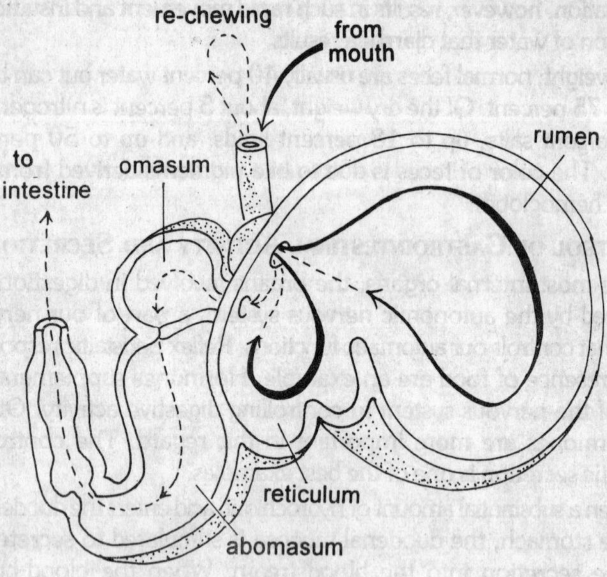

Fig. 5.10. Specialization in Ruminant "Stomachs".

and protozoans. Aided by copious secretions of fluid, body heat, and the churning and kneading caused by the contractions of the muscular wall of the rumen, the food undergoes partial digestion by the microorganisms and is reduced to a pulpy mass. Later the food moves into the reticulum, from which mouthfuls are regurgitated as "cud" and are throughly chewed for the first time. When reswallowed, the

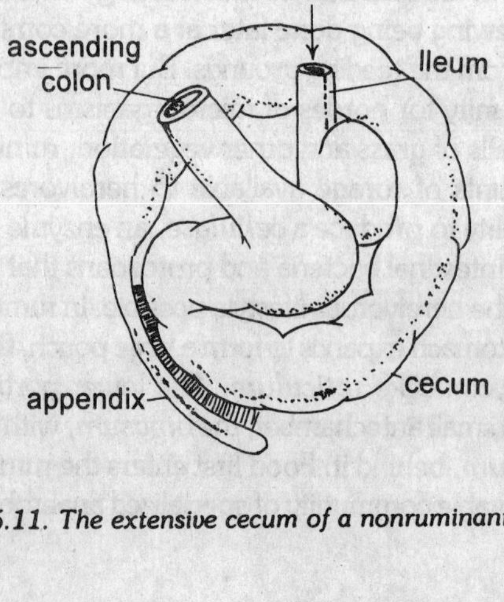

Fig. 5.11. The extensive cecum of a nonruminant herbivore (rabbit).

food usually reenters the rumen. Another cud is then passed up to the mouth to be chewed. This process continues, and the material becomes more completely digested and more liquid in consistency. Only then does it begin to flow out of the reticulum and into the lower parts of the stomach, first the omasum and then the glandular region, the abomasum. It is here that the food first encounters the animal's digestive enzymes, and digestion continues in the usual fashion.

Another remarkable benefit of the ruminant way of life is that nonprotein sources of nitrogen can be utilized as a raw material for the synthesis of protien through the agency of the microorganisms of the rumen. The best example of such a nitrogen source is urea, usually a nitrogenous waste product of mammals. In cattle, urea diffuses from the blood into the rumen where bacteria use it in protein synthesis. Thus recycled, urea becomes a valuable source of "dietary" amino acids. The rumen bacteria, moreover, synthesize all vitamins except A and D, marking their hosts nearly self-sufficient in regard to dietary vitamin requirements.

Whereas the food of ruminants is attacked by bacteria before gastric digestion, in the typical nonruminant herbivore, bacterial action on cellulose occurs after gastric digestion. Rabbits, horses, and rats accomplish this feat by maintaining a population of microorganisms in their unusually large cecum, the blind pouch that extends from the colon. Adding further to nutritional efficiency, a few nonruminant herbivores, such as mice and rabbits, can eat their own feces and process the remaining materials in them, thus having the best of both types of herbivore worlds.

Absorption Surfaces

There are several digestive tract specializations that provide increased surface area for nutrient absorption. Perhaps the simplest way of providing a large surface area for absorption is to have a long absorptive section in the gut. Many animals do indeed have long, coiled guts.

But surface areas are increased in the relatively straight, uncoiled guts of some animals by others means. For example, most earthworms have a longitudinal fold called a *typhlosole* that given their gut nearly twice as much surface area as a simple round tubular gut of similar length. Sharks have a complex folded membrane called *spiral valve* that forces food to follow a route resembling a spiral staircase as it passes through the gut. As a result, the food is exposed to a much larger surface area and more nutrients can be absorbed.

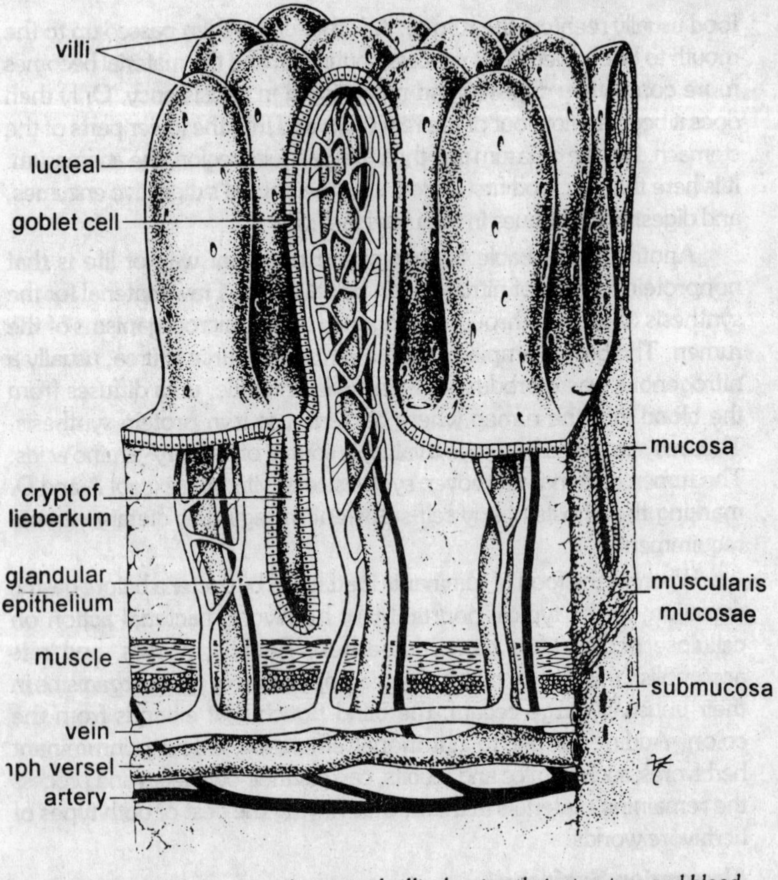

Fig. 5.12. Detailed diagram of intestinal villi, showing their structure and blood and lymph supply.

Large mammals require large absorptive surfaces. They must obtain nutrients to support large body masses and the relatively high metabolic rates required to maintain their body temperatures. Further modification of the layers lining the gut produces even greater absorptive surface areas in many mammals. For example, the human small intestine is not a simple, smoothly lined tube; it has ridges and furrows that give it an almost corrugated appearance. On the surface of these ridges and furrows are small fingerlike projections called *villi.* Cells on the surfaces of villi have minute projections called *microvilli*. These structural features increase the absorptive surface of the human small intestine to about 120 times the area of a simple smooth tube of similar length. A smooth tubular small intestine would have to be five to six hundred meters long to have a surface area comparable to that of the human small intestine. Such a long gut would be hard to carry around, and if the movement rate were the same as it is in a normal human gut, it would require weeks for food to traverse the small intestine.

Control of Digestive Tract Secretions

In order to maintain homeostasis, it is important that all digestive secretions be released when food is present and not at other times.

Salivary gland secretion is controlled by reflex activity of the nervous system. Saliva flows in response to the taste, smell, or sometimes even the thought of food. In fact, you might notice right now that saliva can flow in response to just thinking about saliva flowing.

Neural reflex responses to the sight, smell, and taste of food initiate secretion by glandular cells in the stomach, and then, the arrival of food in the stomach stimulates sensory receptors whose activation causes additional reflex nervous stimulation of gastric secretion.

But a major factor in gastric secretion regulation is a hormone, *gastrin,* which is produced in the stomach itself. When proteins contact the stomach mucosa, gastrin-producing cells are stimulated to release gastrin into the blood-stream. As soon as gastrin circulates through blood vessels and reaches the acid-and enzyme-secreting cells of the stomach lining, these cells respond by secreting large quantities of HCl and pepsinogen. As the stomach empties, both the neural reflexes and gastrin release subside, and less HCl and pepsinogen are secreted.

The duodenum also may produce some gastrin, but this is of minor importance in the overall regulation of gastric secretion. Some duodenal cells produce the hormone *secretin* that stimulates the pancreas to release pancreatic juice, especially the bicarbonate component. Other cells release a hormone called *CCK-PZ* (formerly called *cholecystokinin*) that stimulates the release of bile. Bile, produced by the liver, is stored in the *gallbladder,* a sacklike structure embedded in the surface of the liver. CCK-PZ stimulates the gallbladder to empty its contents through the *common bile duct* into the duodenum. This same hormone also stimulates the pancreas, especially the enzyme secretion of the pancreas, and therefore, it was renamed CCK-PZ, which stands for *cholecystokinin-pancreozymin,* in recognition of this dual role.

A hormone called *enterogastrone* is released into the circulation by duodenal cells in response to the presence of fatty food in the intestine. Enterogastrone inhibits stomach gland secretion and slows muscular movements of the stomach. This slows stomach emptying, thus allowing more time for processing meals that contain large amounts of fat, which is digested more slowly than other nutrients.

Liver and Pancreas Functions

In addition to its contribution to digestion through bile production, the liver is an extremely important and versatile chemical-processing center. Blood leaving the intestinal tract reaches the liver by way of the *hepatic portal system*. This blood, carrying recently absorbed nutrients comes into contact with liver cells, which can make a variety of exchanges with it.

For example, the liver plays an important role in *blood glucose* regulation. Many cells in the body are sensitive to changes in the level of glucose being supplied to them by the circulatory system. Brain cells, for instance, are especially dependent on a stable glucose supply, and they can be seriously damaged by unusual fluctuations in blood glucose level.

When blood sugar level are high, the liver absorbs glucose and uses it to synthesize glycogen for storage. Once the liver has stored

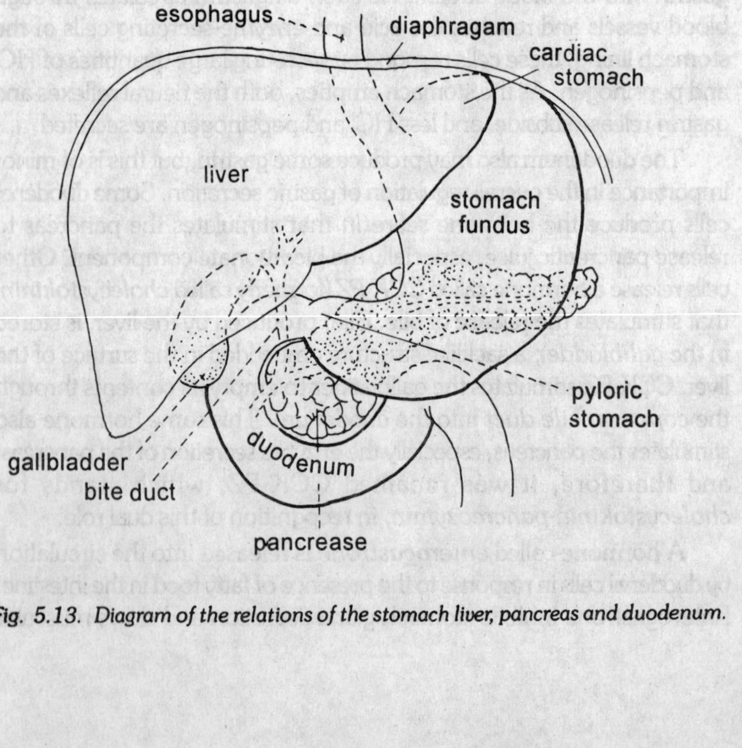

Fig. 5.13. Diagram of the relations of the stomach liver, pancreas and duodenum.

adequate quantities of glycogen to meet the body's needs for a few hours, it stores additional nutrients as fat and also exports fat for storage in tissues elsewhere in the body. When blood sugar levels fall, the liver breaks down glycogen and releases glucose into the blood. This process continues until glycogen reserves have been used up. Then stored fats are used for glucose synthesis. If blood glucose levels remain very low for an extended period, as in starvation, eventually blood glucose is maintained by synthesizing glucose from deaminated amino acids. This depletes body proteins and seriously damages the body if it is continued too long.

The liver also plays an important role in processing the body's extra amino acids. During times of adequate nutrient supply, there may be excesses of particular amino acids. Liver cells deaminate these amino acids to produce usable carbohydrartes. Ammonia, a product of deamination, is quite toxic and must be incorporated into *urea* molecules by liver cells to prevent its buildup.

In addition to these and other metabolic functions, the liver collects and breaks down many toxic substances that enter the body, such as ethanol, many drugs, and other compounds.

The pancreas, like the liver, is a functionally complex organ that no only makes digestive contributions but also contains specialized cells populations, known as the *islets of Langerhans,* that produce the hormones *insulin* and *glucagon.* These pancreatic hormones are involved in the control of blood glucose level and several related aspects of metabolism.

Problems with the Digestive Tract

The gut tube, for all its twists and expansions, is really just a tunnel through the body. In the strict sense, the contents of the gut are outside the body and remain so unless they pass across cell membranes of gut-lining cells. Indeed, the contents of the gut must not come into direct contact with the blood or other interior tissues of the body, because the chyme contains bacteria and other potentially hazardous components. The nature of this risk is graphically illustrated by two rather common digestive ailments—appendicitis and ulcers.

The *vermiform appendix*, or just appendix for short, is a short, blind sac off the beginning of the colon, just past its junction with the small intestine. The human appendix contains lymphoid tissue and might, therefore, be involved in resistance to infection, but its function is poorly understood. The appendix all too often becomes infected and develops an inflammation called *appendicitis.* Once severe appendicitis is diagnosed, standard medical procedure calls for surgical removal of the infected appendix. If untreated, the inflamed appendix can rupture, allowing gut contents to come into contact with the *peritoneum,* the tissue that covers all of the digestive organs and lines the body cavity. The material that escapes from a ruptured appendix causes *peritonitis,* a life-threatening infection.

Ulcers are breaches, of varying depth and position, in the wall of the gut. Most are located in the duodenum and are called *duodenal ulcers.* Somewhat more rarely, *gastric ulcers* may form in the stomach itself. Ulcers are caused by the combined effect of acid and enzymes eroding the protective lining of the gut. However, what is not well understood is why in some individuals the normal mechanisms that protect the linings of the digestive tract function adequately for a lifetime, and in others they do not. It is widely believed that worry, stress, or frustration can contribute to ulcer development because gastric juice secretion is at least partly controlled by the nervous system. In fact, one of the more drastic treatments for serious ulcers is to cut the vagus nerve, the nerve that stimulates stomach secretion.

Most ulcers are rather mild, albeit painful, abrasions in the gut lining. Healing can be rapid because the cells lining the gut tube are quickly replaced, but the causative features (often stress) must be removed and some patients may switch to a bland diet to reduce acid secretion. The real dangers come when ulcers begin to bleed or when they perforate (produce a hole directly through the gut wall). Peritonitis can develop following perforation unless prompt medical care is received. Such serious cases require removal of the source of stomach acid. This can be accomplished by cutting the vagus nerve or by removing part of the stomach. An interesting alternative technique accomplishes the same thing without surgery. The patient swallows a balloon and then alcohol, at a temperature cold enough to freeze and kill stomach lining cells, is poured into the balloon. This destroys a large percentage of the acid-secreting cells and has the same effect as surgical treatments. By the time these cells are replaced, the ulcer has had a chance to heal.

Appendicitis and ulcers are not the only digestive ailments. Problems can also occur in the digestive glands. For example, the liver and its bile-storing structure, the gall-bladder, are subject to several types of problems. One common condition is *gallstones,* which result from the precipitation of bile salts in the gallbladder or in the bile duct. Blockage of the bile duct may cause bile to back up and ultimately lead to jaundice. Surgical removal of the gall-bladder is the usual solution.

Jaundice also may be a symptom of various types of liver failure, a much more serious condition given the overall importance of this organ. Inflammation of the liver is known as *hepatitis*, and it may result from a virus infection (viral hepatitis), poison (toxic hepatitis), or protein deficiency (deficiency hepatitis). The liver has considerable regenerative ability, but when liver cells are destroyed, they are sometimes replaced by scar tissue. This scarring is called *cirrhosis* of liver. All of these liver problems weaken the liver's metabolic capacities and greatly reduce its ability to deal effectively with toxins entering the body. Thus, persons suffering from liver disease, whatever the original cause, are strongly urged not to drink alcohol because alcohol is normally metabolized almost entirely by liver cells. A damaged liver may well succumb to the onslaught of alcohol in the blood, with death as the likely result.

In addition to being susceptible to problems such as appendicitis, ulcers, gallstones, and liver disease, the digestive tract provides an ideal environment for the growth of many disease organisms, from viruses to tapeworms.

Despite all of these potential problems, most human digestive tract function normally, and except for occasional minor upsets, seldom cause much concern. A normally functioning digestive tract is essential if the body is to receive an uninterrupted supply of vital nurtients.

6

RESPIRATION

In unicellular organisms oxygen can diffuse from the watery environment to all parts of the cell. With increasing size and complexity of the animal, however, the distance between the surface and the tissues becomes so great that special means are required to carry oxygen to the individual cells. In mammals the oxygen is taken up in the capillaries of the lungs and is conveyed by the blood to the tissues. At the same time the carbon dioxide produced in the tissues is carried to the lungs and there leaves the body.

Air is moved in and out of the lungs by their 'bellows action' (*ventilation*). It is then distributed within each lungs so that alveoli are adequately ventilated and also perfused by blood (*distribution*). Oxygen and carbon dioxide are the exchanged between the gas mixture in the alveoli and the blood in the pulmonary capillaries (*gas transfer*). The gases are then transported in the blood to the tissues.

VARIOUS RESPIRATORY ORGANS

Gills

The problematical ancestral vertebrates, amphioxus and the tunicates, that are largely sedentary in habit, obtain their microscopic food through ciliary action by creating a current of water which flows into the mouth. Since the current of water also contains oxygen essential to respiration, it is obvious that respiratory organs placed in its pathway will be most favourably located for obtaining oxygen. Gills, lungs, and even more uncommon respiratory devices, such as swim bladders in fishes and parayngoesophageal capillaries in lungless salamanders, therefore, are found colonizing at the anterior end of the alimentary canal. There is in fact and intimate relationship between the respiratory organs of vertebrates and the anterior end of the digestive tube.

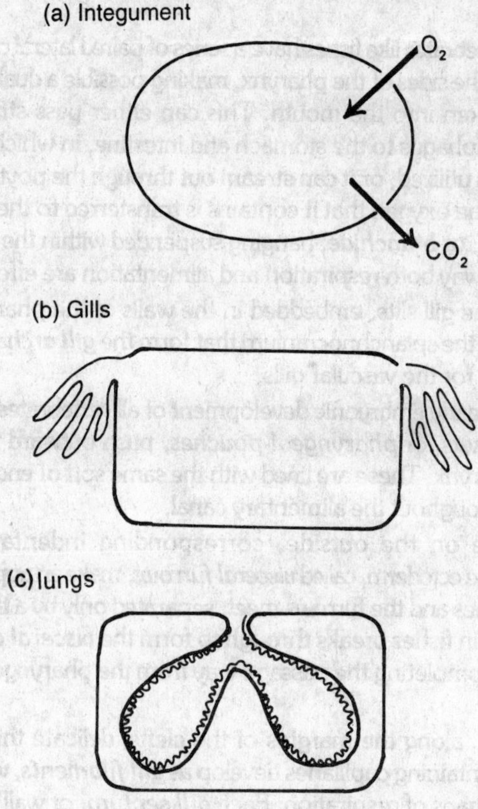

Fig. 6.1. Adaptations for exchanging Gases.

Originally perhaps the entire digestive tube shared in the respiratory function, as suggested in the rectal breathing of certain fishes. Surely in amphioxus as much as the anterior half of the tube is concerned with respiration. In most vertebrates, however, the apparatus for breathing is more restricted, although still closely connected with the pharyngeal region.

Aquatic vertebrates like fishes have a series of paired lateral openings, the *gill slits*, in the sides of the pharynx, making possible a dual disposal of the water taken into the mouth. This can either pass straight on through the esophagus to the stomach and intestine, in which case its blood content is utilized, or it can stream out through the poythole-like gill slits, when the oxygen that it contains is transferred to the delicate respiratory *gills,* or *branchiae*, hanging suspended within the space of the slits. In this way both respiration and alimentation are effected.

Between the gill slits, embedded in the walls of the pharynx, are skeletal parts of the splanchnocranium that form the *gill arches,* which furnish support for the vascular gills.

During the early embryonic development of all vertebrates, a series of internal pockets, or *pharyngeal pouches,* push outward from the sides of the pharynx. These are lined with the same sort of endodermal tissue found throughout the alimentary canal.

Meanwhile on the outside, corresponding indentations or depressions in the ectoderm, called *visceral furrows,* make an appearance. Later the pouches and the furrows meet, separated only by a thin sheet of tissue which in fishes breaks through to form the *visceral clefts,* or *gill slits,* thus completing the passage-way from the pharyngeal cavity to the outside.

In addition, along the margins of the clefts, delicate thin-walled evaginations containing capillaries develop as *gill filaments,* which are the essential organs of respiration. Each *gill septum,* or wall between successive visceral clefts, together with all of the filaments on both of its surfaces makes up a *holobranch,* or *gill.*

The number of pairs of gill slits varies from fourteen in the cyclostome *Bdellostoma polytrema* of the Pacific Coast, to only one pair, between the third and fourth gill arches, in certain salamanders.

The complicated branchial basework of tunicates forms a remarkable exception, as well as the sixth or more pairs of gill slits in the elongated pharynx of amphioxus.

Among reptiles, birds, and mammals gill slits do not normally break through, although both pouches and furrows occur in the embryo. Whenever a pharyngeal fistula, as occasionally reported in medical literature, turns up in a human being, the comparative anatomist, if not himself the possessor or the strange anomaly, is delighted with this remainder to past history.

1. Invertebrate Gills

The gills of invertebrates are not associated with pharyngeal gill slits, as in vertebrates, for there are no gill slits present. Gills may be located anywhere along the outside of the body within easy access of the water. For example, in free-swimming annelids, such as *Nereis*, large leaflike, parapodial appendages extending along either side of the body functions as "gills." In mud-dwelling or tube-inhabiting annelids gills when present crowed together in beautiful feathery tufts at the exposed anterior end, since they would be at a disadvantage buried within the tube or the mud burrow.

The ciliary "gills" of lamellibranch mollusks are concerned primarily with directing a stream of microscopic food toward the mouth opening and they are of doubtful respiratory service, this function being performed instead by the expanded surface of the "mantle."

The *thoracic gills* of crustaceans and the *abdominal gills* of larval aquatic insects are simply thin feathery expansions of the chitinous exoskeleton, enclosing tracheal tubes which in this way extend outside of the body rather than being turned inside, as in land tracheates.

2. Primitive Gills of Amphioxus and Tunicates

The fact that tunicates are enveloped in a thick non-respiratory cellulose tunic makes necessary an enlarged pharyngeal basket, that forms a respiratory structure often exceeding in size all the other organs of the body. The tunicate *Phallusia*, for example, has a branchial basket perforated with several hundred gill slits, while the ghostly transparent Appendicularia, which are microscopic tunicates without a cellulose mantle and consequently with greater capacity for diffuse integumental breathing possess only a single pair of gill slits.

Amphioxus is characterized by an exceptionally generous number of gill slits, and these are bordered by flat primitive vascular tissue that presents much less respiratory surface than is common to the more elaborate gills of fishes and amphibians. No expansive gill filaments, such as appear in higher aquatic vertebrates, are found hanging in the gill slits, so this ancestral animal is obliged to make up in number what it lacks in area of individual gills. The entire pharyngeal region is enclosed in an *atrial cavity,* so that the pharynx opens indirectly to the outside through a single ventral *atrial pore,* instead of directly through the gill slits. The wall of this atrial chamber protect the delicate gill surface from the sand in which amphioxus burrows, while allowing the unobstructed passage of water through the gill slits.

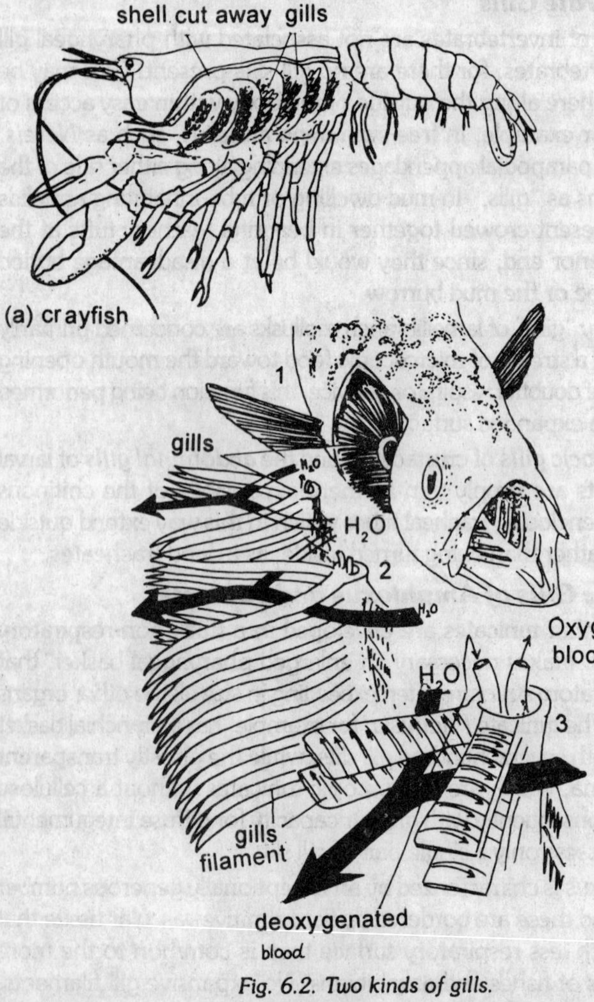

Fig. 6.2. Two kinds of gills.

Since the numerous gill slits of both amphioxus and the tunicates increase the hazard of the escape of food through them, the development of a device for directing bits of food past these lateral openings to the esophagus is necessary. The apparatus accomplishing this end consists of two ciliated glandular grooves, the ventral *endostyle,* and the dorsal *epibranchial groove,* in which food particles become collected together into a sort of mucous rope that spins along continuously into the gullet. Later in vertebrate evolution, when the endostyle is no longer needed for the purpose of collecting and steering food, this pharyngeal groove becomes transformed into the *thyroid gland* with an entirely different function.

The student of comparative anatomy often finds things like this in the morphological junk pile, that have become utilized for purposes entirely different from those to which they were originally put.

3. Cyclostome Gills

The arrangement of gills in cyclostomes is somewhat different from that in typical fishes. In *Petromyzon* a forward extension of the esophagus forms a partition between the alimentary and respiratory parts of the pharynx, leaving the branchial apparatus in a blind pocket ventral to the esophagus. Water in its passage over the gills consequently may *enter* through the anterior gill slits instead of the mouth, which is frequently otherwise occupied on account of the sectorial habits of these animals.

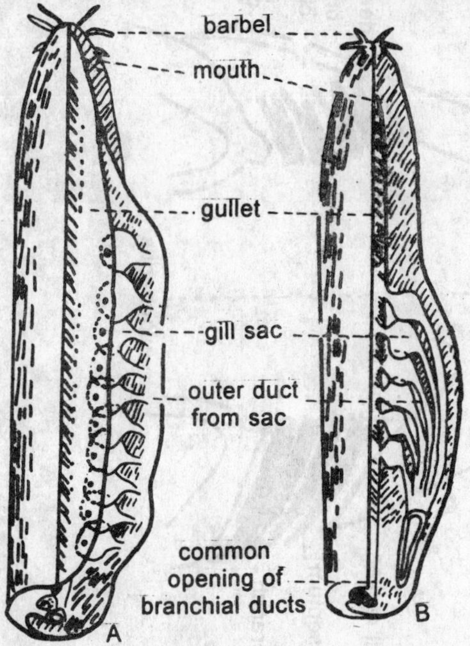

Fig. 6.3. Diagram of the heads of (A) Bdellostoma and (B) Myxine. Dorsal views partly dissected to show arrangement of gills.

The gill slits are also modified by the narrowing of each end of the visceral cleft so that the filaments are enclosed in a rounded sac with a slender duct leading to it from the pharynx and another from it to the surface of the body. Since the gills are located in these pockets, cyclostomes are frequently referred to as *Marsipobranchs,* meaning "pouched gills." In this group different species of *Bdellostoma* have from ten to fourteen pairs of gill slits; *Petromyzon,* eight embryonic pairs but only seven in the adult; and *Myxine,* six pairs, with all the external passage-ways on each side of the body uniting into a common canal opening by a single pore to the outside.

4. Elasmobranch and Holocephalan Gills

Gills of elasmobranch fishes are lateral in position in sharks and dogfishes but ventral in the flattened skates and rays. They open

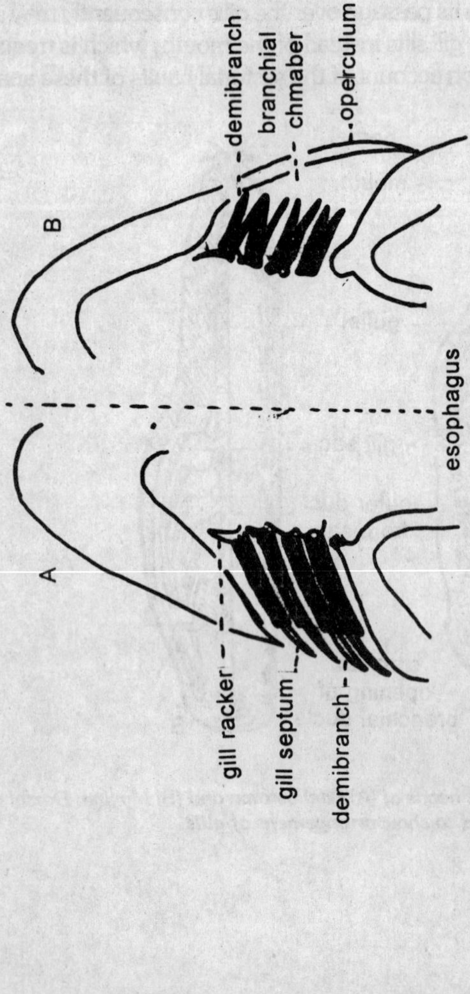

Fig. 6.4. Relation of gills to branchial chamber. A—elasmobranch, with long gill septa; B—teleost, without gill septa.

independently to the outside and are separated from each other not only by skeletal gill arches of cartilage, but also by primitive partitions attached to these arches, called *branchial* septa, on either side of which is located a set of filaments, or *demibranch.* A septum and its two demibranch together make up a *gill*, or *holobranch*. Along the inner margins of the gill arches are fingerlike projections, the *gill rakers,* which not only keep food from entering the slits but also aid in directing it along the straight and narrow esophageal path in which it should go.

The most anterior pair of the pharyngeal portholes in elasmobranchs, between the mandibular and the hyoid arches, develops into the *spiracles,* not far posterior to the eyes in position. The walls of the spiracles are not provided with true gills, but in some cases may support, on one side at least, a "false gill," or *pseudobranch,* so called because its blood supply is not derived directly like tht of true gills from an afferent branchial artery bearing "impure" blood, but from the efferent branchial artery of the following gill arch which, having already given up its load of carbon dioxide and taken in oxygen, carries "pure" blood.

In bottom-feeding skates and rays the spiracles open dorsally instead of ventrally as the other pharyngeal gill slits do. They are useful, therefore, in taking the water of respiration into the branchial cavity when the mouth is otherwise occupied grubbing for food in the mud. No doubt sharks and dogfishes, which swim about freely and gracefully in water, likewise use the spiracles upon occasion, instead of the mouth, as an accessory part of entry for the water of respiration. The cub shark, *Carcharias,* and the mackerel shark, *Lamna*, are without spiracles.

The primitive shark, *Heptanchus,* has seven pairs of gill slits, and the frilled shark, *Chlamydoselachus,* as well as *Hexanchus,* six pairs, while for most elasmobranchs the typical number, aside from the spiracles, is five pairs. On each face of each gill slit, except the posterior side of the last slit, there is demibranch, making a total of nine pairs of demibranchs in a fish with five pairs of slits.

Certain larval elasmobranchs which undergo considerable development within the eggshell before hatching have gill filaments so long that they hang out of the gill slits as temporary "external gills". These unusual structures may serve not only for respiration but also as absorbing organs in connection with the enormous yolks present in these eggs.

The holocephalans, or strange elephant fishes, which have much in common with elasmobranchs, process only four pairs of gill slits and are further differentiated from them by an *operculum.* This is a flap of the integument developed on eithr side and extending backward from the margin of the hyoid arch, until it covers the external openings of the three anterior pairs of gill slits, leaving only the last pair open directly to the outside after the elasmobranch fashion. Possibly the forerunner of the opercular flap is seen in the elasmobranch. *Chamydoselachus,* where the skin on the anterior margin of each gill slit extends backward as a small independent protective fold covering the opening of each gill slit separately, a feature which gives *Chlamydoselachus* the common name of "frilled shark".

5. Ganoid and Teleost Gills

The gill system of ganoids in some ways represents a connecting link between that of elasmobranchs and teleost fishes. Most Chondrostei still have nine pairs of demibranchs but nealry all Holostei lose the most anterior pair so that they are limited to eight pairs of demibranchs, i.e., four pairs of complete gills. In all bony fishes the gills are covered by an operculum that is stifened by flat skeletal plates between the two surfaces of folded integument. Outside and anterior to the operculum on either

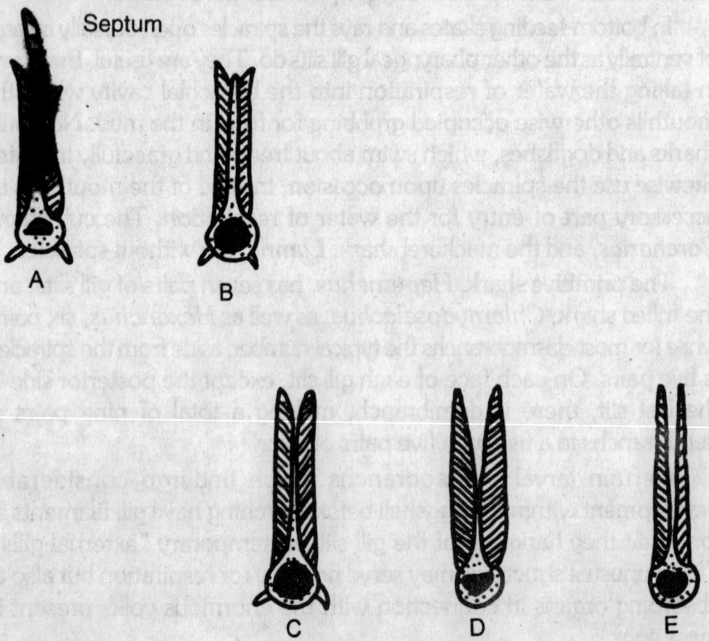

Fig. 6.5. The progressive reduction of gill septa in different fishes. A—elasmobranch; B—Holocephalan; C—Ganoid; D and E—telcosts.

side, there is degenerate spiracle in some of the ganoids, while on the inner surface of the operculum there is attached a small *opercular gill* which is not homolgous either with true gills or with the pseudobranchs of the hyoid arch.

The interbranchial septa in ganoids are reduced so that the demibranchs placed upon them back to back are no longer in separate individual chambers, but occupy a common branchial cavity. The reduction of the interbranchial septa becomes complete in teleost fishes, so that the gills all lie compacted closely together in a common chamber covered by the operculum.

The number of gill arches in both ganoids and teleosts is usually or five pairs, although they may be reduced to three, or even two pairs in some of the bony fishes. Spiracles are not characteristic of teleosts.

The opercular opening becomes much diminished in such fishes as eels, which are thus enabled to retain water in the branchial chamber under unfavourable conditions. Probably the immediate reason why fishes suffocate when removed from a water environment is not because the gills dry up at once with a collapse of the capillaries in the gill filaments, but because when out of water the gills adhere to each other, leaving the exposable respiratory surface reduced beyond the danger point.

The structure of a typical teleost gill, with its relatively great expanse of respiratory surface within a small compass, and the arrangement of its capillaries are indicated in Fig.

In ganoids and teleosts there are additional devices, besides the opercular lid, for protecting the delicate gills. Along the posterior and ventral margins of the operculum beyond the part stiffened by the flat opercular bones, a bordering fringelike flap is sometimes developed, which is supported by fanlike skeletal elements, the *branchiostegal rays,* that help in controlling the passage of water out of the branchial cavity.

On the inner pharyngeal side of the gill slits also there are present in varying degree, a series of stiff comblike projections along the inner margins of the gill arches, the *gill rakers*.

6. Dipnoan Gills

Of the three genera of living lungfishes that are found respectively in Australia, South America, and Africa, *Neoceratodus* has four pairs of gills, *Lepidosiren*, three, and *Protopterus*, two. Spiracles are present in the embryos of these fishes, although not in the adults. In addition to the pharyngeal gills common to this group, four pairs of supplemental *true external gills* of the pinnate (*pinna,* feather) type are present in the larval stages of *Lepidosiren* and *Protopterus.*

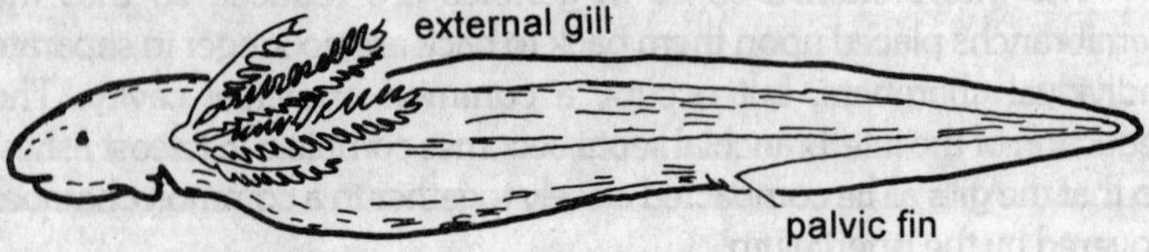

Fig. 6.6. External gills in a larval stage of Lepidosiren.

7. Amphibian Gills

The external gills of amphibians are attached as capillary detours upon the aortic loops in such a way that blood can go either directly around the vascular loops or round-about through the gills. This is quite different from the arrangement of the internal gills of fishes, which offers no alternative for the circulating blood except to pass through the gills themselves.

As compared with internal gills, external gills are only weakly supported by a skeletal framework, and although always present during the tadpole stages of amphibians, they persist throughout life only in the perennibranchiate urodeles, as the name indicates. Such external gills occur in the larval forms of a few fishes, for example, in the dipnoans as already mentioned, in the cartilaginous ganoids, *Polypterus* and *Calamoichthys,* and in the teleosts, *Gymnarachus* and *Heterotis.*

The urodele amphibians have been separated into three groups according to the character of their gills, as follows: (1) *Perennibranchiata,* retaining both gills and gill slits throughout life; (2) *Derotremata,* losing gills and all the gill slits except one pair; and (3) *Myctodera*, or true salamanders, having neither gills nor gill slits when adult.

Although perennibranchiates preserve throughout life their tadpole-like external gills with two or three pairs of gill slits, these pharyngeal openings are no longer useful for their original purpose, since the water of respiration does not pass through the pharynx. Instead a fresh supply of dissolved oxygen is brought into contact with the gill filaments as the gills wave to and fro in the water, by means of muscles attached to the base of each gill. Five pairs of gill pouches form embryonically in the pharyngeal cavity, but the first and the fifth no longer break through. *Necturus* and *Proteus* be have in this group.

In derotremes only the gill slit between the third and the fourth gill arches becomes complete, while the external gills vanish during larval life *Amphiuma, Cryptobranchus Siren* are derotremes.

The myctodere salamanders are well as frogs and toads, have no gills slits, although embryonic pouches and furrows develop. The temporary external gills of these forms are sacrificed with the development of pulmonary and cutaneous respiration. The newts and true salamanders, including *Amblystoma* and *Triton,* are myctoderes.

The external gills of frog tadpoles become enclosed during metamorphosis by a fold of the skin without skeletal support, while the protective peribranchial chamber thus formed usually has single opening to the outside, more rarely two, as in *Pipa* and *Xenopus*, corresponding physiologically to the artial pore in the peribranchial chamber of amphioxus.

One of the tropical limbless amphibians, *Caecilia*, is exceptional, having larval external gills of peculiar crumpled leaflike structure, with a relatively large respiratory surface.

8. Gill Structures in Land Vertebrates

The persistence of branchial remains in land forms, that have no use for gills even in embryonic life, is striking evidence of the common ancestry of all vertebrates.

Although gills are never present in reptiles, birds, or mammals, there are five pairs of embryonic visceral pouches and furrows in reptiles and mammals and four pairs in birds. Ordinarily these break through only briefly in reptiles and birds but not at all in most mammals. Only the anteriormost, or hyomandibular, pair remains well developed in adult mammals. The hyomandibular pouches become the *Eustachian* tubes and the *middle ear cavity,* which ordinarily remains separated by the *ear drum* from the *external ear canal,* derived from the ectodermal furrows.

Air Breathing Organs

Breathing is an essential activity of life and involves the movement of air into lungs through a set of structures which together constitute the respiratory system. This system consists of two portions, viz. (i) the *conducting portion* which furnishes respiratory passages or airways for the transport of air from the outside of the body into all parts of the lung. These include the nasal cavity, pharynx, larynx, trachea and bronchial passages. Most of these passage are lined by a special respiratory mucosa, of which these are three types: nasal mucosa, olfactory mucosa and buccopharyngeal mucosa; and (ii) *the respiratory portion* which commences where the conducting portion ends, i.e., at the terminal bronchiole. The conducting and the respiratory portions are connected by a *transitional portion* represented by several respiratory bronchioles or alveolar bronchioles. The respiratory portion of the lung consists of many small alveolar ducts, alveolar sacs, and alveolar. Exchange of gases takes place at the alveolar surface through an air-blood interface or barrier.

The Conduction Portion

A. Nasal cavity

The nasal cavity is the gateway of the respiratory system and represents the first segment of the conducting portion of the respiratory tract. It has several distinctly different regions, viz. voslibule. respiratory region. Olfactory region, and paranasal sinuses. Its functions include, (1) cleaning of air: sweeping of particulate matter trapped in mucus towards the pharynx by ciliary action; (ii) air-warming; (iii) air-humidifying; and (iv) olfaction.

1. Vestibule

This is a narrow zone restricted within the nostrils and constitutes the first or upper one-third part of the nasal cavity. The facial skin continues through the anterior nares into the vestibule of the nose. It is lined by a thin, stratified squamous and moderately keratinized epithelium. It is provided with sweat and usually large sebaceous glands and small but coarse/stiff hairs (vibrissae) which are believed to help in excluding dust particles from the inspired air.

2. Respiratory Region (Respiratory Mucosa)

The remaining two-third part of the nasal cavity is lined by a respiratory mucous membrane (mucosa) with a lamina propria attached directly to the underlying cartilage and bone. The submucosa and muscularis mucosae are not present. The *mucosa* lines the *nasal septum*, the floor of the nasal cavity and its lateral walls. Protruding into the cavity from the lateral walls are three scroll-like turbinal bones (conchae), viz., superior, middle and inferior. The inferior concha is the largest and covered by a thicker mucosa membrane. The mucosa is composed of a mucus-secreting epithelium resting on a basement membrane and a lamina propria.

(a) *Epithelium*. It is composed of pseudostratified, ciliated columnar cells richly interspersed with goblet cells. This epithelium, with minor modifications, is also present throughout the conducting portion of the respiratory system. The basement membrane is formed of a thin basal lamina and network of reticular and collagenous fibrils.

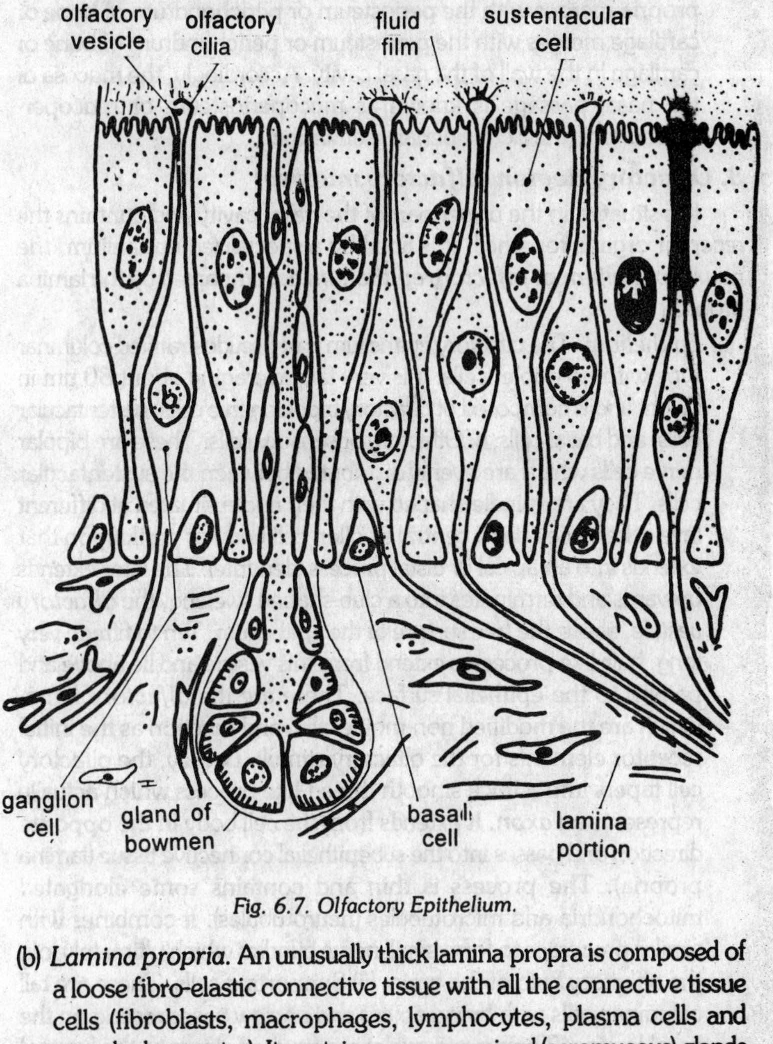

Fig. 6.7. Olfactory Epithelium.

(b) *Lamina propria.* An unusually thick lamina propra is composed of a loose fibro-elastic connective tissue with all the connective tissue cells (fibroblasts, macrophages, lymphocytes, plasma cells and granular leucocytes). It contains numeous mixed (seromucous) glands and a rich vasculature, nerves and lymphatics. The mucus from these glands as well as goblet cells keeps the lining of the nasal cavity moist. The cilia of the respiratory epithelium constantly move this toward the nasopharynx whence it is either swallowed or expectorated. The mucous layer or blanket spread over the epithelium serves to pick up the particulate matter (foreign particles, bacteria and pollen grains) from the incoming air. The deeper layer of lamina propria merges with the periosteum or perichondrum of bone or cartilage merges with the periosteum or perichondrum of bone or cartilage in the wall of the nasal cavity. Accordingly, the mucosa of the nasal passage is known as mucoperiosteum or mucoperichondrium (the Schneiderian membrane).

3. Olfactory Region (olfactory mucosa)

It is situated in the upper part of the nasal cavity and contains the receptor organs for smell. It is lined by a specialized epithelium, *the neuroepithelium* or *olfactory epithelium* accompanied by the lamina propria.

(a) *Epithelium.* The olfactory epithelium is of pseudostratified columnar type with no goblet cells. It is very tall, averaging about 60 um in height and is composed of olfactory bipolar nerve cells, sustentacular cells, and basal cells; (i) Olfactory or sensory cells. These are bipolar nerve cells which are evenly distributed between the sustentacular cells. They are spindle-shaped with their nuclei situated at different levels. Each cell has a central swollen cell body or perikaryon that extends into an apical or distal process (dendrite). The later extends upwards and terminates into a club-shaped swelling, the *olfactory vesicle*, above the free surface of the epithelium. Ten to fifteen very long, hair-like processes extend from this vesicle and lie above and parallel to the epithelial surface. These are the *olfactory hairs,* which are the modified non-motile cilia, and function as the initial receptor elements for the olfactory stimuli. Basally, the olfactory cell tapers into a thick smooth thread-like process which actually represents an *axon*. It extends from the cell body in the opposite direction and passes into the subepithelial connective tissue (lamina propria). The process is thin and contains some elongated mitochondria and microtubules (neurotubules). It combines with similar processes to form small nerve bundles which ultimately join the olfactory bulb of the brain; (ii) Supporting cells. These are tall columnar cells with broad apices and narrow bases resting on the basal lamina. Their ovoid nuclei are situated closer to the luminal surface. The apical surface has a fine, striated border formed by long, slender microvilli. These microvilli are submerged in a blanket of mucus and are intermingled with cilia. Mucous droplets fill the apical region of the cell. The cells also contain a few lipofuchsin (pigment) granules of a yellowish-brown colour, and (c) *Basal Cells*. The basal cells are small and in number forming a single layer in the basal region of the epithelium. They are spherical or cone-shaped with dark ovoid nuclei and stellate or branching cytoplasmic processes. They may be responsible for giving rise to other cells including the sustentacular cells.

(b) *Lamina propria*. The lamina propria contains a loose network of fibroelastic connective tissue. Also present within the lamina propria are less prominent lymph and venous plexuses. Many large branched tubulo-alveolar olfactory Bowman's glands are also present in this region. The glands are characterized by two types of cells — (i) The light pyramidal cells, characterized by large central nuclei. They secrete a serous or aqueous solution. Odoriferous substances could only be registered in solution and the serous fluid provides this medium, and (ii) The dark cells, containing secretory granules of low density. These cells provide the lipid which acts as a solvent for the odoriferous substances.

The Bowman's glands are serous glands from which a watery secretion is carried to the epithelial surface by narrow ducts. It moistens the surface of the olfactory regions and serves to dissolve the odour-producing chemicals. The continuous flow of secretion from these glands cleans the epithelial portion of the olfactory cells. In this manner, compounds which stimulate the sense of olfactory are constantly being removed, thus keeping the receptors in a state of readiness to respond to new stimuli. This prevents repetition of stimulation of the olfactory hairs by a single odour.

4. Paranasal Sinuses

These are air-filled blind pockets contained within the bones of the skull and named as maxillary, frontal, sphenoidal and ethmoidal sinuses. They reach the nasal cavity through tiny apertures. These pockets or cavities are lined by a thin mucous membrane which is continuous with the nasal mucosa and is similar in type, i.e., pseudostratified ciliated columnar type. However, in man, much of this epithelium is thinner and is of simple, ciliated columnar type with a few goblet cells. The basal lamina is poorly developed. It is very thin and indistinguishable from the periosteum of the bone. It contains fewer and smaller seromucous glands than that of the nasal mucosa. The sinuses contribute to the humidification of the nasal cavity.

B. Nasopharynx

The nasal cavities lead into the uppermost part of the pharynx, the *nasopharynx*. The structure of the nasopharyngeal lining is similar to that of the nasal respiratory mucous membrane, i.e., it is a pseudostratified ciliated columnar epithelium, which changes to stratified squamous epithelium in the oral part of the pharynx (oropharyx) that is subjected to friction, e.g. where the free edge of the soft palate and posterior pharyngeal wall come in contact during deglutition. A thick lamina propria contains fibroelastic connective tissue and lymphocytes, many of which are aggregated to form pharyngeal and tubal tonsils on the surface of the nasopharynx. Seromucous glands are also present in the lamina propria. There is no muscularis mucosae in the nasopharynx, but a loose submucosa is present only in its lateral part. A thick layer of skeletal muscle fibres constitutes a muscularis externa. It is composed of an inner longitudinal layer (elevator) and an outer circular layer (constrictor).

C. Larynx

The larynx is that part of the respiratory system which is interposed between the pharynx and trachea. Besides acting as a part of the airway of conducting portion of the respiratory system, it has become specialized for vocalization (production of speech). It helps prevent swallowed food from entering the lower respiratory tract. It is situated atop the trachea. Its wall is supported by a framework of hyaline and elastic cartilages, some connective tissue, mucous glands and a number of small skeletal muscles, serving as a part of the speech mechanism. The laryngeal cavity is lined by a mucous membrane (mucosa) of respiratory type. It forms two pairs of prominent folds, which hang into the laryngeal cavity. The upper pair forms the *false vocal cords* or *vestibular folds,* while the lower pair constitutes the *true vocal cords*. These vocal cords contain two cores of elastic tissue (constituting the *vocal ligaments*) to which skeletal muscle fibres of thyroerytenoid muscles are attached. Contraction of muscles relaxes the vocal ligaments, modifies the opening of the vocal cords and changes the shape of the laryngeal cavity, resulting in the production of sounds with different tones, when air passes through the larynx.

The *epithelial lining* of the mucous membrane is not of the same type of different locations of the larynx. It is subject to friction on the epiglottis and superior surface of the vocal cord and hence is of the stratified squamous, nonkeratinized type. Over the remainder of the larynx, it is of pseudostratified ciliated columnar type interspersed with goblet cells, thus representing the typical respiratory tract epithelium. Its cilia move surface secretion towards the pharynx. A well-defined mucosa is not present but the lamina propria is quite thick. it contains an unusually rich amount of elastic fibres, mast cells and some lymphatic tissue. Small lymphoid nodules and migratory leucocytes frequently occur here. Numerous *laryngeal glands* found in the *Lamina propria* are small, branched, tubuloalveolar invaginations of the epithelium. They, however, are not present in the true vocal cords. They are of mixed type and their secretion protects the vocal cords from desiccation. The pseudostratified epithelium and its associated glands are akin to those of the of the trachea and bronchi and serve a similar function, i.e., conditioning the inspired air and in providing protective coating of muscus spread over the epithelial surface. Considerable amount of this mucus is swept to the larynx from the lower part of the respiratory tract and in turn is driven into the pharynx by local ciliary action.

(b) *Trachea*. From the nasal cavity air passes back through the *nasopharynx* to the *oropharynx*, where it crosses the food route on the way to the esophagus, thence entering the trachea through a slit-like opening, the *glottis*.

The trachea, or "windpipe," is a rigid tube, very short in frogs and toads whose lungs are far anterior in the body cavity. It is somewhat longer but still insignificant in the urodeles. In lizards it is relatively shorter than in other reptiles, although unmistakably present, while in turtles and crocodiles it is frequently so long that it becomes convoluted or even spiral in form. The lengthened trachea in the chelonians is an accommodation to the accordion-like movements of the head and neck.

Birds with long necks are of course provided with a long trachea, but frequently the trachea is even longer than the neck itself, so that it cannot remain straight but loops about. In swans these extra loops are stored within the hollow breastbone, while in some birds they lie coiled under the skin, or may even extend into the body cavity. Extra long tracheae make it possible for their possessors to stretch out the neck without pulling the lungs out "by the roots."

Usually the windpipe is nearly cylindrical, but sometimes, as in the little vocal wall lizards, or "geckos," and also certain bucks, it may show a bulblike enlargement that acts as a resonance sac when air is expelled.

In mammals the trachea is practically straight, with a length directly dependent upon that of the neck, except in the three-toed sloth, *Bradypus* whose trachea is so elongated that it extends down as far as the diaphragm and back before entering the lungs. It will be remembered that the upsidedown sloth, while hanging from the limbs of tropical trees, feeds lazily upon leaves without scrambling about. It therefore has a very stretchable neck.

At all times the elastic walls of the trachea are kept mechanically distended for the passage of air by encircling rings of cartilage, resembling the metal rings embedded in a garden hose to give it flexibility and durability, and at the same time to keep it uncollapsed and open. In the case of mammals these skeletal tracheal rings are usually in incomplete on the dorsal side that is, on the side liable to press against the esophagus that lies parallel to it, thus minimizing the 'corduroy road' effect that might otherwise be encountered by a bolus of food when swallowed.

Among reptiles, birds, and pinniped mammals, the tracheal rings are entire, while cetaceans present the unusual case of tracheal rings, incomplete on the ventral rather than the dorsal side.

Camel and giraffes are noteworthy in having upwards of one hundred separate tracheal rings, and whales and sea-cows in having these skeletal structures spirally arranged.

Although usually of hyaline cartilage, tracheal rings become bony in the python *Agama,* and also in many birds.

The Respiratory Portion

Respiratory Unit

As stated in the lung, it is composed of one respiratory bronchiole (alveolar bronchiole) and all its subsequent divisions such as alveolar ducts, alveolar sacs and alveoli.

1. Respiratory Bronchioles

These are very short and their width from 0.5 mm to 0.1 mm. They are not always present in man, as this is a region of transition between the conducting and respiratory portions of the respiratory system. The respiratory bronchiole can be considered as a terminal bronchiole when a part of the wall outpockets into minute saccules called the alveoli, which are quite abundant at the distal end of the respiratory bronchiole. At this level the epithelium is of low cuboidal variety and largely made of nonciliated Clara cells and some ciliated cells. Elastic fibres and smooth muscle cells form a thin and an incomplete layer.

2. Alveolar Ducts

Near the terminal, the respiratory bronchioles flare out and give rise to two or more long sinuous alveolar ducs, each bearing several dilated sacs called alveoli. The openings of these ducts are ringed by

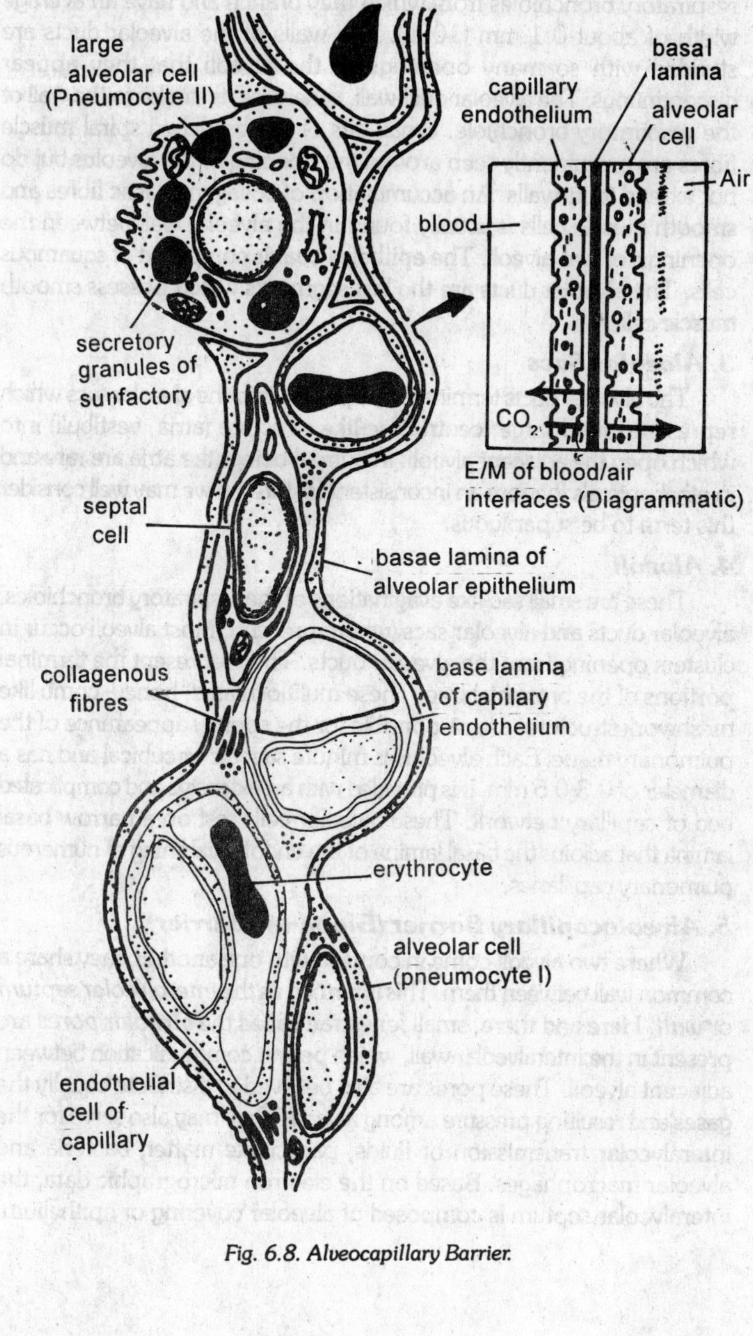

Fig. 6.8. Alveocapillary Barrier.

smooth muscles. The alveolar ducts are structurally similar to the respiratory bronchioles from which they branch and have an average width of about 0.1 mm (100u). The walls of the alveolar ducts are studded with so many openings of the alveoli that they appear discontinuous. The alveolar duct wall, otherwise, is similar to the wall of the respiratory bronchiole. Small bits of the branched spiral muscle fibres are prominently seen around the opening of the alveolus but do not extend to its walls. An accumulation of collagen, elastic fibres and smooth muscle cells is usually found in the alveolar wall between the openings of two alveoli. The epithelial coat is composed of squamous cells. The alveolar ducts are the last segments which possess smooth muscle cells.

3. Alveolar Sacs

The alveolar ducts terminate by opening into the alveolar sacs which represent rather large, central sac-like structure (atria, vestibuli) into which open the adjacent alveoli. In human beings the atria are rare and in other animals they are an inconsistent feature, so we may well consider this term to be superfluous.

4. Alveoli

These are small sac-like evaginations of the respiratory bronchioles, alveolar ducts and alveolar sacs (where present). Most alveoli occur in clusters opening from the alveolar ducts. They represent the terminal portions of the bronchial tree. These mulitloculated, honey- comb like meshwork structures are responsible for the spongy appearance of the pulmonary tissue. Each alveolus is minute saccule or cubical and has a diameter of 0.3-0.5 mm. It is provided with an extensive and complicated bed of capillary network. These alveolar cells rest on a narrow basal lamina that adjoins the basal lamina of the endothelial lining of numerous pulmonary capillaries.

5. Alveolocapillary Barrier (Blood-air Barrier)

Where two alveoli come in contact with one another, they share a common wall between them. This is termed as the *interalveolar septum* or *wall*. Here and there, small fenestrae called the *alveolar pores* are present in the interalveolar wall, which permit communication between adjacent alveoil. These pores are also believed to distribute equally the gases and resulting pressure among alveoli. They may also serve for the interalveolar transmission of fluids, particulate matter, bacteria and alveolar macrophages. Based on the electron micrographic data, the interalveolar septum is composed of alveolar covering or epithelium, basal lamina, and capillary endothelium. The interalveolar septum is supported by a loose connective tissue that constitutes its interstitium.

(a) *Alveolar covering.* Both surfaces of the septum which face the air contained in the alveoli are lined by a single but continuous layer of extremely attenuated (thin) and flattened epithelial cells not easily identified by light microscopy. The alveolar epithelium is covered by a stabilizing fluid film (*surfactant*) and lies on a *basal lamina* which sometimes merges with, and sometimes is separated from the basal lamina of the endothelial lining of the adjacent capillaries.

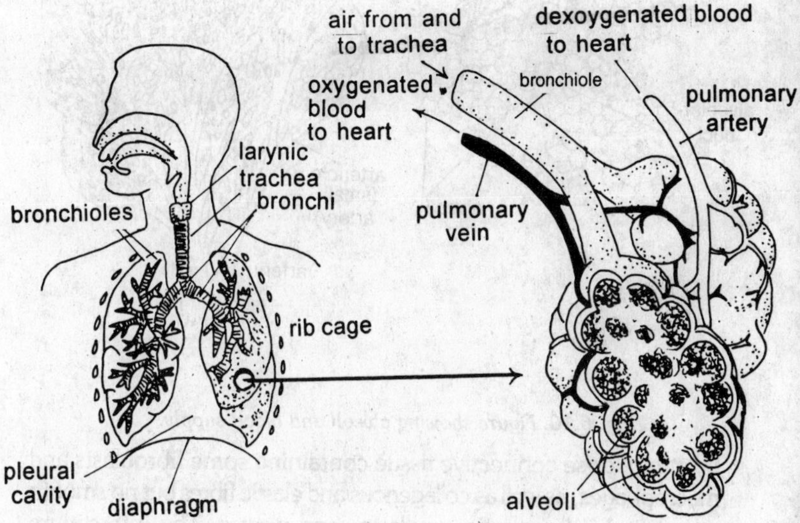

Fig. 6.9. Human respiratory system.

(b) *Pulmonary capillarion*. The interalveolar capillaries are lined by a continuous unfenestrated squamous endothelium which is surrounded by a continuous basal lamina.

(c) *Interstitium* (*Interstitial connective tissue space*). Where the two basal laminae are separated, the space, called *zona diffusa,* is taken

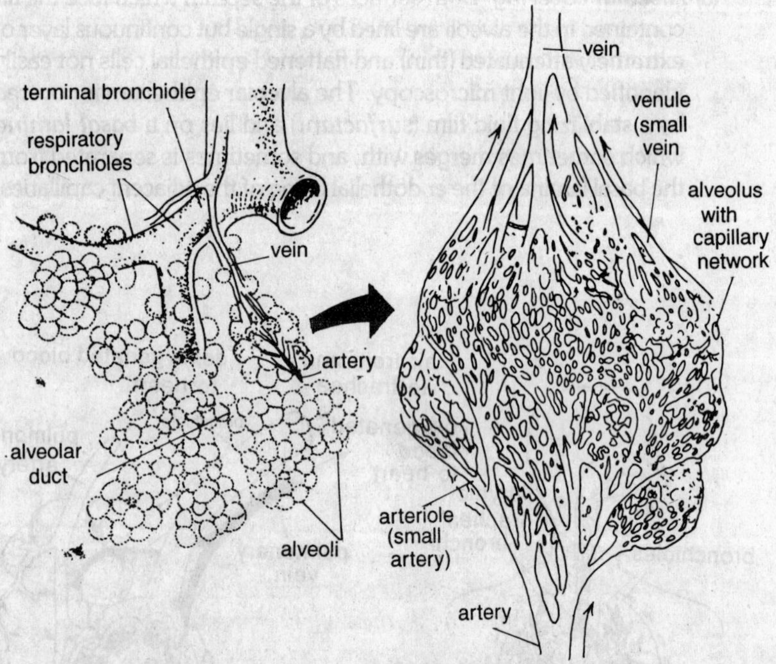

Fig. 6.10. Figure showing alveoli and blood supply.

by a thin loose connective tissue containing some fibroblasts and macrophages, as well as collagenous and elastic fibres but no smooth muscle cells, lymphatic capillaries and nerves. The interstitium provides a supporting framework for the alveoli and their surrounding capillary networks and resists overexpansion of the alveoli.

Based on electron microscopy, three types of cells can be made out in the alveolar or epithelial lining,—(i) Great alveolar or Septal cells, (ii) Squamous or small alveolar cells, and (iii) Alveolar phagocytes.

(i) *Septal cells.* The ultrastructure of the great alveolar cells is quite interesting. They are present side with the epithelial lining cells and have a rounded cuboidal shape. Their free or luminal surfaces have short microvilli. Their cytoplasm contains numerous mitochondria, ribosomes, well-developed Golgi complex, lysosomes and granular endoplasmic reticulum. The most peculiar feature of these cells is the presence in their cytoplasm of round, oval bodies termed as "*cytosomes*" or "*multilamellar*" bodies. They consist of concentric or parallel lamellae in their interior and are bounded by a trilaminar

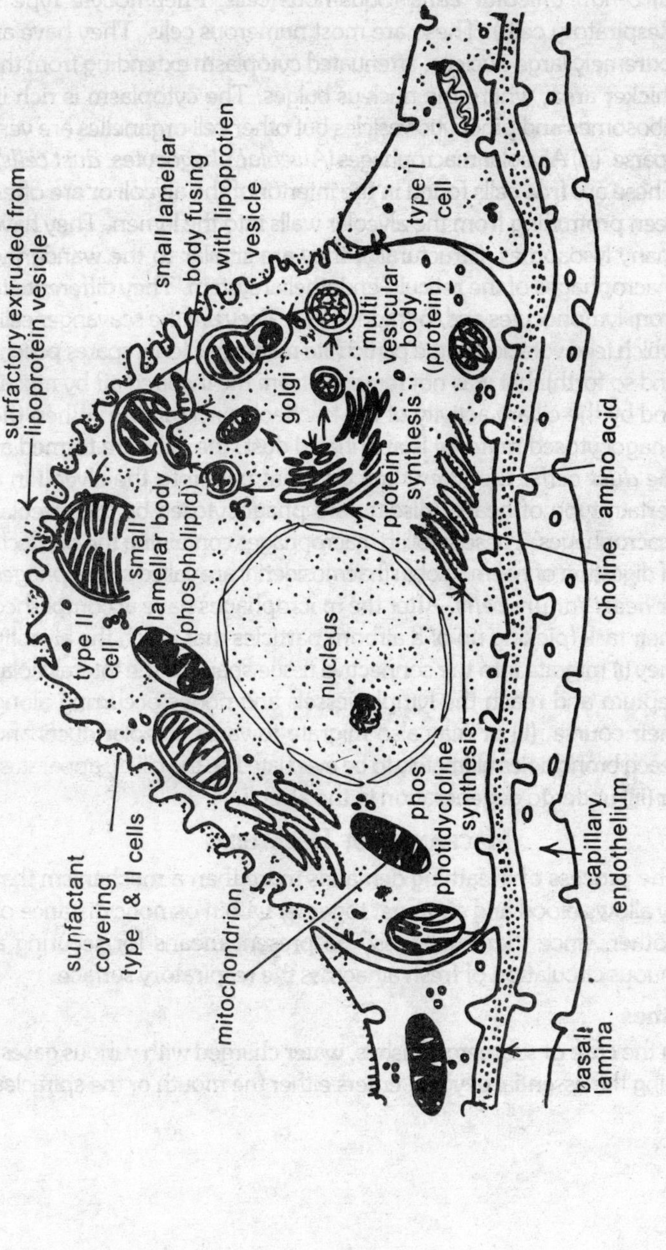

Fig. 6.11. Structure and secretory process of a septal cell.

membrane. They contain a surface active phospholipid termed as the *aurfactant*. The surfactant is lipoprotein which spreads on the surface of the alveoli forming a thin stabilizing fluid film of surface active material that lowers surface tension and is essential for maintaining the shape and stability of the alveoli during respiration. (ii) *Small alveolar cells* (Squamous cells, Pneumocyte type I, Respiratory cells). These are most numerous cells. They have an extremely large nucleus, attenuated cytoplasm extending from the thicker area, where the nucleus bulges. The cytoplasm is rich in ribosomes and pinocytic vesicles but other cell organelles are very sparse. (iii) Alveolar macrophages (*Alveolar phagocytes, dust cells*). These are free cells found in the interior of the alveoli or are often seen protruding from the alveolar walls into the lumen. They have many lysosomes. Structurally, they are similar to the wandering macrophages of the reticulo-endothelial system. They differentiate from lymphocytes and/or monocytes. They are the scavanger cells which feed voraciously the particulate matter (bacteria, spares pollens and so forth) that was not removed from the inspired air by mucus and by the ciliary activity of the tracheobronchial tree. When the phagocytosed material is an inhaled dust, the cells are termed as the *dust cells*. Blood erythrocytes that pass into the alveoli in a certain type of heart—disease are phagócytosed by the alveolar macrophages. These swollen macrophages containing the products of digestion of haemoglobin (haemosiderin) are called *siderophages* or *heart-failure cells*. After the macrophages have accomplished their task (picking up of a airborn particles that reach the alveoli), they (i) migrate into the connective tissue space of the interalveolar septum and reach the lymph vessels and nodes occurring along their course, (ii) or may also migrate towards alveolar ducts and reach bronchioles ultimately to be evacuated by the ciliary apparatus, or (iii) undergo degeneration in the alveoli.

Mechanism of Breathing

The process of breathing demands more than a mechanism that simply allows blood and air to get together within osmotic distance of each other, since there must also be present means for securing a continuous circulation of fresh air across the respiratory surface.

1. Fishes

In the case of submerged fishes, water charged with various gases, including the essential oxygen, enters either the mouth or the spiracles and passes out through the gill slits. It is forwarded and directed in its course, not only by muscular movements which alternately expand and contract the walls of the orobranchial chamber, but also by a system of valves that prevents the water from going the wrong way.

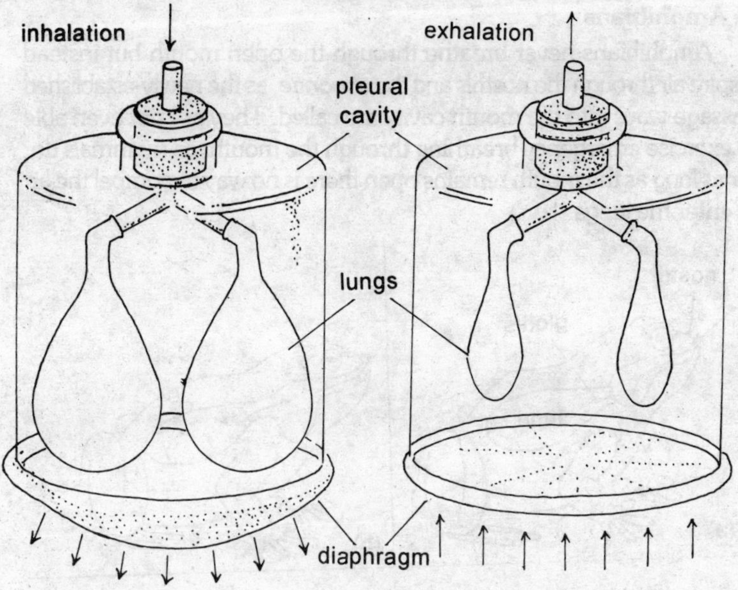

Fig. 6.12. The Diaphragm ventilates the lungs.

The anterior set of these valves are collapsible folds along the inner edge of the mouth opening, those of the upper edge being called *maxillary*, and of the lower, *mandibular*. The reach their greatest differentiation in teleost fishes which have a well-developed opercular apparatus. The posterior, or *branchiostegal,* set of valves are membranes along the free margins of the opercular flaps. A freely moving current of water is produced in the following way. First, the mouth remains open as a narrow slit, while the anterior maxillary and mandibular valves lie flat, or open, and the posterior branchiostegal valves close. Next the walls of the orobranchial chamber spread apart by muscular action, thus pulling water into the mouth to occupy the increased space. Then the valves reverse, that is the anterior ones close the slitlike mouth aperture and the posterior ones open the opercular slit, while the walls of the orobranchial cavity squeeze together, forcing the water backward over the gills and out of the opercular openings.

In some teleosts, particularly those that feed on microscopic plankton, the branchiostegal valves play the major role in this process, but in others, for example the Percidae and the Sciaenidae, the opercular flaps take the most prominent part.

2. Amphibians

Amphibians never breathe through the open mouth but instead inspire air through the nostrils and the *choanae*, as the newly-established passage-ways into the mouth cavity are called. They are not even able to exercise emergency breathing through the mouth, as mammals do, for as long as the mouth remains open there is no way to compel the air to enter the lungs.

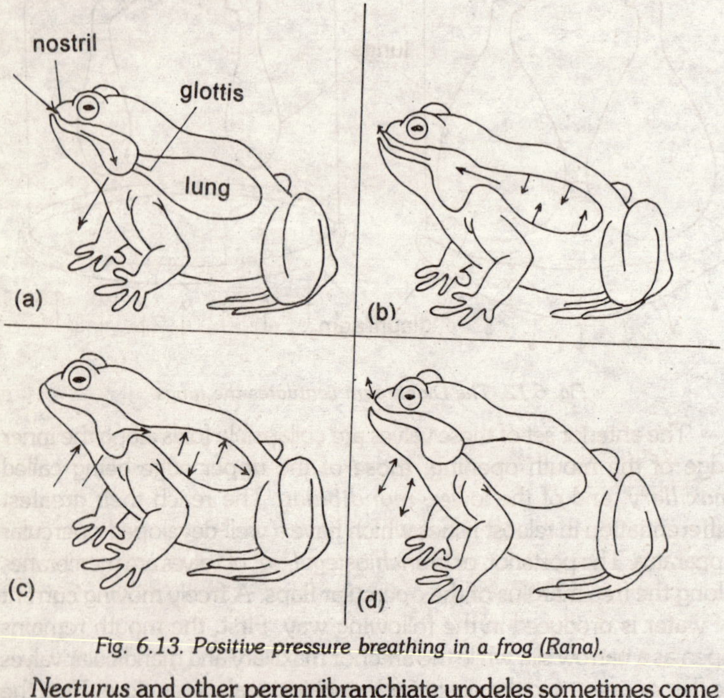

Fig. 6.13. Positive pressure breathing in a frog (Rana).

Necturus and other perennibranchiate urodeles sometimes come to the surface of the water and gulp air through the mouth, which may soon be seen escaping in the form of bubbles through the gill slit, although it is doubtful whether much of it reaches the lungs. This occasional air-gulping behaviour does not furnish fresh air for the external gills that lung outside the gill slits, since by means of muscles that cause them to wave back and forth, these animals obtain their suppy of oxygen dissolved in the water.

The intake of air in the frog, which may be taken as a representative to the anurans is accomplished by a combination of pumplike throat muscles and nostril valves. It will be seen that when the nostril valves open in the manner of lidgs and the throat muscles draw down, the oral cavity within is enlarged and air is necessarily inhaled. With the closure of the nostril valves and the contraction of the throat muscles, the lungs automatically become filled by the mouthful of air that is backward. The expiration of air alternates with inhalation and is accomplished by means of the contraction of body muscles.

3. Reptiles

The problem of getting air into the lungs of reptiles is much like that in the case of amphibians, although some improvement is seen since ribs and rib muscles furnish a mechanical means for admitting air that is not present in the practically ribless amphibians. This improvement, however, is ineffective in turtles, whose ribs form a boxlike armour of uncompromising rigidity. These animals still resort to the amphibian method of utilizing throat muscles and nostril valves, swallowing the air by "working the throat." No doubt the in-and-out movements of the turtle's head and neck aid in pumping air into the lungs while the pectoral muscles, which are *inside of the rib* in the bizarre retpiles instead of outside as in other vertebrates, are aided by the addominal muscles in bringing about the expulsion of air from the lungs. The usefulness of rib muscles in pumping air in and out of the lungs is very apparent in the panting snakes, lizards, and alligators.

The latter have an exceptionally elongated nasal passage-way with a curtain, or *velum,* that close off the inner choanal openings from the mouth cavity. This device makes it possible for the alligator to breathe with the mouth open under water while holding the drowning prey between the cavernous jaws, only the tip of the snout with the opening of the external nares being above the water line.

4. Birds

So long as a bird is not in flight it breathes by means of its rib muscles after the typical reptilian manner. When a bird flies, however, as already explained, the powerful pectoral muscles, on which flight depends, require and secure anchorage upon a rigid thoracic basket that does nto change shape with every breath. The bellows-like air sacs, which are filled and emptied by the action of the flying muscles rather than the rib muscles, furnish an effective means for irrigating the lungs of a flying bird with air, while the inactive rib muscles remain temporarily fixed and rigid.

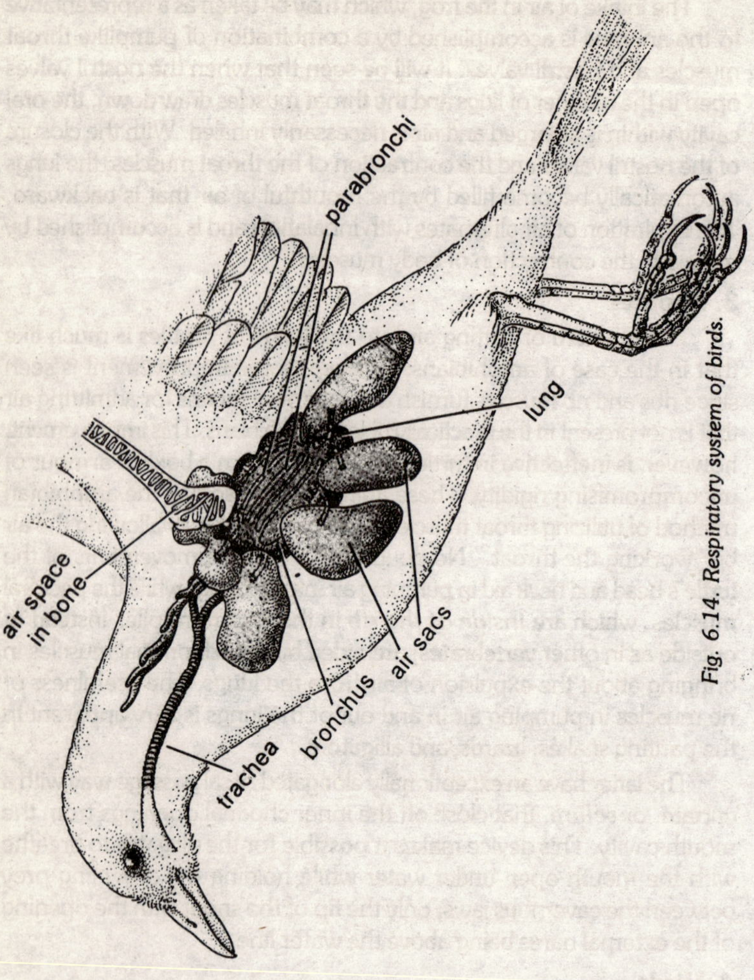

Fig. 6.14. Respiratory system of birds.

5. Mammals

In mammals both nasal and oral breathing are made possible by the backward migration of the glottis to a position in the posterior region in the throat. Nasal breathing, however, with the greater facilities thus provided for warming and moistening the inhaled air, and the added advantage of testing its quality by means of its passsage over the sensitive olfactory surfaces in the nasal chamber, is the better and more favoured method among mammals generally.

The outstanding advance in the breathing mechanism of mammals is furnished by the *muscular diaphragm*, which in lower vertebrates is fore-shadowed by the transverse septum that separates pleural and pericardial chambers from the body cavity. The diaphragm consists of a central tendinous component, from which extend radiating muscle fibres dervied from mid-cervical myotomes. The diaphragm when relaxed is shaped somewhat like an arched valut, and is perforated by the dorsal aorta, esophagus, azygos vein, thoracic duct, postcava, and vagus nerve. As its radiating fibres shorten by contraction, the valut of the cuplike diaphragm lowers or flattens, thus increasing the space within the thoracic cavity. Consequently the atmospheric pressure from without forces air into the lungs. At the same time the viscera within the body cavity are crowded down so that the abdominal wall bulges out. The muscular opponents of the diaphragm are the strong walls of the abdomen.

In addition to abdominal diaphragmic breathing, mammals also utilize the reptilian method of rib muscles to enlarge the thoracic cavity when inspiring air. The ribs are bent, like jointed leaves at an oblique

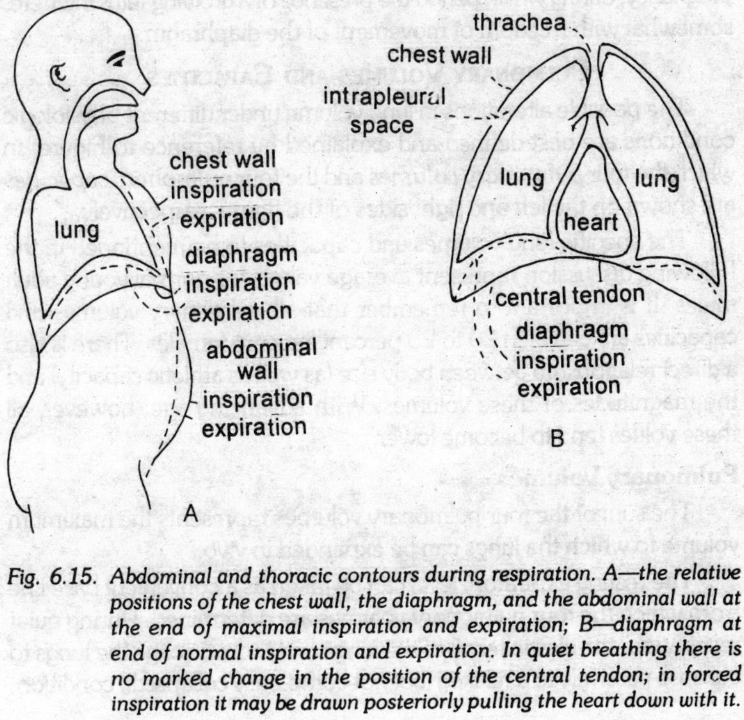

Fig. 6.15. Abdominal and thoracic contours during respiration. A—the relative positions of the chest wall, the diaphragm, and the abdominal wall at the end of maximal inspiration and expiration; B—diaphragm at end of normal inspiration and expiration. In quiet breathing there is no marked change in the position of the central tendon; in forced inspiration it may be drawn posteriorly pulling the heart down with it.

angle to the vertebral column and if acted upon by the intercostal muscles the movable sternum, to which they are attached ventrally, moves farther away from the relatively stationary backbone, thus enlarging the thoracic cavity in which the lungs are located. So it comes about that inspiration is effected not only by the depression of the diaphragm but also by the elevation of the ribs, both efforts calling for muscular activity.

Expiration, on the other hand, is to a large extent automatic through the elasticity of the stretched body walls, the taut cartilaginous ends of the but ribs, and the tensity of the expanded lung tissues.

In big heavy animals, abdominal or diaphragmic breathing predominates over rib breathing. Jumping animals, like kangaroos and monkeys, utilize rib muscles rather move than the diaphragm in respiration.

Breathing by means of the ribs is also more pronounced in human females than in males in whom abdominal breathing predominates. The reason for the sexual difference in the respiratory mechanism may be an evolutionary adaptation brought about in connection with pregnancy, during which period the presence of a growing fetus interferes somewhat with freedom of movement of the diaphragm.

Pulmonary Volumes and Capacities

The possible alterations in lung volume under different physiologic conditions are best defined and explained by reference to Figure, in which the four *pulmonary volumes* and the four *pulmonary capacities* are shown on the left and right sides of the figure, respectively.

The specific lung volumes and capacities to be mentioned in the following discussion represent average values for normal young adult males. It is important to remember that all pulmonary volumes and capacities are between 20 to 25 percent lower in females. There is also a direct relationship between body size (as well as athletic capacity) and the magnitudes of these volumes. With advancing age, however, all these volues tend to become lower.

Pulmonary Volumes

The sum of the four pulmonary volumes represents the maximum volume to which the lungs can be expanded in vivo.

The resting expiratory level can be taken as a convenient base line from which the four pulmonary volumes are determined. During quiet respiration, the elastic recoil of the lungs and thoracic cage the lungs to collapse passively to a relaxed (but not completely collapsed!) condition, and the volume of air then present within these organs is known as the resting expiratory level.

The tidal volume (*TV*) is the total volume of air inhaled and exhaled during quiet breathing. This volume usually is around 400 to 500 ml, but it can be considerably greater in large (or athletic) individuals. This fraction of the total pulmonary volume is so named because of its resemblance of oceanic tides, which move back and forth along the same path.

The inspiratory reserve volume (*IRV*) is the volume of air that can be inhaled above the normal tidal volume, usually about 3,000 ml.

The expiratory reserve volume (*ERV*) is that volume of air which can be expelled by forceful expiration following a normal tidal expiration. This volume is usually around 1,100 ml.

The residual volume (*RV*) is the quantity of air remaining in the lungs following a forceful expiration. This volume is approximately 1,200 ml.

Pulmonary Capacities

For convenience is discussing certain events pertaining to respiratory physiology, it is often useful to consider simultaneously two or more of the volume described above. Such combinations of respiratory volumes are known as pulmonary capacities.

The Functional Residual Capacity (*FRC*) is equal to the expiratory reserve volume plus the residual volume, i.e., (*ERV* + *RV*). This is the volume of air remaining in the lungs at the resting expiratory level as described above, and following a normal tidal expiration. The *FRC* volume is around 2,300 ml.

The Inspiratory Capacity (*IC*) is equal to the tidal volume plus the inspiratory reserve volume (i.e, *TV* + *IRV*). This is the total quantity of air an individual can inspire starting at the normal resting expiratory level, followed by a maximal inhalation. This volume is about 3,500 ml.

The Vital Capacity (*VC*) is the sum of the inspiratory reserve volume, the tidal volume, and the expiratory reserve volume (i.e., *IRV* + *TV* + *ERV*). The vital capacity thus represents the maximum volume of air that an individual can exhale during a maximal forced expiration following a maximal forced inspiration. This volume is approximately 4,600 ml.

The Total Lung Capacity (*TLC*) represents the sum of the four individual lung volumes, that is, the inspiratory reserve volume, the tidal volume, the expiratory reserve volume, and the residual volume (i.e., *IRV* + *TV* + *ERV* + *RV*). Therefore, the total lung capacity is the total volume of gas that is contained in the lungs at the end of a forced inspiration around 5,800 ml.

Anatomic and Physiologic Dead Space

As noted earlier in this chapter, the air that enters the respiratory tract is "insert" (insofar as respiratory gas exchange is concerned) until it has passed through the conducting airway (the nasal passage, mouth, pharynx, larynx, trachea, bronchi, and bronchioles). The total internal volume of these passageways is known as the *anatomic dead space*. this volume is difficult to determine in the laboratory, but tables have been prepared for estimating this value on the basis of age, sex, and tidal volumes as well as lung volumes. A useful rule of thumb is that the anatomic dead space in an adult is roughly equal in (ml) to the "ideal weight" of that individual in pounds.

The anatomic dead space is subject to change, depending upon the body position of the subject. In a healthy young man lying supine, the approximate value for the anatomic dead space is about 115 ml, and this value increases to around 156 ml in the semirecumbent position. During maximal expiration, the value falls to around 110 ml, whereas during maximal inspiration the anatomic dead space increases to about 230 ml, because of the increased that volume of the conducting airways as the pulmonary structures dilate, hence increase in overall volume capacity. As noted earlier, the values for the anatomic dead space are proportionally lower in women.

7

CIRCULATION

THE BLOOD

Blood, the red fluid of the blood vessels with which we are all familiar, is the transport medium of the body. As we have seen in the General Function Design of the body, it is the medium by which all living tissues are related to their external environment, that is the outside world. It is pumped to and fro between the tissues and the lungs and between the tissues and the alimentary canal to take up oxygen and nourishment respectively, while by its aid the products of oxidation and metabolism in the tissues are returned to the outisde world by the lungs, the skin, and the kidney. It is also the medium by which growth and repair substances are transported to the tissues and by which the various controlling glands of the body can distribute their chemical messengers. In spite of having all these functions which are constantly tending to change its composition, this remains remarkably constant.

Blood is a somewhat viscid fluid, and in man and in all other vertebrate animals, with the exception of two, is red colour. It consists of a yellowish fluid, called *plasma,* in which are suspended numerous *blood-corpuscles* or *cells,* the majority of which are coloured, and it is to their presence that the red colour of the blood is due. They form 45 percent of the blood volume. It also contains smaller bodies known as platelets.

Even when examined in very thin layers, blood is *opaque*, on account of the refractive power possessed by its two constituents, the plasma and the corpuscles. On treatment with ether, water, and other reagents, however, it becomes transparent and assumes a lake colour, in consequence of the colouring matter of the corpuscles having been discharged into the plasma. The average *specific gravity* of blood at 15°C. (60°F.) varies from 1055 to 1062. A rapid and useful method of estimating the specific gravity of blood was invented by Roy. Drops of blood are taken and allowed to fall into fluids of known specific gravity. When the drop neither rises nor sinks in the fluid it is taken to be of the same specific gravity as that of the standard fluid. In Van Slyke's (1942) method a series of appropriate solutions of copper sulphate are made up and a drop of blood added to each. With approximately normal plasma the result really depends on the haemoglobin concentration. The *taste* is saltish. Its *temperature* varies slightly, the average being 37.8°C. (100°F.). The blood stream is warmed by passing through the muscles and glands, but it is somewhat cooled on traversing the capillaries of the skin. Recently drawn blood has a distinct *odour*, which in many cases it characteristic of the animal from which it has been taken; it may be further developed by adding to blood a mixture of equal parts of sulphuric acid and water. The *reaction* of the blood is faintly alkaline and is dealt with in later chapter.

Blood Volume

The quantity of blood in an animal may be estimated in the following manner; a small quantity of blood is taken from an animal by venesection; it is defibrinated and measured, and used to make standard solutions of blood. As much blood as possible is removed from the animal by bleeding, and the blood-vessels washed out with saline solution until the washings are no longer coloured. The amounts of blood in the washings are then determined by reference to the standards. The application of the method to decapitated criminals has given values from 1/8 to 1/14. In man during life it is evident that other methods are necessary.

Those commonly in use involve the injection of dyes, usually "vital red" or "Evans blue."

The subject is of some practical interest as it is found that it may vary in disease. In talking of blood volume it shold be realised that we really usually mean *plasma volume,* as the dye does not enter the corpuscles.

The following method is that now most extensively used. Two samples of blood (5 c.c.) are withdrawn and oxalated. A 1.5 percent solution of the dye is injected into an arm vein—after five minutes, samples of blood are withdrawn and centrifuged. The colour is compared, the tint of the dye in the plasma compared with standards consisting of uncoloured plasma and dye solutions of known concentrations. This method gives a total blood volume of from 1/11 to 1/14 of body-weight. In an average man the amount is then about 6 *litres,* but this may be increased as in anaemia or at high altitudes. Similar results have been obtained by studying the change in blood concentration which occurs when plasma is injected.

It is possible also to withdraw a number of the subject's blood corpuscles, to label them with radio-active phosphorus, then to reinject them and determine their dilution with a Geiger counter. This method gives a slightly lower value than the dye method.

All methods depending on the dilution of an injected substance have the disadvantage that they measure only the circulating volume of blood and not necessarily that available in the blood-depots, which may or may not be full at the time of the estimation. Any state which causes them to contract, such as excitement and some diseases, may upset results when they are most needed, but in the same individual at rest repeated examinations under uniform conditions give reasonably constant results, the experimental error being quite small. The resting blood volume is relatively much greater in children that in adults and those adults which take active exercise, but mere rest in bed is sufficient to cause a decrease.

The Regulation of Blood Volume

The blood volume obviously depends ultimately on the water content of the body as a whole; but even when conditions arise which tend to change it, it keeps approximately constant. This is made possible by the fluid passing through the capillary walls in either direction according to requirements, while excess is lost by the kidney and skin. Thus the injection of small quantities of fluid changes the blood volume for only a short time, for a lowering of the blood pressure allows water to leave the vessels while excessive loss of water has the opposite effect. The subject is closely related to the water-balance of the body and to the formation of lymph, which are dealt with in separate sections.

The circulating blood volume is increased in a number of states, notably mental stress, muscular exercise, and high temperature. This is no doubt due in part to the constriction of the blood-depots, for there is an increase of cells as well as plasma. More rapidly moving food also tends to sweep static corpuscles into the circulation. It has been calculated that five-sixths of the total blood is in the heart and larger blood-vessels at any one time.

A reduction of circulating blood volume is brought about by a number of conditions. It may occur in any generalised anhydraemia (*i.e.* loss of body water), in haemorrhage, or it the blood pools in the capillaries, as may occur when there is extensive damage to tissues. Prolonged standing has a similar effect from the pooling of blood and lymph in the legs (see "The Effect of Gravity on the Circulation"). A reduction is also produced by exposure to cold, apparently because of the vasconstriction, of the skin brought about by the nervous system, for it does not occur if the spinal cord is cut (Barbour). The skin vasoconstriction, by raising capillary pressure, presumably increase the amount of tissue fluid and excretion by the kidney. It seems possible, too, that continued mental stress acting similarly may eventually reduce the blood volume, and hence the effect of a holiday in causing a subject to "look better" may be due to additional blood filling the superficial vessels more fully. There are, however, as yet no experimental data on the subject.

Functions of Blood

Mephistopheles, in Goethe's *Faust,* is biologically quite right when he exclaims: *"Blut ist ein ganz besonderer Saft!"* (Blood is a very peculiar fluid!)

In the first place it is not a lifeless fluid incapable of metabolic changes, but a *circulating tissue,* made up of detached living cells floating in a liquid plasma. The fact that it moves about marks it off distinctly from all other tissues. It is conspicuously kaleidoscopic in character, changing constantly in its intimate composition as it passes through the different parts of the body, for it is not only the common carrier to the various tissues of everything needful for their maintenance, but also the collector *from* the tissues of the products of metabolism, such as carbon dioxide and soluble nitrogen compounds.

In the last analysis cells of every kind within the animal body sooner or later contribute something to the blood and receive something in return, so that the blood reflects the entire metabolism of the body. In spite of this fact its component parts are normally kept at nearly a constant level. When serious deviations occur, as in the excessive loss of water from the blood during cholera, or the diminution of the number of blood cells in anaemia or hemorrhage, pathological consequences are sure to follow.

In general the following functions may be performed by the blood.

(a) Equilibration of Water Content

Water is the fluid universal that facilitates the internal transport of materials, making good loses by evaporation and otherwise, as well as preventing the drying up of tissues. It is also the great solvent of all sorts of substances in the food and throughout the tissues. The degree of activity exhibited by any tissues is directly dependent upon the fluidity of the cytoplasm within its cellular units, which in turn is ultimately a matter of the amount of water supplied by the blood.

(b) Liberation of Energy

Tissues "burn" in the presence of oxygen thus releasing energy, which is what constitutes "living." Certain softer tissues, like the muscles for example, lend themselves particularly to this oxidative process, while others, such as skeletal tissues, resemble more the iron girders of a "fire-proof" building, and do not burn as readily. The oxygen necessary for the release of stored energy is delivered to the tissues of vertebrates by the *haemoglobin* in the red corpuscles of the blood.

(c) Distribution of Food

From one point of view blood is a solution or an emulsion of food substances carried in the plasma. To some extent also it transports food as solid undissolved particles engulfed in the white blood cells. It is thus the grocery delivery boy for the cellular community.

(d) Regulation of Temperature

Body temperature, which results from the oxidation of the tissues, is equalized by means of the circulation of the blood, much as is the temperature of a building by the hot-water pipes of its heating plant. Such equalization is necessary because of the unequal production of heat-energy by different tissues of the body.

In so-called "warm-blooded" animals the body temperature is maintained at a practically constant level, regardless of the temperature of the environment, thus enhancing the animal's independence. "Cold-blooded" animals, on the other hand, owing partly to the low oxygen-carrying capacity of their blood, have a body temperature which fluctuates in response to that of the surroundings. A cold-blooded animal is consequently a thermal slave to the environment in which it finds itself.

It has been demonstrated, however, that even cold-blooded fishes when ill may show fever-like fluctuations in temperature.

(e) Transmission of Chemical Substances

Hormones, the chemical messengers from endocrinal glands, frequently perform metabolic feats at some distance from their point of origin, after travelling along the blood highways. Drugs and poisons introduced into the organism likewise gain ready disposal over the body by means of the circulating blood. This is why a person with malaria, for example, feels "sick all over," since the blood carrying the organisms producing malarial poison literally goes over the whole body.

(f) Defense against Parasitic Invasion

Troublesome foreign invaders, such as infective bacteria and protozoans, are regularly combatted by white blood cells which capture and devour them. The cures of most "catching" diseases depend upon the successful outcome of this function of the blood.

(g) Disposal of Cell Wreckage

The blood is a continual funeral cortege, in which are being borne away the "ashes" of dead cells, foreign bodies, bacterial products, and wastes of metabolism generally.

GENERAL PROPERTIES OF THE BLOOD

Composition

If something is added to a sample of blood to keep it from clotting and it is permitted to stand—or better, because it goes faster, if the blood is centrifuged—it separates into two approximately equal fractions. The rapidly rotating centrifuge drives the heavier elements, e.g., the red blood cells, to the outer (lower) part of the tilted test tube. The upper, lighter half is the liquid portion of the blood, the blood plasma; it is pale yellow, some what vicious, and sticky. The *plasma* usually constitutes about *55 percent* of the total, the cells 45 percent. The instrument used for this determination is the *hematocrit,* and the measurement is called the "hematocrit reading." Thus if a certain blood sample gave a hematocrit reading of 40, it would mean that the sample consisted of 40 per cent cells, 60 per cent plasma.

Colour

Everyone knows that blood is red, but it is not, necessarily. The colour of the blood depends on the nature of the hemoglobin. Much will be said about hemoglobin later. Let us only say now that it is the coloured pigment within the red cells of the blood, that it is the agent which transports oxygen, and that the colour of hemoglobin varies, depending on whether or not it is carrying oxygen. As the blood passes through the lungs, the hemoglobin picks up oxygen and is then called *oxyhemoglobin*. Oxyhemoglobin gives blood a bright scarlet colour. However, if the hemoglobin contains no oxygen whatever (as is the case after death), the blood is dark blue, nearly black. In passing through the living body, the blood gives up some, but by no means all, of its oxygen. Therefore blood which has passed through the tissues (*venous blood*) has less redness and more blueness than fully oxygenated (arterial) blood. As blood flows from a cut it is red, because it has not had a chance to give up much of its oxygen. Hemoglobin may take on other colours as well. For instance, in case of carbon monoxide poisoning, it is a distinct cherry red.

Density

A moment ago we said that 6.4 kg. of blood was *roughly* 6.4 liters. It would have been *exactly* if the density of blood were 1. However, blood is not only thicker than water; it is also slightly heavier. To determine the density of blood, it is necessary to know the volume and the weight. A droplet from a skin prick would give the weight accurately enough, but it is difficult to measure small volumes. It is apparently even more difficult to withdraw larger volumes of blood, except by nurses trained for the task. Therefore, for ordinary office procedure, and indirect method is used.

If one had a fluid of known density with which the blood did not mix and placed a droplet of blood in it, if the droplet neither rose nor sank, it would have the same density as the fluid. And if instead of a single fluid, a mixture of two fluids, one lighter and one heavier than blood, were used, the density of the solution could be adjusted to the density of the droplet of blood. Now, benzene has a density of 0.9 and chloroform has a density of 1.5. The two mix together, but blood does not mix with them. It happens that 10 cc. of the mixture, containing 7.3 cc. of benzene and 2.7 cc. of chloroform, weighs 10.6 gm; the density of the mixture is therefore *10.6, the density of normal blood.* If one places a drop of blood having a normal density of 1.06 in the mixture, it will neither rise nor fall. However, suppose the blood is abnormal and falls, then known volumes of chloroform can be added to adjust the density. *If 0.5 cc. of chloroform must be added, what is the density of a sample of blood?*

Density equals weight divided by volume. The *volume* of the sample is the standard 10 cc. plus 0.5, or *10.5 cc.* The total *weight* of the sample is 7.3 cc. of benzene multiplied by 0.9 (density of benzene), or 6.57; plus 3.2 cc. of chloroform multiplied by 1.5 (density of chloroform), or 4.8; this equals *11.37 gm.* The density of one cubic centimeter of the sample would be 11.37 (weight) divided by 10.5 (volume), or 1.08.

Osmotic Pressure

Whole blood has practically the same osmotic pressure as that of plasma. This is not only because the red cells do not behave as molecules, being relatively much too large, but because there are so few of them compared to the number of molecules. It will be recalled that a molar solution ("one containing a molecular weight of a substance in one liter of solution") will exert an osmotic pressure of 22.4 atm. This solution will contain 6.06 x 10^{23} molecules. In a liter of blood there are 5,000,000 red cells per cu. mm., or 5 x 10^{12} red cells which is only one one-billionth as many particles as in a molar solution.

The osmotic pressure of blood is rarely determined but can be calculated from the freezing point. The freezing point of plasma is – 0.56°C. A molar solution has a freezing point of –1.86° and exert an osmotic pressure of 22.4 atm. Therefore, 0.56/1.86 X 22.4 equals 6.7 atm. This is the osmotic pressure of blood, a considerable pressure, equivalent to column of water 69,251 mm., or 230 ft. in height.

The pH of the Blood

The term "pH" was introduced by S.P.L. Sorenson in 1909 as a handy way to express the degree of acidity or alkalinity of a solution. If a solution is neutral, i.e., neither acid nor alkaline (e.g., distilled water), it is said to have a pH of "7." Going upward on the scale, "8-9-10" indicate constantly rising *alkalinity*; going downward, "6-5-4" indicate constantly increasing *acidity*. Thus, a solution with a pH of 8.5 would be fairly alkaline, but one whose pH is 10.5 is more strongly alkaline. A solution whose pH is 6.3 is weakly acid; one with a pH of 2.5 is much more strongly acid. The *pH of the blood is 7.35,* which means that it is a very weak alkaline solution. The cells of the body cannot stand much departure from this figure. A pH of 8 on the one hand or much below 7 would certainly be fatal. Shifts towards the acid side occurs in diabetes, toward the alkaline side, in hyperventilation (over-breathing) and sometimes after taking large doses of sodium bicarbonate.

Choice of the number 7 to express neutrality was not because 7 is presumably a lucky number. Although ordinarily water is not thought of as ionizing-for one thing, it does not conduct the electric current-nevertheless, to a small extent it does, thus: HOH produces (H^+) plus (OH^-). The extent to which this happens is only one part hydrogen ions in 10,000,000 parts of water, or 10^7. So Sorenson choose 7, the power that expresses the ionization of water. A solution with pH of 6 contains one part hydrogen ions per 1,000,000 parts of water, or ten times as many hydrogen ions as water. It is therefore an acid.

Measurement of the pH may be done colorimetrically. There are a large number of dyes, called "indicators," which change colour at various pH's. Or it may be done electrically by means of a meter such as "Beckman's pH meter."

Blood is a highly complex extracellular fluid which is composed of two parts-a faintly yellow transparent fluid known as the *plasma* and floating in this fluid are the *blood corpuscles.* The plasma constitutes about 55 percent, and the cells about 45 percent of the total volume of human blood.

PLASMA

The plasma is a non-lining fluid of much more chemical complexity than appears in the test tube. It contains a constantly changing variety of substances in solution, chief among which are dissolved food materials on the way to cellular delivery, and waste products that are being collected for elimination. There are also present *enzymes* of diverse sorts which activate chemical changes: *opsonins*, that prepare trespassing bacteria for phagocytosis; *hormones,* the chemical messengers from endocrine glands on their way to the performance of tasks of internal regulation; *anti bodies,* and other problematical substances engaged in constant warfare against harmful invasion.

Plasma is made up of about 90 per cent water and 10 per cent solids. The solids include about 7 per cent proteins, and 0.9 per cent inorganic matter; the remainder is non-protein organic matter.

Albumin (4.7%) is the major component of the plasma proteins. Its molecular weight is about 69,000. It is soluble in distilled water. It is precipitated by full saturation with ammonium sulphate. It is heat coagulable. It is made up of a single polypeptide chain containing 575 aminoacids. It is synthesized in liver. A healthy liver is able to synthesize about 10-15 gm. of albumin per day. The molecules of this protein do not readily pass through vessel walls, so they aid in keeping fluid in the blood vessels by increasing the osmotic pressure of the blood. Its isoelectric pH is 4.7.

Albumin performs several functions as stated below:

(i) The albumin released from the liver is circulated to all tissue which can utilize the amino acids of albumin for the synthesis of their own proteins. Thus, it serves as a source of amino acids.

(ii) Nearly half of the calcium present in blood is loosely bound to albumin at different sites.

(iii) It helps in the transport of thyroxine and steroid hormones.

(iv) It binds the toxic products of metabolism and carries to the sites of excretion

(v) It plays a major role in the osmotic regulation of the fluid distribution.

Globulins (1.5%). It has a molecular weight varying from 90,000 to 1,300,000. Because of the high contents of carbohydrates it is highly soluble, and is not coagulated by heat. It is a mixture of several, globulins (1) α_1, and α_2 *globulin*-mol. weight ranges from 41,000 to 200,000 with isoelectric pH 5.1. α_1 globulin is present in small amount. It is said to have an inhibitory effect on trypsin. It consists of two fractions: (a) one fraction combine with bilirubin, (b) another fraction helps in the carriage of lipids, steroids and glycoproteins. α_2 globulin consists of α_2 macroglobulins, mucoproteins, ceruloplasmin and hepatoglobulins—the latter combines with free haemoglobin in the plasma. (ii) β-*globulin*. It has molecular weight varying from 90,000 to 1,300,000. Its isoelectric pH is 5.6. The β globulin is primarily concerned with the carriage of the insoluble lipid along with fat soluble vitamins. It is also concerned with the iron transport. Prothrombin is a β-globulin. (iii) γ *globulin*. Its molecular weight ranges from 150,000 to 190,000. Its isoelectric pH is 6.0. Antibodies belong to this class.

Fibrinogens (0.3%). Fibrinogen is another glycoprotein or is globulin in nature. It has molecular weight of 340,000 daltons. It has an isoelectric pH of 5-8. It is made up of 6 peptide chains held by S-S-linkage. It is synthesized by the liver at a rate of 4 to 5 gm. per day.

Fibrinogen is primarily concerned with blood clotting during which it is converted to insoluble fibrin by the action of thrombin.

After depletion of plasma proteins, such as by severe haemorrhage or after blood donation, the plasma proteins come to the normal level in about fourteen days. Fibrinogen is regenerated first, then comes globulin and last of all serum albumin.

The inorganic material of blood includes various anions and cations such as chloride, bicarbonate, phosphate, sulphate, sodium, potassium, calcium and magnesium. Minute amounts of iron and iodine are also found. Some of these are essential for cell metabolism, and some function as buffers to maintain the pH of the blood within a normal range.

The formed elements of the blood consist of erythrocytes, leucocytes and platelets.

Erythrocytes or Red Blood Corpuscles

First described by the Duch microscopist *Leeuwenhock* in 1674 as small round globules.

Erythrocytes are found in all vertebrates except the larvae of eels ('leptocephalus' larvae) and a few antarctic fish of the family chaenichtyidae. Red corpuscles are for the most peculiar to vertebrates, although a few invertebrates, including the worms *Glycera* and *Phoronis,* the "blood clam" *Arca* and the holothurian *Thyone,* also possess red blood cells.

Shape

In all vertebrates except mammals the erythrocytes are oval shaped, biconvex and nucleated. In mammals except camel, the mature erythrocytes are circular, biconcave and non-nucleated. Camel has oval shaped erythrocytes resembling non-mammalian red blood cells in form. The biconcave shape increases the surface area, while the absence of nucleus increases the haemoglobin contents.

Size

In size erythrocytes, measured from dried smears of blood, range from 75 micra in the caudate *Amphiuma* (Salamander) to 2.5 micra in the musk deer *Tragulus.* Other measurements in micra for the largest diameter of blood corpuscles from dried smears have been reported as follows:

Proteus, 58; frog, 22; turtle, 20; carp, 17; pigeon, 14; lamprey, 13; chick, 12; conger eel, 11; elephant (biggest among mammals), 9.4; man, 7.5; dog, 7.3; rabbit, 6.9; cat, 6.5; horse, 4.6; goat, 4.1.

Human erythrocytes average 7.5 µu in in diameter 2.4 µu in thickness at the edge, but only 1 µu thick in the centre, and a volume of 90 Cu µu. One erythrocytes contains 30 µu gm. of haemoglobin which

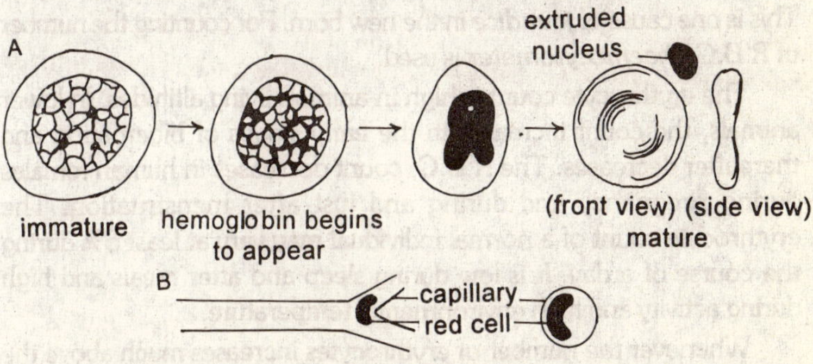

Fig. 7.1. Red blood cells. A—Stages in the formation of human red cells; B—bending and twisting of red cells as they pass through capillaries. Their supporting elastic framework enables them to resume their normal disc shape after the tension is released.

constitutes 90% of the dry weight. The erythrocytes smaller than 6 microns are called *macrocytes*. The erythrocytes of variable shape and size are called *poikilocytes*. The total area of an individual's erythrocytes amounts to about 3000 sq. m. which is 1500 times the whole surface area of the body. If all the red cells of an individual were put in a straight line, its length would almost equal one half the distance from earth to moon.

The erythrocytes are acidophilic and therefore stained with eosin. The erythrocytes are very flexible; they are able to squeeze through capillaries of only half their diameter and regain their orignal shape. Mammalian mature erythrocytes lack mitochondria, Golgi body, endoplasmic reticulum and other cell organelles. The human erythrocytes has a highly oriented structure of lipids, proteins and haemoglobin consisting of approximately 64% water, 28% haemoglobin (approximately 100 million haemoglobin molecules), 7% lipids or fatty materials, and the remaining 3% sugars, salts, enzymes and other proteins. Under the microscope, a single red cell seems to have a light brown or yellowish colour. But when seen in bulk, the erythrocytes appear to be red.

Number

The number of erythrocytes differs in different animals as well as the physiological state of an individual. There are about 5 million red blood cells per cubic millimeter of blood in the adult human male and 4.5 million per cubic millimeter of blood in the adult female. In infants, the count is 6 to 7 million, whereas in foetus, 7.8 million. In the first ten days of the postnatal life, a large number of red cells are destroyed. This is one cause of *jaundice* in the new born. For counting the number of R.B.C., *haemocytometer* is used.

The erythrocyte count is high in animals living altitudes. In lower animals, the count increases in the early stages of hibernation and thereafter decreases. The R.B.C. count decreases in human females during pregnancy and during and just after menstruation. The erythrocyte count of a normal individual may vary at least 5% during the course of a day. It is low during sleep and after meals and high during activity and high environmental temperature.

Whenever the number of erythrocytes increases much above the normal value, the stage is called *polycythemia*. The persons living at high altitude (Kashmir, Tibet etc.) have rosy cheeks.

Life Span

Within a period of 100-120 days (maximum 124 days in dog), red blood cells become very fragile. Worn out red cells are removed from circulation and destroyed by spleen and liver. In one second, there is a destruction of about 2.5 million of erythrocytes (5×10^{11} per day). This is the rate of its production. This is why the number is maintained in the body. Each red blood takes about 50,000 rounds of the body before its death. The destruction of red blood cells in healthy person produces about 250 mg. of bile salt per day.

Toble 7.1. Erythrocyts count in a Various Vertebrates

Name of animal	Number per mm^3 of blood
Fish	
Lamprey	1,33,000
Ray	2,30,000
Flounder (Pleuronectus)	2,000,000
Amphibians	
Salamander (Amphiuma)	95,000
Frog	4,00,000
Reptiles	
Turtle	6,30,000
Birds	
Humming bird	6,500,000
Ostrich	1,600,000
Mammals	
Human-being	5,000,000
Rabbit	7,000,000
Sheep	10,000,000
Llama	11,000,000

Erythrocyte sedimentation rate

E.S.R. When blood to which an anti coagulant has been added is allowed to stand in a narrow tube the red blood corpuscles form rouleaux in which the cells are arranged face to face like a pile of coins and the aggregates gradually settle down (sediment) to leave a clear zone of plasma above. The sedimentation rate is measured as the length of the column of clear plasma above the sedimentatating erythrocytes at the end of one hour. The E.S.R. depends almost entirely on the relative concentration in the plasma of fibrinogen, and globulins. The asymmetry of the molecules promotes the aggregates of red cells into rouleaux and clumps. Almost all infections, whether acute or chronic, are accompanied by a rise in α_2- and γ-globulins and most wasting and destructive disease by an increase in fibrinogen. The E.S.R. is thus a simple, rapid procedure for revealing such changes. It is useful in detecting the presence and activity of disease but it is quite unspecific. The sedimenation rate also varies according to the red cell count, the lower the red cell count the greater the sedimentation rate.

The metabolism of erythrocytes

Since the red cell can be bisected without its contents escaping its seems likely that it consists of a lose network or of a jelly which is none the less flexible enought to allow the cells to assume the roughly paraboloid form in which they pass through capillaries as a succession to tiny plugs separated by plasma. The usual biconcave disc shape is regained immediately when the cells pass into a larger vessel. The elongated paroaboloid form of the erythrocyte when passing through a capillary presents a much larger surface area next to the capillary wall for the transference of oxygen.

The stroma or framework of the erythrocyte consists of some 50 percent protein and 10 percent lipid, mainly phospholipid arranged in three layers. The outer layer of sialic acid determines the cell's electrical charge. It is traversed by pores about 0.4 mm wide which allow the passage of ions and molecules not exceeding a molecular weight of about 100.

Although erythrocytes have no nuclei and normally contain no glycogen they are metabolically quite active utilizing some 1.5 to 2.2 mmols of glucose per litre per hour equivalent to the liberation of 100 J (25 cal). This energy is required to pump out sodium and pump in potassium against electrochemical gradients and to reduce methaemoglobin to haemoglobin.

The metabolism of glucose in red blood cell, which is essential for its survival, takes place along the two common pathways, glycolysis and the pentose phosphate pathway, 80 percent of the glucose being broken down by the former and 20 percent by the latter. The red cell has no mitochondria and therefore no citric acid cycle.

In the pentose phosphate pathway, the first enzyme involved is the *glucose-6-phosphate dehydrogenase* (G-6-PD) in the operation of which large amounts of reduced NADP is utlized by the red blood cell to maintain glutathione in the reduced state.

$$\text{glucose-6-phosphate} + NADP^+ + H_2O \xrightarrow{\textit{G-6-PD}} \text{6-phosphogluconic acid} + NADPH + H^+$$

$$GSSG + NADPH + H^+ \xrightarrow[\textit{reductase}]{\textit{glutathione}} 2\,GSH + NADP^+$$

The function of the reduced glutathione within the erythrocyte is not clearly understood, but it probably prevents the oxidatión of haemoglobin to methaemoglobin and its sulphydryl groups play a part in maintaining the structure of the red cell and also of the globin part of the haemoglobin molecule.

In the population of many parts of the world, there is a deficiency of the (G-6-PD) in the red blood cells. This is an inherited condition which affects more than 100,000,000 people in the world at the present time. This deficiency is a common cause of jaundice in babies, it also is responsbile for heamolysis after the administration of certain drugs, particularly the antimalarials such as primaquine, and certain antibacterial drugs. Haemolysis also occurs in G-6-PD-deficient persons after the consumption of the bean of *Vicia faba* which is a common foodstuff in many countries. Many enzyme deficiencies have been discovered but only a few have serious effects.

Hydrogen peroxide is formed in small amounts in normal red cells during metabolic processes and also is generated within the cells as a result of the action of certain drugs, including those which cause haemolysis in G-6-PD deficiency. If not destroyed as rapidly as it is produced, hydrogen peroxide would oxidize haemoglobin to methaemoglobin and then, irreversibly, to oxidized degradation products; it is probable that other substances essential for the integrity of the red cell would also be oxidized, and haemolysis would follow.

There are two protective mechanisms against hydrogen peroxide within the red cell. Glutathione peroxidase uses hydrogen peroxide to peroxide to oxidize reduced glutathione. Normal activity of glutathione peroxidase depends on the presence of adequate amounts of reduced glutathione which requires normal functioning of the pentose phosphate pathway, including G-6-PD. The other protective agent is the enzyme catalase which decomposes hydrogen peroxide to water and oxygen. Chemically, catalase resembles haemoglobin in being an iron-containing haemoprotein. The relative importance of glutathione peroxidase and of catalase is not fully understood.

At least one hundred enzyme systems have been identified in the red cell. But because it has no nucleus it cannot synthesize proteins (including enzymes).

HEMOGLOBIN

The most important protein of the red cell is hemoglobin. This protein finds an important place among the other proteins due to its ability to combine with oxygen reversibly. It carries oxygen from the lungs to the tissues, and CO_2 from tissues to lungs. Normally hemoglobin is solely confined to the red blood cells.

It is a conjugated protein, possessing heme as is prosthetic group and globin as its protein portion. The heme part of the hemoglobulins is similar but the globin portion shows considerable variations in the amino acid composition and sequence. Each molecule of hemoglobin contains four molecules of heme attached to one molecule of globin.

Heme is a chelate compound in which iron is bound to protoporphyrin nucleus. The protoporphyrin nucleus itself is made up of four pyrrole nuclei, attached to form a ring structure. Iron is found at the middle of this ring. The ring structure is designated as tetra-pyrolic ring; or protoporphyrin nucleus. The heme nucleus remains attached with the globin molecules by its coordinate valancy at the histidine residue. Since, each hemoglobin molecule contains four heme molecules, four such attachments are found in each of the hemoglobin molecule.

As regards the globin portion of hemoglobin, it is usually formed of various combinations of four different peptide chains which are called as alpha (α), Beta (β), gamma (γ), and delta (δ) chains. Sometime another chain known as epsilon (e) chain is also found in the hemoglobin molecules. Most of the human hemoglobin contain two alpha chains and two other chains. Normal adult hemoglobin which is represented as (HbA) consists of two alpha and two beta chains as is shown as α_2A β_2A or more simply as α_2 β_2. Approximately 90 percent of the hemoglobin found in the adult man belongs to this group. Another hemoglobin known as human foetal hemoglobin which is represented as $\alpha_2\gamma_2$. A phemoglobin which forms only a mines Component of adult hemoglobin is enoun as HbA_2 and it contains two alpha chains and two gamma chains and is therefore represented as Hb. As Hb_2A and contains two alpha and two delta chains. It is represented as $\alpha_2\delta_2$. One more form of hemoglobin which is alleged to be embryonic hemoglobin contains two alpha and two epsilon chains and is represented as $\alpha_2\varepsilon_2$.

The four peptide chains of globin are believed to be held together by non-covalent forces. The structure of globin portions of hemoglobin has been well studied now. The alpha chains, are made up of 141 amino acid, the beta and gamma chains of 146 amino acids each. The δ chains are also made up of 146 amino acids. There are differences in the amino acid sequence in these chains, however, there do also exist some striking similarities in these chains. For example there exists twenty four consecutive identical amino acid in the beta and gamma chains. Also, there occurs 65 identical amino acids in the alpha and beta chains, and 108 in the beta and gamma chains. Subsequent studies have revealed that beta and delta chains are even more similar showing 138 identical amino acids. The sequences of the N-terminal and C-terminal ends of the alpha, beta and gamma chains in the globin portions has been shown below:

	N-terminal residue	C-terminal residue
Alpha chains	Val-lew-ser-	—lys-Tyr-arg
Beta chains	Val-Hist-lew	—lys-Tyr-hist
Gamma chains	Gly-Hist-Phe	—Arg-Tyr-Pha

Most of the mammalian T hemoglobins show a molecule weight of the order of 67,000 and their shape is nearly spherical.

Chemical Properties of Hemoglobins

The chemical properties of hemoglobins have been discussed as follows:

1. Combination with Oxygen

Reduced hemoglobin can loosely combine with Oxygen forming Oxyhemoglobin. The attachment of oxygen, occurs at iron in the heme portion. Iron is present in the ferrous form both in hemoglobin and oxyhemoglobin. Oxygen remains attached with the unpaired electrons of iron. Each heme can bind only one molecule of oxygen. Since, each molecules of hemoglobin contains 4 molecules of heme, hence, one molecule of hemoglobin can maximally combine with four molecules of O_2. The combination of hemoglobin with oxygen is governed by partial pressure of oxygen (pO_2) and partial pressure of Co_2 (pCo_2). The former favours combination where the later favours dissociation.

$$\text{Reduced Hemoglobin} + 4o_2$$
$$\uparrow\downarrow$$
$$\text{Oxyhemoglobin (Hb.}4o_2\text{)}$$

2. Combination with Carbon Monoxide

The heme portion of hemoglobin can also combine with carbon mono-oxide forming Carbonyl hemoglobin whch is a very stable compound. Affinity of hemoglobin for carbon monoxide is approximately 210 times more as compared to oxygen. In the presence of Co_2 therefore, hemoglobin loses capacity to combine with O_2 and asphyxiation starts. In cases of Co poisioning, the patient must be removed immediately from the poisoned atmosphere to fresh air and should be given artificial respiration using a mixture of Co_2 and Oxygen. In serious cases of such poisoning, it might also become necessary to give a blood transfusion.

$$\text{Hb} + \text{Co} \longrightarrow \text{Hb.Co}$$

Carbonyl hemoglobin.

3. Combination of Carbon Dioxide

Hemoglobin does also combine with Co_2 through its amino groups present on the globin portion forming carbamino-hemoglobin compound. Formation of this complex is a reversible process as has been shown below.

$$\text{Hb.NH}_2 + \text{Co}_2 \rightarrow \text{Hb.NH.COOH} \rightarrow \text{HbNH.COO-HbNHCOO}^- = \text{H}^+$$

Normally, about 10 percent of the total CO_2 is carried by the blood in this form.

Hemoglobin cannot combine with the hydrogen cyanide, or other cyanides.

Methemoglobins

Administration of certain drugs, or exposure of hemoglobin to some poisons leads to oxidation of hemoglobin to methemoglobin in which iron remains present in ferric form. Methomoglobins therefore contain one extra positive charge per heme molecule and four extra positive charge in all. Each heme can therefore bind one negatively charged OH^- ion.

The drugs which can lead to the conversion of hemoglobin to methemoglobin are various sulphonamides, sulphonals, antipyrine nitrites and nitrobenzenes.

Methemoglobin can combine with HCN without any difficulty. Advantage is therefore taken of this property of methemoglobin in removing cyanide poisoning. A small amount of hemoglobin is allowed to be converted into methemoglobin by giving a mixture of sodium nitrite and sodium thiosulphate to the patient with cyanide toxicity.

Methemoglobin is formed and it readily takes up cyanide forming cynamethemoglobin which is non toxic. The residual cyanide is changed by sodium thiosulphate into thiocyanide which is also non-toxic, and can be readily excreted.

Abnormal Hemoglobins

Sometimes, due to faulty DNA formation or Mutational changes in the DNA structures abnormal hemoglobins are formed which can be separated by electrophoresis. Some of the amino acid in these hemoglobins differ with those of the normal hemoglobins. Although, most of such abnormal hemoglobins are harmless but some of them may create problems during conditions of stress. Their oxygen carrying capacity remains unaltered.

Some of the abnormal hemoglobin have been mentioned below.

In hemoglobin C, the glutanic acid in the beta chain at position 6 is replaced by lysine. In hemoglobin D sparagine at position 68 in the alpha chain is replaced by lysine. In hemoglobin E, glycine in the beta chain at position 26 is replaced by lysine. In hemoglobin-G, glumatic acid in the alpha chain at position 36 is replaced by glutamine. In hemoglobin-M, histidine in the alpha chain at position 58 is replaced by tyrosine. In hemoglobin-Q, glutamine in the beta chain at position 121 is replaced by lysine. In Hb-S glutamine in the beta chain at position 6 is replaced by valine.

Anaemias

Normally, men contain 13 to 16 g hemoglobin and women contain 12 to 15 g hemoglobin per 100 ml blood. In man, a value of 15 g per 100 ml is taken to be the average hemoglobin. In women this average is about 14 g per 100 ml. In the blood hemoglobin falls below about 11.9 g per 100 ml. The consequences become undesirable. The state of an individual whose hemoglobin content is reduced below the normal range, or the number of red blood cells is reduced below the normal range is known as anemia.

There may be many factors responsible for development of anaemic conditions, for example, excessive loss of blood, increased rate of destruction of red cells defective formation of the red cells or blood loss or blood may be lost by excessive bleeding from wounds, or in women due to menstruation. The rate of destruction of the red cells may be increased by certain hemolytic agents of metabolic, drug or bacterial origin formation of hemoglobin may be reduced during the deficiency of iron, copper, vitamin B_{12} in the diet, or due to lack of intrinsic factor in the gastic juice, or due to toxic inhibition of the bone marrow. Development of tumours, or fibrous or leukemic growth in the bone marrow also lead to anaemic conditions.

The classical examples of anemia are hypochronic anaemia pernicious anaemia and sickle cell anaemia. A hypochronic anaemia is that in which red cells contain less hemoglobin than the normal. It may develop due to lack of iron and copper in the diet. The proportion of hemoglobin is reduced in comparison to the number of red cells. The size of the red cells is reduced. In the prematurely born children who have low internal store of iron, such a type of anaemia is very common. In the foetus, iron is mostly stored from mother towards the latter part of gestation. In the young women too, this type of anaemia is very common due to dietary deficiency of the said elements, and due to excessive menstrual bleeding.

In pernicious anemia, the number of the red blood cells as well as haemoglobin content are reduced followed by a rise in the blood plasma iron. This type of anaemia is believed to develop due to deficiency of intrinsic factor in the gastric juice or due to deficiency of vitamin B_{12} in the diet. Vitamin B_{12} cannot be absorbed in the absence of intrinsic factor, therefore, presence of both the factors is essential for prevention of the development of this type of anaemia.

In sickle cell anaemia, the red cells become crescent shaped and also abnormality is incorporated in the globin portion of hemoglobin. Such red cells become fragile & undergo destruction at a very low rate resulting into anaemic state. Sickle cell anaemia is hereditary in nature and mostly occurs in the Negro races.

Fate of Hemoglobin in the Body

After completing a life span of about 100-120 days, the red blood cells are phagocytised by the macrophages of reticulo endothelial system mostly in the liver, tissue, spleen and bone marrow. Simultaneously, hemoglobin is also broken down. Iron of hemoglobin is retained for further use. The globin portion is also separated from the protoporphyrin and metabolized in a manner analogous to other proteins. The prophyrin nucleous is cleaved and ultimately converted into bile-pigments. There is still controversy on the point whether protoporphyrin ring is separated from globin before or after the cleavage of the portoporphyrin nucleus itself. Whatever may be the case, biliverdin is the first pigment which is transformed. In the reticulo-endothelial cells, biliverdin is the first pigment which is formed. In the reticulo-endotheu at cells, biliverdin is the first pigment which is formed. In the reticulo-endotheu at cells, biliverdin is readily reduced to bilirubin. Bilirubin thus is synthesized in the other tissues is transported to liner in the form of bilirubin albumin complex. One molecule of albumin able to bind two molecules of bilirubin. After reaching in the liver, bilirubin is conjugated with glucuronic acid in the presence of a microsomal enzyme uridine-diphosphate-glucuronyl transferase with the formation of bilirubin mono and diglucuronides. These conjugates are fairly water solu ble and are readily excreted from the liver into digestive tract through bile and further reduced in the G.I.T.

Normally a small amount of bilirubin is found in circualtion in the plasma. Under certain pathological conditions, however, circulatory bilirubin level is found to be raised. All these conditions leading to increased red blood cell destruction increase the bilirubin level e.g., hemolytic jaundice, pernicious anaemia and other hemolytic infections. Any obstruction in the billary passage, as in obstructive jaundice, also, increases circulatory bilirubin level. Decreased hepatic function as in atrophy of liver, typhoid fever, liver poisoning by carbon tetra-chloride, arsenic or other heavy metals and infective hepatits also lead to increased serum bilirubin level because of the decreased capacity of liver to conjugate bile pigments with glucuronic acid.

Biosynthesis of Hemoglobins

The globin portion of hemoglobin is synthesized from amino acids in a manner similar to other body proteins. Biosynthesis of hememoiety occur in the following manner.

Succlnyl coenzyme A reacts with glycine in the presence of a microsomal enzyme delta-aminolevulinic anid synthetase and pyridoxal phosphate to give alpha-amino beta-ketoadipic acid which is decarboxylated to give delta-aminomoleculinic acid. Another enzyme delta aminolievulnic acid dehydrase brings about condensation of two molecules of delta amino levulinic acid to form prosphobilinogen. In the next step, three molecules of prophobilinogen condense to form tripyrryl-methano which is splitted in the presence of the third enzyme at position (A) and (B) to form two types of dipyrryl-methane. Subsequently, one molecule of type (A) dipyrryl-methane and another molecule of type (B) condense to form uroporphyrinogen III. These reactions are believed to be catalysed by the enzyme prophobilinogen deaminase. In the next step, uroporphyrinogen III is decarboxylated eliminating the molecules of Co_2 per molecule of uroporphyrinogen III forming Coproporphyrinogen III. This step is catalysed by the enzyme uroporphyrinogen decarboxylase coproroporphringoen III is oxidatively dicarboxylated to protoporphyrin III. This reaction is catalyzed by two enzymes, namely coproporphyrinogenosidase and protoporphyrinogen oxidase. At this stage, iron is incorporated into the protoporphyrin nucleus forming hemenucleus. In the nucleated avian erythrocytes, synthesis of hemogobin from apohemoglobin and heme has been shown in vitro. An enzyme heme synthetase has been shown to be involved in this coupling.

Iron for hemoglobin synthesis is supplied by transferrin or ferritin as has been shown in the bone marrow cells. Furthermore, it has also been shown in the rabbit bone marrow that heme and globin synthesis proceed at almost parallel rates, and simultaneously, these get attachment to form new hemoglobin molecules. In the developing red blood cells however, the rate of protoporphyrin synthesis slightly exceeds the rate of hemoglobin synthesis. In addition to the red blood cells, all the aerobic cells can independently synthesize heme-proteins in the mitochondria. The heme synthesized by such cells can be incorporated into various types of cytochromes, myohemoglobin as well as hemoglobin.

Amount of Haemoglobin

To measure the haemoglobin contents *haemometer* is used. In average human adult, irrespective of sex, the haemoglobin is 14.5 gm. per 100 ml. of blood. Total haemoglobin can be calculated by multiplying the concentration of haemoglobin (gm./ml.) with blood volume (ml.) as follows:

Total haemoglobin = conc. of haemoglobin (gm./ml.) x blood volume (ml)

$$= 0.15 \times 5000 = 750 \text{ gm.}$$

Similarly the rate of its production can also be detected in a manner given below:

$$\text{Rate of production per day} = \frac{\text{Total haemoglobin}}{\text{Life span in days}}$$

$$= \frac{750}{120} = 6.25 \text{ gm.}$$

The amount of haemoglobin varies under different physiological conditions as shown in the following lines:

(i) *Age.* In the foetus the concentration is highest. At birth the average concentration is about 23 gm. per 100 ml. By the end of 3rd month it falls below normal, probably, because of deficiency of iron in milk. The gradual recovery takes place at the end of the first year.

(ii) *Sex.* In females the haemoglobin amounts from 13.5 to 14.5 percent whereas in males from 14.5 to 15.5 percent.

(iii) *Altitude.* At higher altitude, haemoglobin percentage rises.

(iv) Diurnal variation, exercise and adrenalin injections increase the amount of haemoglobin.

Chart Showing R.B.C. Content and Haemoglobin Content in Different Animals

Name of Animal	*R.B.C content in million cu mm.*	*Haemoglobin contents in gm. %*
1. *Pisces*		
Carp	0.84	10.5
Rock cod	1.55	5.9
American fresh water eel	2.48	9.0
White cat fish	2.65	9.2
Atlantic mackerel	3.94	14.9
2. *Amphibians*		
Mud puppy	0.02	4.6
Two toed amphiuma	0.03	9.4
Hell bender	0.07	13.3
American bull frog	0.44	7.8
Tiger salamander	1.63	8.6
3. *Reptiles*		
Turtles	0.62	7.3
Tortoise	0.63	7.4
Water snake	0.77	10.0
Greater snake	1.18	8.5
4. *Birds*		
Turkey	2.3	11.2
Crow	2.5	—
Chicken	2.8	10.3
Duck	2.8	14.8
Street pigeon	3.2	12.8
5. *Mammals*		
Elephant	2.02	—
Human female	4.5	14.5
Human male	5.0	15.5

Rhesus monkey	5.2	12.6
Guinea pig	5.6	14.4
European rabbit	5.7	11.9
Dog	6.3	14.8
Cow	6.5	9—14
Swine	6.5	10.5—16.5
Cat	8.0	11.2
Cattle	8.1	11.5
Horse	8.2	9.0—14
House mouse	9.3	14.8
Camel	10.4	13.0
Sheep	10.3	10.9
Goat	16.0	10.5

Haemoglobin is a conjugated protein, prossessing haeme as its prosthetic group and globin as its protein portion. The haeme part is similar in most of the case but the globin part shows considerable variations in the amino acid composition and sequence. Each molecule of haemoglobin contains four molecules of haeme attached to one molecule of globin.

Haeme is a protoporphyrin compound and consists essentially of four pyrrole groups joined together forming a ring structure. Iron is found at the middle of this ring. Thus haeme is a metalloporphyrin where the metal is iron. The iron content of haemoglobin is about 0.34% and about 3-4 gm. of iron is present as haemoglobin in the total amount of blood of an adult having 70 kg body weight. Small amount of iron is excreted through urine. It is about 0.5-1 mg daily in adult male. The loss is normal menstruation is about .1-2 mg. daily and it is 1.5-2.5 mg. per day in pregnant woman. Iron remains in ferrous (Fe^{++}) form.

The globin part of haemoglobin is usually made up of various combinations of four different peptide chains which are called the alpha (α), and beta (β), gamma (γ), and delta (δ) chains. The alpha chains are made up of 141 amino acids, the beta, gamma and delta chains of 146 amino acids each. The difference lies in the sequence of amino acids, however, they show some striking similarities. There exist twenty four consecutive identical amino acids in the beta and gamma chains, 65 identical amino acids in the alpha and beta chains, and 138 in beta and delta chains.

Most of the human haemoglobins have two alpha chains and two beta chains. This category is represented as HbA and is shown as $\alpha_2\beta_2$.

This forms about 90 percent of the human haemoglobin. HbF is another haemoglobin known as human *foetal* haemoglobin. It contains two alpha chains and two gamma chains ($\delta_2\gamma_2$). A haemoglobin which forms only a minor component of adult haemoglobin is known as HbA_2. It is made up of two alpha and two delta chains and is represented as $\alpha_2\delta_2$.

The approximate chemical formula is $(C_{712}H_{1130}O_{245}N_{214}S2Fe)_4$. The molecular weight of haemoglobin is about 68,000.

The Blood Pigments of Haemochromes

In both the plant and animal worlds pigments play an important part in the supply of oxygen.

Haemoglobin and Oxyhaemoglobin

Pigments are by far the most abundant and important of the constituents of the red cells. These are conjugated proteins, a compound of the protein globin with the complex iron-containing pigment *haem,* which belongs to the class of pigments known as *porphyrins* which contains also the green pigments of plants -chlorophyll-and cytochrome, the pigment which is associated with oxidation in cells. These pigments, together with the bile pigments and other animal pigments which are derived from them, are sometimes known as the *pyrrole pitments*, as they are derived chemically from the simple heteroring compound pyrrole, C_4H_4N. Some respiratory pigments, such as the green chlorocruorin of certain worms, contain iron, but the haemocyanin of molluscs and crustacea contains copper. In the invertebrate animals the haemoglobin is not in the cells but free in the plasma.

Haemoglobin is the pigment which gives the red colour to the blood. When combined with oxygen (but not oxidised) it becomes the familiar bright red oxyhaemoglobin which has already been studied in relation to Respiration. The actual shade of red which blood assumes depends on the proportion of haemoglobin to oxyhaemoglobin, and, as we have seen, even venous blood contains a considerable amount of the latter.

Crystals of oxyhaemoglobin may be obtained with readiness from the blood of such animals as the rat, guinea-pig, or dog; with difficulty from the other animals, such as man, ape, and most of the common mammals. The following methods are the best:

1. Mix a drop of defibrinated blood of the rat on a slide with a drop of water or Canada balsam; put on a cover-glass; in a few minutes the cells are rendered colourless, and then the oxyhaemoglobin crystallises out from the solution so formed.

2. On a larger scale, crystals may be obtained by mixing the blood with one-sixteenth of its volume of ether; the cells dissolve, and the blood assumes a *laky* appearance. After a period varying from a few minutes to days, abundant crystals are deposited.

The shape of the oxhyaemoglobin crystals in different animals varies somewhat, probably owing to the varying amounts of water of crystallisation they contain.

Haem, the parent pigment chemically, is chiefly of academic interest but has the property of being easily oxidised or reduced. It exists in both forms (see schema). The term haem usually refers to reduced haem. In nature, however, it does not seem likely that it ever occurs free from globin. The globin prevents the oxidation of the haem without interfering with its power to carry oxygen in loose combination. If the globin is thrown out of action in blood the haem becomes oxidised and methaemoglobin is formed (see below). This occurs to haemoglobin if they are stored.

Haematin

On adding an acid or alkali to haemoglobin, it is broken up into two parts-a brown pigment called *haematin,* which contains all the iron of the original substance, and *globin*.

Haemoglobin disintegrating in the circulation is first split into globin and reduced haematin (ferrous) and is immediately oxidised to haematin (ferric), which in man combines with serum albumin to form methamalbumin.

Haematin has the formula $C_{34}H_{33}$ or $_{35}O_5N_4Fe$. It presents different sepectroscopic appearances in acid and alkaline solutions (see accompaying plate). It may be reduced in alkaline solution by adding a reducing agent, and the well-marked absorption spectrum of reduced haematin or haemochromogen forms the most delicate of the spectroscopic tests for blood pigment. Haemochromogen is of red colour; haematin is brownish.

Haemochromogen is neutral solution undergoes spontaneous oxidation to the ferric state of parahaematin. It was fact that suggested that the function of the neutral globin in haemoglobin is to prevent this oxidation. The reducing agent destroys this protective action of the globin by denaturing it.

Haemin is of great importance, as the obtaining of this substance forms the best chemical test for blood. The test is much used in medicolegal work. Haemin crystals maybe prepared for microscopical examination by boiling a fragment of dried blood with a drop of glacial acetic acid on a slide; on cooling, triclinic plates and prisms of a dark brown colour, often is star-shaped clusters and with rounded angles, separate out. In the case of an old blood stain it is necessary to add crystal of sodium chloride. Fresh blood contains sufficient sodium chloride in itself.

The action of the acetic acid is to split the haemoglobin into haematin and globin; a hydroxyl group of the haematin is then replaced by chlorine.

The structure of haemin has now been confirmed by synthesis and has been shown to consist of four pyrrole rings (dark in figure) linked together. The various haem derivatives differ in having different side-chains, while in the case of the bile pigments the iron is removed and the large ring breaks up to form a chain.

From haemin it is possible to produce the parent pigment haem.

Haematoporphyrin, $C_{34}H_{38}O_4N_6$, is iron-free haematin; it may be prepared by mixing blood with strong sulphuric acid; the iron is taken out as ferrous sulphate. It is also found sometimes in nature; it occurs in certain invertebrate pigments, and may also be found in certain forms of pathological urine. Even normal urine contains traces of it. It is isomeric with the bilirubin of bile. It presents different spectroscopic appearances according as it is dissolved in acid or alkaline media. The absorption spectrum figure is that of acid haematoporphyrin.

Haemopyrrol is formed by reduction from haematoporphyrin and proves to be a mixture of several pyrrol derivatives. Similar derivatives are obtained from chlorophyll, a fact which illustrates the near relationship of the principal animal and vegetable pigments.

Haematoidin

This substance is found in the form of yellowish red crystals in old blood extravasation, and is derived from the haemoglobin. Its crystalline form and the reaction it gives with fuming nitric acid show it to be closely allied to *billirubin*, the chief colouring matter of the bile, and on analysis it is found to be identical with it. (See Bile.)

Haematoidin, like haematoporphyrin, is free from iron, but differs from it in showing no absorption bands in the spectrum.

Compounds of Haemoglobin

Haemoglobin forms at least four compounds with gases:

With oxygen	1. Oxyhaemoglobin 2. Methaemoglobin
With carbon monoxide	3. Carbon monoxide haemoglobin
With nitric oxide	4. Nitric oxide haemoglobin

These compounds have similar crystalline forms with the exception of methaemoglobin; each consists of a molecule of haemoglobin combined with one molecule of the gas in question. They part with the combined gas somewhat readily; they are arranged in order of stability in the above list, the least stable first.

Oxyhaemoglobin is the compound that exists in arterial blood. Many of its properties have already been described in relation to Respiration.

We have seen that the oxygen may be removed by exposing the blood to a vaccum. The blood may also be reduced by passing hydrogen through it or by addition of other reducing agents, such as ammonium sulphide or Stokes' reagent (an ammoniacal solution of ferrous tartrate), or, best of all, sodium hydrosulphite. One gramme of haemoglobin will combine with 1.34 c.c. of oxygen.

If any of these methods for reducing oxyhaemoglobin is used, the bright red (arterial) colour of oxyhaemoglobin changes to the darker (venous) tint of haemoglobin. On once more allowing oxygen to come into contact with the haemoglobin, as by shaking the solution with the air, the bright arterial colour returns. These colour changes may be more accurately studied with the spectroscope, and the constant position of the absorption bands seen constitutes as important test for blood pigment. It will be first necessary to describe briefly the instrument used.

The Spectroscope

When a ray of white light is passed through a prism, it is refracted or bent at each surface of the prism; the whole ray is however, not equally bent, but it is split into its constituent colours which may be allowed to fall on a screen. The bands of colours beginning with the red, passing through orange, yellow, green, blue, and ending with violet, is called a *spectrum*: this is seen in nature in the rainbow.

The spectrum of sunlight is interrupted by numerous dark lines crossing it vertically, called Frauenhofer's lines. These are perfectly constant in position and serve as landmarks in the spectrum. The more prominent are A, B and C, in the red; D in the yellow; E, b, and F, in the green; G and H, in the violet. These lines are due to certain volatile substances in the solar atmosphere. If the light from burning sodium or its compounds is examined spectroscopically, it will be found to give a bright yellow line, or, rather, two bright yellow lines very close together. Potassium gives two bright red lines and one violet line and the other elements, when incandescent, give characteristic lines, but none so simple as sodium.

A convenient form of small spectroscope is the direct vision spectroscope, in which, by an arrangement of alternating prisms of crown and flint glass, the spectrum is observed by the eye in the same line as the tube furnished with the slit—indeed, slit and prisms are both contained in the same tube.

If we interpose between the source of light and the slit a piece of coloured glass, or a solution of a coloured substance contained in a vessel with parallel sides, the spectrum is found to be no longer continuous, but is interrupted by a number of dark shadows, or absorption bands corresponding to the light absorbed by the coloured medium. Thus a solution of oxyhaemoglobin of a certain strength gives two bands between the D and E lines; reduced haemoglobin gives only one; and other red solutions; though to the naked eye similar to oxyhaemoglobin, will give characteristic bands in other positions.

The Properties of Haemoglobin

We may now summarise its properties and it will be seen how well it is adapted to perform its functions of transporting oxygen and assisting in the carriage of carbon dioxide.

It has a speculiar affinity for oxygen, as shown by its dissociation curve, and just as readily gives it up.

It is amphoteric, that is, it unites with acids and alkalies. When oxidised, i.e., oxyhaemoglobin, in the red blood cells its reaction is 7.4, and it acts as an acid and combines with base.

It takes up acids especially when reduced. This property, as we have seen, plays an important part in the maintenance of the reaction of the body generally.

It is precursor of the bile pigment

Methaemoglobin

As has been said, the globin prevents the oxidation of haem by the oxygen of the air, but this occurs if the globin is thrown out of action by adding a variety of agents, especially protein precipitants or oxidising agents. Methaemoglobin, a very dark brown pigment, is thus formed. It is commonly produced by adding potassium ferricyanide or amyl nitrite ot blood, and as it also may occur in certain diseased conditions and pass into the urine, it is of considerable practical importance.

Potassium ferricyanide produces another effect as well, namely it causes an evolution of oxygen, if the blood had been previously laked.

In relation to respiration it will be seen that use is made of this fact in the estimation of the oxygen content of the blood. After the oxygen is discharged from oxyhaemoglobin, fresh oxygen, due to the oxidising action of the reagent added, takes its place; this new oxygen, however, is combined in some way different from that which was previously united to the haemoglobin.

This oxygen is not removable by the air-pump, nor by a stream of neutral gas such as hydrogen. It can, however, by reducing agents such as ammonium sulphide, be made to yield reduced haemoglobin. Methaemoglobin is of a brownish-red colour, and gives a characteristic absorption band in the red between the C and D lines (spectrum 7 in coloured plate). In dilute solutions other bands can be seen.

Carboxyhaemoglobin may be readily prepared by passing a stream of carbon monoxide or coal-gas through blood or through a solution of oxyhaemoglobin. It has a peculiar cherry-red colour. Its absorption spectrum is very like that of oxyhaemoglobin, but the two bands are slightly nearer the violet end of the spectrum (spectrum 4 in coloured plate). Reducing agents, such as ammonium sulphide, do not change it, the gas is more firmly combined than the oxygen in oxyhaemoglobin. Co-haemoglobin forms crystals like those of oxyhaemoglobin. It resists putrefaction for a very long time.

Carbon monoxide is given off during the imperfect combustion of carbon such as occurs in charcoal stoves or during the explosions that occurs in coal mines; it acts as a powerful poison, by combining with the haemoglobin of the blood, and thus interferes with normal respiratory processes, and a fatal anoxaemia or lack of oxygen is produced. The bright colour of the blood in both arteries and veins and its resistance to reducing agents are in such cases characteristic.

The Origin of Haemoglobin

As has been said, haemoglobin belongs to the class of pigments known as porphyrins and contains iron; so also does the green chlorophyll of plants. Both are widely disposed in nature but not always in a form that can be utilised; but there is no evidence of this, for the body cannot break down the chlorophyll or even the iron-containing porphyrin, haem. Articles of diet such as hare soup and black pudding made from blood are of no value as source of iron to the body. The exact source of the pyrrole remains unknown but the globin is probably synthesised.

It animals are bled, however, and given different food stuffs we get an idea of the origin of the constituents of the blood generally, and from such studies it becomes evident that a meat diet is best for restoring haemoglobin. For this, milk and its derivatives are of little use and carbohydrates valueless. Apricots and spinach are the best foods of vegetable origin, although it is obvious that the herbivorous animals must obtain their haemoglobin from more usual sources. Most of such investigation has been carried out by Whipple and his colleagues in America on a standard anaemic dog which has had its total haemoglobin reduced to about a quarter of its nromal by haemorrhage, and which is given a standard anaemic diet largely of carbohydrate to which the food under investigation is added. To a considerable extent the value of foods in this respect depends on the amount of iron they contain, although it is evident from the beneficial effect of drugs that the iron need not to be in organic form. (See Iron.) Haemoglobin formation is restricted by infections and this is one reason why if prolonged they lead to pallor of the face.

Factors affecting the Haemoglobin Content

These are the number of red blood cells and the amount of haemoglobin in each. A fall in either will obviously bring about a fall of total haemoglobin, although when there is a deficiency in cell formation, as in pernicious anaemia, there may be a compensatory rise of colour index and the fall of haemoglobin is not then so low as the cell count would lead one to expect.

By far the most important cause of low haemoglobin values is a deficiency of iron in the diet, and this is particularly liable to occur in women whose requirements are greater than those of men, because of their periodic loss of blood. Small amounts of copper are also needed.

THE HEART

Gross Morphology

The adult human heart is essentially a complex pulsatile blood vessel which has a large lumen and extremely muscular walls. This vessel has its embryologic origin partly in the fusion of simple paired endothelial tubes of mesodermal origin. These embryonic vessels in turn form the lining of the adult heart, the *endocardium*. Cells of the embryonic mesoderm external to the endothelial primordia in turn give rise to the *myocardium,* or heart muscle proper, as well as the *epicardium,* the outermost serous layer of the heart. The complex form and structural relationships found within the adult heart are direct consequences of the differential growth of the embryonic tubular vessels resulting in their flexion as well as a concomitant reversal of their primitive cephalocaudal positions within the body. Thus, in the adult, the sinus venosus region of the primitive embryonic heart lies cephalad (anterior) to its original site of development during early intrauterine life. In addition, the development of partitions or septa between the two atria and the two ventricles, as well as the formation of certain values, further compartments the heart into the four discrete chambers found in the adult form of this organ.

The adult human heart is a blunt cone-shaped organ enclosed in a dense fibrous sac called the *pericardium*. It is located within the mediastinal space of the thoracic cavity between the lungs, and lies anterior to the trachea and the thoracic aorta. The heart rests obliquely upon the superior surface of the diaphragm, so that about two-thrids of the organ is located to the left of the midline of the body. In size, the heart weighs approximately 320 gm in the adult male and 260 gm in the adult female; its dimensions are about 12 cm in length, 9 cm in width, and 6 cm in its anteroposterior diameter.

The two thin-walled upper chambers, the right and left atria, are separated by the interatrial septum, and serve principally to convey blood to the right and left ventricles, respectively; the latter structures are relatively thick-walled, and are divided by the interventricular septum. The two thick-walled ventricles perform the major pumping functions of the heart.

The right atrium receives venous blood from the superior and inferior venae cava; the blood then passes to the right ventricle, and hence passes via the pulmonary artery to the lungs in the so-called *lesser circulation*. Oxygenated blood from the lungs then enters the left atrium via the four (two left and two right) pulmonary veins, from whence it passes to the heavy-walled left ventricle. It is next pumped into the aorta and from there to the *greater*, or *systemic*, *circulation* for distribution throughout the entire body.

Size

The size of heart varies from animal to animal among vertebrates. The heart of a bird is proportionately larger than that of another vertebrate. Among mammals small species have relatively larger hearts than large forms. Fishes have smaller hearts, about 0.2% of their weight. Amphibians have a heart size of about 0.46% whereas reptiles about 0.51% of body weight. A one kilogram bird can be expected to have a heart weight of 8.2 gm., while for a mammal of the same size, the expectation is a heart weight of 5.9 gm. In adult human male the heart is about 0.43% of the body weight and in the female about 0.4% but in active animals like the deer or greyhound the heart is about 1% of the body weight.

The heart is a highly specialized overgrown portion of the blood vessel, the muscular wall of which has become differentiated for intermittent contraction. Structurally, the cardiac walls (atrial and ventricular walls) are a kin to the arteries, being formed of three layers: inner endocardium, middle myocardium and outer epicardium which match correspondingly with the tunica intima, tunica media and tunica adventitia of arteries.

Endocardium

It forms a continuous lining that covers all internal surfaces of the heart, and is composed of: (i) Endothelium: This is the innermost component of the endocardium that lines the cardiac cavity and is continuous with that of the blood vessels entering and leaving the heart. It consists of a single file of flat, polygonal *endothelial cells* which rest on a thin basement membrane. (ii) Basement membrane: It is formed of a thin basal lamina and a layer of reticular fibrils. (iii) Subendothelial

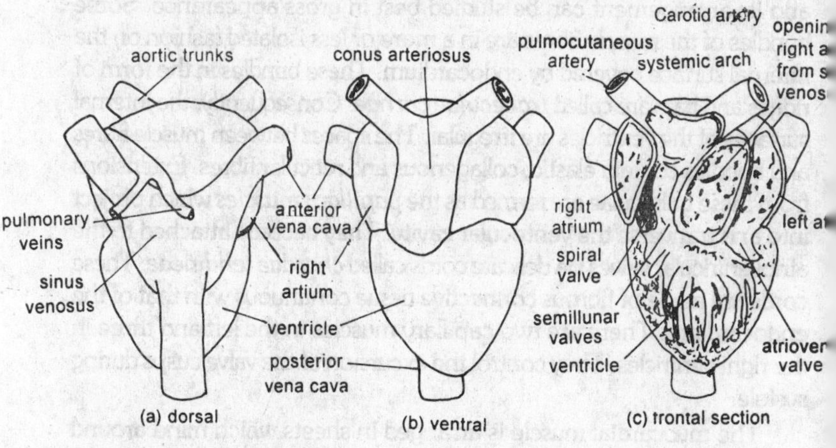

Fig. 7.2. The frog heart. Note the single ventricle.

layer: It is a fibroelastic layer beneath the basement mambrane. This narrow zone is composed of variable amount of collagen elastic fibres and fibroblasts. Smooth muscle cell may be present in this layer. This layer accounts for the variation in thickness between atrial (thick, milky and less transparent) and ventricular (thin and transparent) endocardium. (iv) Subendocardial layer: This is a layer of losse connective tissue mixed with some adipose tissue. It lies farthest from the cardiac cavity and binds the endocardium proper to the underlying myocardium. Small blood vessels, nerves and branches of the special impulse-conducting systems of the heart are located in the Subendocardial layer.

Myocardium

It is thickest of the three coat and is mainly composed of cardiac muscle. It varies in its degree of development from chamber to chamber of the heart and is quite thin in the atria, of medium thickness in the right ventricle and very thick in the left ventricle. Individual cardiac muscle cells are arranged in bundles and would in spiralling sheets in a loose connective tissue meshwork which carries numerous small blood capillaries, lymphatic vessels and autonomic nerve fibres. Small numbers of histocytes are present in the interstitial tissue.

The ventricular myocardium is more complex than that of atria, and its arrangement can be studied best in gross appearance. Some bundles of the muscle fibres are in a more or less isolated fashion on the internal surface covered by endocardium. These bundles in the form of ridges and bars are called *trabeculae carnae.* Consequently, the internal surfaces of the ventricles are irregular. The spaces between muscle fibres and bundles contain elastic, collagenous and reticular fibres. Extensions from these trabeculae are termed as the *papillary muscles* which project into and tranverse the ventricular cavity. They become attached to the atrioventricular valves via delicate cords called *chordae tendineae.* These cords are made of fibrous connective tissue continuous with that of the endocardium. There are two capillary muscles in the left and three in the right ventricles. They control the excursion of the valve cusps during systole.

The myocardial muscle is arranged in sheets which mind around the ventricles in complex, spiralling fashion. The spiral arrangement of muscle bundles makes contraction peculiarly efficient. Each heartbeat virtually brings out the blood contained in the heart. The structure and the ultrastructure of the cardiac muscle fibres is considered in details in the chapter on Muscular Tissue.

Epicardium

The epicardium is the visceral layer of the pericardium. It is a thin, transparent serious membrane that covers the exterior of the heart. It is composed of a single layer of mesothelial cells lying on a basal lamina and a thin layer of loose connective tissue containing numerous elastic fibres. A subepicardial layer, consisting of areolar tissue binds the epicardium to the myocardium. Adipose cells, coronary blood vessels and nerve fibres are embedded and course within it. The parietal layer of the pericardium is a connective tissue membrane of fibres supporting the mesothelium. It faces the epicardium across the pericardial cavity containing a small amount of lubricating fluid.

CARDIAC SKELETON

For the heart to function properly, it requires a central supporting structure of dense fibrous connective tissue which prevents overdistension of AV, aortic and pulmonary orifiices and provides attachment for sheets of cardiac muscle. The cardiac skeleton includes cardiac valves, chordae tendineae, annuli fibrosi, trigona fibrosa, and septum membranaceum or membranosum.

(a) Caridac Valves

Two such valves are *atrioventricular valves* positioned between atria and ventricles, (i) Tricuspid valve lies at the right atrioventricular opening and has three cusps, and (ii) Bicuspid or Mitral Valve is situated at the left atrioventricular opening and has two cusps.

Histologically, each cusp of the AV valve is a fold of endocardium containing a core of dense fibroelastic connective tissue fused to the supporting valve rings, the *annuli fibrosi.* Numerous elastic fibres are concentrated on the ventricular aspect of the cusp. Cardiac and smooth muscle cells may be present near the base (atrial side) of the cusps of the chordae tendineae: atrioventricular openings. The valves are largely avascular, and are connected to the papillary muscles of the ventricles by chordae tendineae (essentially fibrous cords). The chordae tandinae have a centre of parallel, dense connective tissue bundles of collagenous fibrils mixed with some elastic elements covered by thin endocardium, both layers (C.T. and endothelium) being continuous with the corresponding layers of the atrioventricular valves. They serve to restrain the valves and prevent eversion of valves when the ventricles contact.

(b) Semilunar Valves

These lie at the commencement of the aorta and pulmonary artery and guard the aortic and pulmonary orifices. Each valve has three

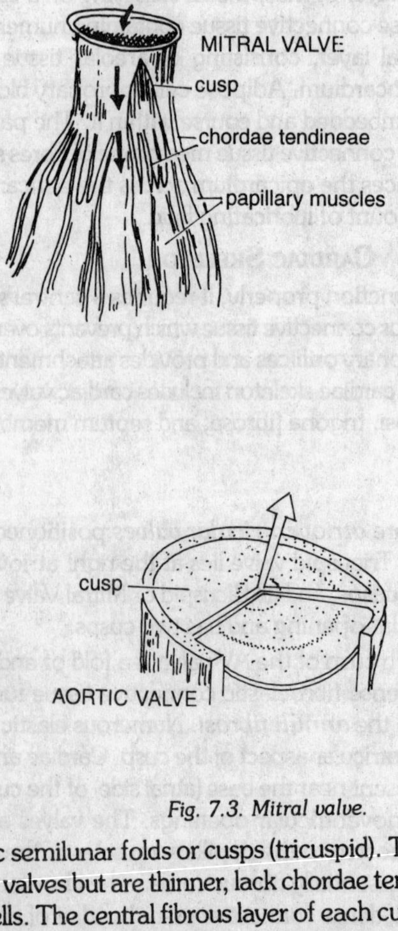

Fig. 7.3. Mitral valve.

crescentic semilunar folds or cusps (tricuspid). These valves are similar to the AV valves but are thinner, lack chordae tendineae and contain no muscle cells. The central fibrous layer of each cusp is thickened to form a nodule of Arantius close to the free margin of the cusp.

(c) Annuli Fibrosi

These are heavy rings of fibrous connective tissue that encircle the AV opening. To the edges of these rings are attached the flaps of the right and left atrioventricular valves. The aortic and pulmonary openings are surrounded by fibrous cuffs which also constitute part of the cardiac skeleton and serve as attachments for the semilunar flaps.

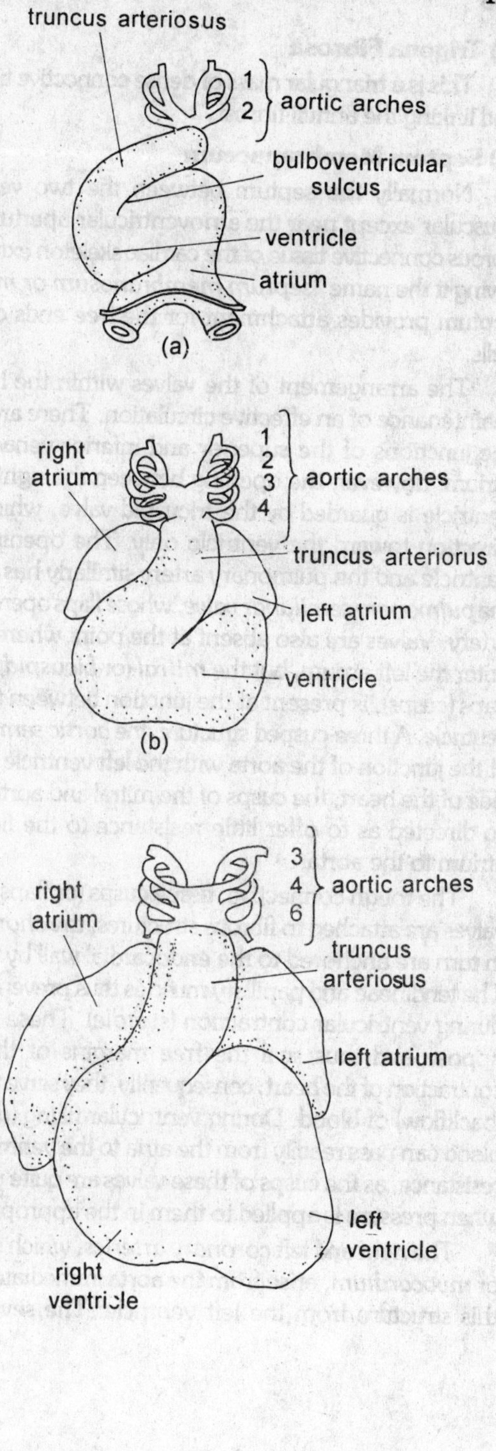

Fig. 7.4. Stages in embryonic development of the human heart (ventral aspect). (a) Three millimeter embryo. (b) Five millimeter embryo. (c) Nine millimeter embryo. Note changing position of the atrium.

(d) Trigona Fibrosa

This is a triangular mass of dense connective tissue located between and linking the annuli fibrosi.

(e) Septum Membranaceum

Normally the septum between the two ventricles is thick and muscular except near the atrioventricular apertures, where the dense fibrous connective tissue of the cardiac skeleton extends into this septum, giving it the name "*septum membranosum or membranosum.*" This septum provides attachment for the free ends of myocardial muscle cells.

The arrangement of the valves within the heart is critical to the maintenance of an effective circulation. There are no valves present at the junctions of the superior and inferior venae cava with the right atrium; however, the opening between the right atrium and the right ventricle is guarded by the tricuspid valve, which opens readily in a direction toward the ventricle only. The opening between the right ventricle and the pulmonary artery similarly has a three-cusped valve, the *pulmonary semilunar valve,* whose flaps open toward the pulmonary artery. Valves are also absent at the point where the pulmonary veins enter the left atrium, but the *mitral* (or *bicuspid*) *valve,* which has two flaps (cusps), is present at the junction between the left atrium and left ventricle. A three-cusped structure, the *aortic semilunar valve,* is located at the junction of the aorta with the left ventricle. Similarly, to the right side of the heart, the cusps of the mitral and aortic semilunar valves are so directed as to offer little resistance to the flow of blood from left atrium to the aorta.

The tough connective, tissue cusps (or flaps) of the several cardiac valves are attached to fibrous structures, the *chordae tendineae,* which in turn are anchored to the endocardial wall by the *papillary muscle.* The tendineae and papillary muscles thus prevent eversion of the valves during ventricular contraction (*systole*). These structures also help to appose and thus seal the free margins of the valve cusps during contraction of the heart; consequently, they serve to prevent regurgitation (backflow) of blood. During ventricular filling (*diastole*), however, the blood can pass readily from the atria to the ventricles without significant resistance, as the cusps of these valves are quite pliant, and open readily when pressure is applied to them in the appropriate direction.

The right and left *coronary arteries*, which supply the heart muscle or *myocardium,* arise from the aorta immediately beyond the origin of this structure from the left ventricle. The several coronary arteries, after branching to supply the myocardium with blood, form a number of *cardiac veins,* which empty into the *coronary sinus*; this vessel in turn empties into the right atrium between the atrioventricular aperture and the opening of the inferior vena cava. The orifice between the coronary sinus and right atrium is guarded by the *valve of the coronary sinus* (*valve of Thebesius*). This semilunar valve is incompetent; hence, regurgitation of some blood from atrium to sinus can occur normally.

Microscopic Anatomy of the Atrial and Ventricular Musculature

The contractile myocardium of the human heart consists of densely packed striated muscle fibres which differ in three important respects from skeletal muscle. These major differences are summarized below:

1. Cardiac muscle fibres are not simple cylindrical units; rather, they bifurcate and anastomose freely with adjacent fibres, thereby forming an intricate three-dimensional network. This morphologic arrangement has important functional implications.
2. The myocardial fibres do not anatomic syncytium (i.e., one huge multinucleate cell), as was believed for many years. Rather, the fibres are made up of individual cellular units that are joined together end to end by specializations of the cell surfaces called *intercalated discs;* these structures, which run transversely across the individual fibres, have two important physiologic properties: (1) They provide low-resistance bridges or pathways for the rapid conduction of electric impulses within the myocardium. Thus, the entire atrial and ventricular musculatures form two separate and distinct functional, but not anatomic, syncytia. (2) In addition, the intercalated discs, by maintaining a firm union between adjacent myocardial cellular units, serve to transmit myofibrillar tension from one cellular unit to the next, and thus link the entire myocardial structure into a single contractile unit.
3. The elongated nuclei of the myocardial cellular units generally are located deep within the fibre rather than just beneath the sarcolemma, as is the case with skeletal muscle.

The cytology of cardiac muscle is quite similar in certain respects to that of skeletal muscle. The pattern of the myofibrillar corss-striations and the nomenclature of the bands as A, I, M, H, and Z are identical in both types of muscle. The thin sarcolemma of cardiac muscle is also quite similar to that of skeletal muscle; however, the mitochondria are far more numerous and the sarcoplasm is relatively more abundant in the former. The contractile material in heart muscle is divided into fascicles by the rows of mitochondria within the interfibrillar sarcoplasm; and around the centrally located nuclei, the myofibrils diverge. Thus, a fusiform-shaped region of sarcoplasm is delineated about each nucleus. A small Golgi apparatus is located near one pole of each nucleus, and glycogen droplets are quite abundant in the sarcoplasm, for more or than in skeletal muscle. Fat droplets are also found in the conical sarcoplasmic regions at the poles of each nucleus.

The *intercalated discs* appear in suitably stained histologic preparations of cardiac muscle as heavy transverse lines which occur at fairly regular intervals and are found only in the region of the bands. The transverse regions of the intercalated discs may show a stepwise appearance, rather than traverse the individual fibre completely, as the apposing ends of these structures interdigitate freely with each other by means of projections, ridges, and papilla-shaped prominences. The individual cell membranes of the adjacent intercalated discs of two individual cardiac cells thus follow an irregular parallel course separated by a cleft some 200Å in width.

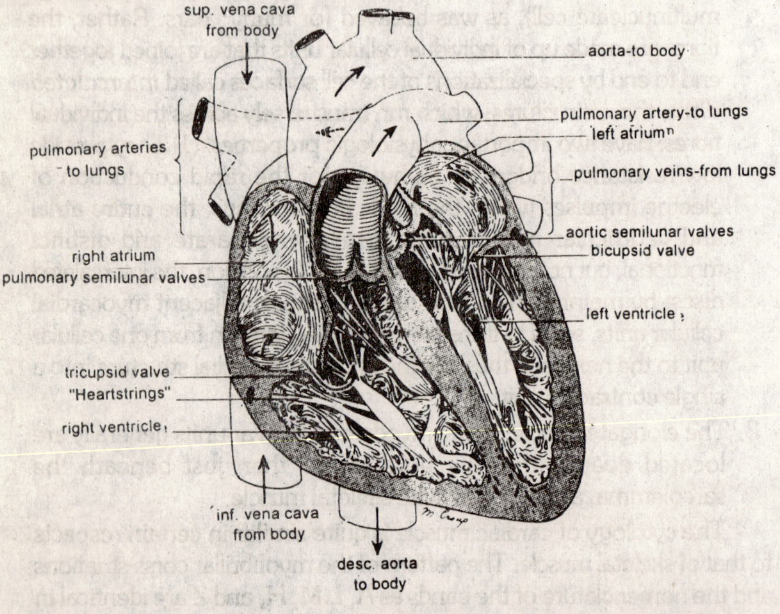

Fig. 7.5. Diagram of the heart, showing chambers, valves and connecting vessels.

Electron microscopy has revealed a pattern of myofibrillar structure within cardiac muscle quite similar to that found in skeletal muscle. A regular alternating arrangement of thick (*myosin*) and thin (*actomyosin*) filaments is observed at extremely high magnification. Hence, the contractile machinery of cardiac muscle is identical with that found in skeletal muscle.

The *transverse tubular system* (or *T-system*) or caridac muscle is relatively much larger than that of skeletal muscle. In human heart muscle, however, these transverse sarcolemmal invaginations are found at the level of the Z lines instead of the A-I junctions. Consequently, as the sarcomeres of cardiac muscle have a resting length of about 2.2 μ, no point in a cardiac muscle cell is more than 2 or 3μ from the extracellular fluid space, either within or at the surface of the fibre. This anatomic fact has important implications insofar as rapid excitation-contraction coupling is concerned, as well as far the similarly rapid exchange of materials between the interior of the cell and interstitial fluid.

The *sarcoplasmic reticulum* of cardiac muscle fibres is relatively more simple than that of skeletal muscle, and it consists of a plexiform network of tubules which occupy clefts among the myofilaments. Within this reticulum, there are no continuous tranverse elements comparable to the cisternae fo skeletal muscle; hence, no well-defined triads are present.

Specialized Tissues of the Heart

The principal function of the atrial ventricular myocardial cells discussed above is to contract, hence to pump the blood throughout the pulmonary and systematic vascular circuits. There is, however, within the heart a distinct system of specialized muscle cells whose primary function is not to contract, but rather to generate the rhythmic bioelectric stimuli that are necessary to ensure the repetitive contractions of the myocardium proper. There is also present a system of specialized tissues whose primary function is to conduct the rhythmic impulses in such a fashion as to ensure a smoothly coordinated sequence of activation of auricles and ventricles so that the heart as a whole behaves as an efficient pump. The elements of the impulse-generating and-conducting systems are illustrated in Figure and will be discussed individually.

The Sinoatrial Node (Keith-Flack Node)

The sinoatrial (SA) node is small, being about 3 mm wide by 10 mm in length in the human heart, and it is located under the epicardium within the posterior wall of the right atrium at the junction of the superior vena cava with the atrium. The individual cells comprising the sinoatrial node are fusiform in shape, distinctly smaller than typical myocardial cells, have a single nucleus, and are arranged in a network embedded within a fairly dense connective tissue framework. Lacking intercalated discs, sinoatrial nodal cells possess abundant sarcoplasm and relatively few, although typical, myofibrillae having the usual cross-striations. This latter characteristic enables these cardiac elements to be distinguished from the fibroblasts with which they are closely associated and which they markedly resemble.

The sinoatrial nodal cells are capable of spontaneously generating rhythmic impulses at a rate that is faster than any other tissue in the heart; this node is referred to as the *cardiac pacemaker.*

Since the sinoatrial nodal cells, which actually are specialized cardiac muscle fibres, are themselves capable of generating rhythmic impulses independently of any external influences, the origin of the human heartbeat is said to be *myogenic* in character; i.e., the beat originates in muscular tissue.

The sinoatrial node is also richly innervated by terminal axons of both the sympathetic and parasympathetic (vagal) branches of the autonomic nervous system; these nerves serve to modify the inherent rate of discharge of impulses by the sinoatrial node and thus alter the basic heart rate. Peripheral ganglia of the parasympathetic nervous system are also associated closely with the tissue of this node.

The sinoatrial nodal fibres are continuous with typical atrial myocardial fibres.

The Atrioventricular Node (Node of Tawara)

In mammals, including human beings, no specialized conducting tissue has been demonstrated so far that links the sinoatrial with the atrioventricular node. Therefore, functional impulse conduction between the two regions of the heart takes place directly through the myocardial fibres of the atria.

The *atrioventricular node* is situated beneath the endocardium in the lower portion of the interatrial septum; it is located between the point of origin of the septal leaf of the tricuspid valve and the orifice of the coronary sinus.

Fibres of the atrioventricular node are small and similar morphologically to those of the sinoatrial node. There is gradual transition between the atrioventricular node and the common atrioventricular bundle, which is a continuation of this portion of the conducting system.

The Common Atrioventricular Bundle (Bundle of His)

The bundle of His originates from the anterior part of the atrioventricular node with which it continuous, and enters the interventricular septum within which it divides into the right and left bundle branches; these branches in turn ramify to supply both the right and left ventricle, respectively. Beneath the endocardium of each of these ventricular chambers, the bundle branches divide considerably, thereby forming extensive systems of delicate fibres which ultimately terminate in close contact with myocardial fibres each ventricle.

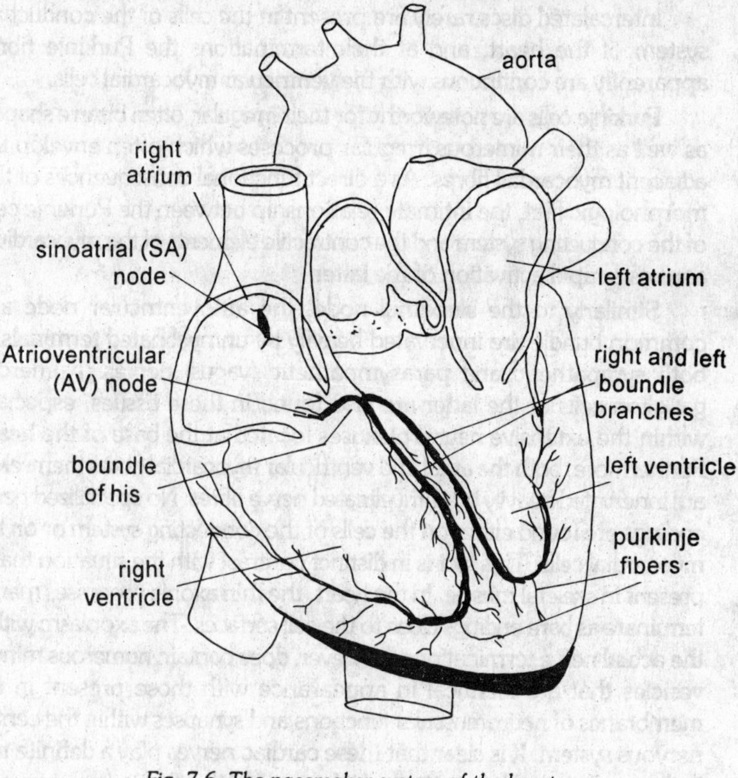

Fig. 7.6. The pacemaker system of the heart.

The cells of the common atrioventricular bundle (of His) are small and similar in appearance to those of the atrioventricular node, but as they proceed downward into the two bundle branches, the cell gradually become considerably larger, have a quite different appearance, and are now called Purkinje fibres.

The Purkinje cells of the left and right bundle branches have one or two spherical or ovoid nuclei located in an abundant clear sarcoplasm situated centrally within the cells. Both glycogen droplets and mitochondria are abundant within this sarcoplasm; the scanty and typically striated myofibrillae are located peripherally around the sarcoplasm, as indicated in Figure. The myofibrils that are present in Purkinje fibres are present singly or in small groups; and unlike these contractile elements are found within true myocardial cells, they are oriented in a random pattern. In general, however, the myofibrils are usually found to lie roughly parallel to the long axis of the Purkinje cells.

Intercalated discs rarely are present in the cells of the conducting system of the heart, and at their terminations the Purkinje fibres apparently are continuous with the ventricular myocardial cells.

Purkinje cells are noteworthy for their irregular, often bizarre shapes, as well as their numerous irregular processes which often envelop the adjacent myocardial fibres. As a direct functional consequences of this morphologic fact, the intimate relationship between the Purkinje cells of the conducting system and the contractile elements of the myocardium assures a rapid activation of the latter.

Similarly to the sinoatrial node, the atrioventricular node and common bundle are innervated heavily by unmyelinated terminals of both sympathetic and parasympathetic (vagus) nerves. Numerous ganglion cells of the latter are also found in these tissues, especially within the extensive neural plexuses located at the base of the heart. Furthermore, both the atrial and ventricular myocardial fibres themselves are innervated heavily by unmyelinated nerve fibres. No specialized nerve endings are found either on the cells of the conducting system or on the myocardial cells. This fact is in distinct contrast with the situation that is present in skeletal muscle. In the heart, the thin axonic processes merely terminate as bare endings close to the cell surfaces. The axoplasm within the actual nerve terminations, however, does contain numerous minute vesicles that are identical in appearance with those present in the membranes of neuromuscular junctions and synapses within the central nervous system. It is clear that these cardiac nerves play a definite role in the regulation of heart rate, and it is assumed that the vesicles represent the excitatory or inhibitory neuromuscular transmitters that are responsible for modifying the inherent cardiac activity in vivo.

What initiates the heart beat?

We might here attempt to answer the question. What is it that acts as a stimulus to the S-A node and thereby initiates the heart beat? It must be admitted that at present the question cannot be definitely answered, but studies on the perfused heart have indicated some of the factors whicha are necessary for the heart to be able to beat at all.

The importance of *potassium ions* and *calcium ions* is unquestioned. Too much potassium will cause the heart to weaken and relax until it stops in the relaxed state. Calcium has the opposite effect; too much of it stimulates contraction, the period of relaxation shortens, and the heart stops in the contracted state. So a proper balance between the two ions is essential and it may not be too far out of line to say that such a balance represents the stimulus. One could assume that the proper balance of ions is a continuous stimulation, the intermittent response of the S-A node being due to a property something like the refractory period in muscle. At one time it was believed that the weakly radioactive property of potassium was the stimulus. However, in view of the much higher doses of radioactive substances now being used therapeutically (e.g., radioactive phosphorus in leukemia; radioactive iodine in thyroid disease), all without effect on the heart, this theory has been abandoned. All that can be said is that if the heart beat does represent "spontaneous activity," it is situation unique in the body.

Heart Sounds

As the direct consequence of the mechanical activity of the heart and valves, as well as the attendant sudden acceleration and deceleration of the masses of blood passing through its chambers, which result in the production of eddy currents, several distinctly audible vibrations are produced within the heart during certain phases of the cardiac cycle. The sound waves produced by the heart during a single cycle exhibit a wide frequency range. Although sophisticated electronic visualization and recording of the heart sounds is possible, from a practical standpoint the technique of direct *auscultation* with a stethoscope gives the most information of the trained ear of an examining physician. The statement holds true despite the fact that the stethoscope neither amplifies sounds accurately nor transmits them faithfully to the ear.

Direct ausculation reveals two distinct heart sounds; occasionally, a third sound may be heard, whereas a fourth sound is revealed by graphic recording. The first two sounds take place primarily during ventricular systole, the latter two during ventricular diastole. The individual sounds now may be discussed, both as to their characteristics and their etiology.

The *first heart sound* commences with the rapid rise in intraventricular tension during the isometric phase of ventricular systole, and it reaches a maximum at the onset of ventricular ejection. This sound is of relatively long duration, low in pitch, and soft in quality.

There are three major factors involved in the production of the first heart sound: (1) A valvular components is produced by closure of the atrioventricular valves combined with the tension developed within the chordae tendineae and papillary muscles as the intraventricular pressure rises. (2) A vascular components is caused by the sudden turbulent surge of blood from the ventricles as well as the shock wave which is transmitted to the aortic and pulmonary arterial walls. (3) Contraction of the ventricular muscle *per se* induces vibrations of the myocardial mass.

When recorded electronically, the first heart sound is composed principally of a series of vibrations having in total a duration that ranges between 90 to 160 milliseconds and a frequency that ranges between 33 and 110 cycles/sec. These vibrations are of low amplitude at the outset, rise to a maximum at the end of the isometric period, and then diminish with a time course to that of their development.

The first sound may be heard most clearly and at maximum intensity when a stethoscope bell is placed over the fifth left intercostal space. At this location, the mitral component of the sound predominates; listening with the stethoscope bell placed over the inferior tip of the sternum is of use in determining abnormal sounds produced by the tricuspid valve.

The *second heart sound* coincides in time with the dip in the incisura of the aortic pressure curve, and this sound is shorter in duration, sharper, and of a higher pitch than the first sound. Verbally described as "lub" and "dup", separated by a brief pause, the first two heart sound demarcate the onset and termination of ventricular systole. The interval between the beginning of each of these sounds, recorded phonocardiographically, gives an accurate determination of the length of ventricular systole in man.

The second heart sound is caused by vibrations that are induced in the blood columns and arterial walls when the semilunar valves develop tension rapidly following their sudden closure.

The second heart sound lasts for about 100 msec, and it commences approximately 90 msec following the peak of the electrocardiographic T wave. The frequency of this sound averages around 50 cycles/sec.

The second sound normally is "split" into two components when the stethoscope bell is placed in the pulmonary area, i.e., along the upper left sternal margin. Phonocardiographic recording indicate that the first component of the second sound is synchronous with the dicrotic notch of the carotid arterial pressure pulse wave; hence, this sound represents closure of the aortic valve. The second component of the split second sound is produced by closure of the pulmonary arterial valve.

The relative delay between the closure of the aortic and pulmonary valves is augmented during inspiration when the increased venous return of blood to the right heart prolongs the ejection time of the right ventricle. During expiration, on the other hand, the situation is reversed; ie., the slightly greater mass of blood which was ejected from the right ventricle to the pulmonary circuit now reaches the left side of the heart and this factor now prolongs left ventricular ejection slightly. Thus, pulmonary and aortic valve closures now become closer together in time. Inspiration widens the split normally present in the second heart sound to around 50 milliseconds, whereas this time interval narrows to around 20 milliseconds, or even disappears, during expiration.

The *third heart sound* sometimes may be heard in normal hearts. This sound follows the second sound by around 80 milliseconds, and its duration is about 40 milliseconds. Commonly heard in young adults, the third sound may be heard best at the apex of the heart.

The third heart sound can be induced to appear, or to become intensified, by manipulations which increase blood flow into the atria, such as is accomplished by exercise or by assuming a supine position.

It is believed that the third sound is caused by vibrations of the ventricular walls that are induced by the surge of blood that fills them toward the end of the rapid diastolic filling phase of the cardiac cycle.

The *fourth heart sound* generally cannot be heard in normal individuals. The phonocardiogram, however, sometimes reveals several small presystolic vibrations, usually having two components. The first component is associated with the atrial a wave; hence it is caused by atrial systole. The second component of the fourth sound occurs after the peak of atrial systole has been reached, and this component can be obtained by intracardiac phonocardiographic recording more often from the ventricles than the atria. It is believed that this second component of the fourth sound is produced by vibrations induced directly in the ventricular wall by the jet of blood expelled from the atria.

The accurate interpretation of both normal and pathologic auscultatory findings, as with ECGs, is a skill requiring considerable practice and experience to develop. It is recommended that the student not only practice auscultations by listening to various subjects directly with a stethoscope, but also obtain a phonographic recording such as are commerically available in order to gain some basic familiarity with both normal and abnormal heart sounds.

The Cardiac Output

It must be noted that this term refers to the output of the left ventricle only, and that the total output is twice as much. The reason that the output of the left ventricle is considered so important that it is given this special name is that it supplies the entire body (except the lungs) with blood. Brain, muscle, the heart itself, depend on it for their blood supply. Another reason is that it does a much greater amount of work than does the right, and, consequently, is more likely so fail. It may seem contradictory to say that the left ventricle does more work than the right, just after saying that both, over a period of time, put out the same amount of blood, but the work that the pressure against which they must beat, and that pressure is about six times as great in the ventricles do depends not only on the volume of blood, but upon the aorta as in the pulmonary artery—about 80 mm in the first and only about 15 min in the second. Much more will be said on the subject of these pressure later. To repeat the definition—*cardiac output* refers to the volume of blood which the left ventricle forces into the aorta per minute of time.

Various means are used to determine the cardiac output; the one most commonly used is based on the *Fick Principle*. In this method, three things must be determined: the subject's oxygen consumption over a fixed period of time; the concentration of oxygen in the arterial blood; and the concentration of oxygen in the *mixed* venous blood.

If, for instance, the oxygen consumption were 250 cc. per minute, the oxygen concentration of arterial blood were 20 cc. oxygen per 100 cc. blood, and the oxygen concentration of venous blood were 15 cc. oxygen per 100 cc. blood, that would mean that every 100 cc. blood left 5 cc. oxygen behind in the tissues. Then every liter of blood leave behind 50 cc. oxygen (1 litre = 1,000 cc.) and to supply the 250 cc. oxygen actually consumed would require 250/50 cc. or 5 litres of blood.

It is as if a factory, which used 250 tons of coal a day, was being supplied by trains of 10 cars each. Each car carried 20 tons of coal out to the factory, unloaded five tons and ame back carrying 15 tons. This is not recommended as the way to run a railroad, but if it cost no more to run a full train than an empty one and if the factory had no way of storing coal, it would not be such a bad idea. Then, if the factory needed more coal, each car could unload a larger amount.

This is the way it is in the body. It is no more work for the heart to move blood containing oxygen than to move deoxygenated blood, and since the body has no other means of storing extra oxygen, the blood carries it around, unloading it as needed.

In making the determination, the oxygen consumption is easily measured; the patient breathes into a tank containing oxygen, which is equipped with a device to absorb the carbon dioxide. Determining the concentration of oxygen in blood is also not difficult; for this a so-called "Van Slyke apparatus" is used. Arterial blood is obtained from any small artery since the concentration of oxygen in the blood of all arteries is the same. This blood has been to the lungs and has been oxygenated and is now being distributed to all parts of the body over the arterial system. Since no oxygen leaves the blood while it is in the arteries *the oxygen content of all arterial blood is the same.*

This is not true of venous blood, however, and it is in getting the sample of *mixed* venous blood that the difficulty with this method arises. The blood entering a vein from active tissue, such a skeletal muscle, will have given up much oxygen and contain little; blood from an inactive tissue, such as the skin or the fatty tissue, will have given up little and retained much. In short, *the oxygen concentration indifferent vains varies* and it would, of course, be impractical to get samples from all and attempt to average them. So the place to get mixed venous blood is the place in the body where it is mixed, namely, the right ventricle of the heart. In order to do this, a catheter (a thin-walled rubber tube) is inserted into the basilar vein of the forearm and pushed upward through the subclavian vein and into the heart. It is this difficult in obtaining mixed venous blood that militates against wide—spread use of this method.

Other methods have been tried. In one—the *ballistocardiograph* - a bed, suspended by very sensitive springs, is used. The recoil of each beat of the heart causes movements of the bed, which are recorded. By using complicated mathematical formulate, the cardiac output can be calculated; it must be admitted, however, that the results are not always very accurate.

Nervous Control of Heart Rate

As was true of the pump purchased a few pages back, cardiac output is the product of two factors: heart rate (the number of beats per minute) and stroke-volume (the volume expelled per beat:

Rate x Stroke-volume = Cardiac output.

We will first consider the factors that affect the rate. Since the chief among these are the *cardiac nerves*—the vagus and the sympathetics—we will have to digress briefly to learn something about these nerves.

A preview of the autonomic nervous system

There are two main divisions of the nervous system, the *somatic*, which controls the activity of skeletal muscles, and the *autonomic*, which controls the activity of involuntary muscles, the heart and the glands. It is with the autonomic division that we now deal. This division consists of two subdivisions, the *parasympathetic* and *sympathetic*. These differ anatomically and functionally.

Anatomically, the parasympathetics leave the central nervous system along with other fibres of certain cranial nerves. These will be discussed later; it is only the tenth cranial nerve, the *vagus,* that concerns us here. It departs from the lower region of the brain (the medulla) and courses downward; it is a large nerve and supplies many organs of the body in addition to the heart; we will encounter it frequently. The sympathetics leave the spinal cord from only one region—the region of the twelve thoracic and the upper three lumbar spinal nerves. However, they extend to all parts of the body. The heart receives three sympathetic nerve trunks on each side, but, being pretty mixed-up by this time, they are simply referred to as *sympathetics to the heart*.

Functionally, too, the parasympathetics and sympathetics differ. As a general rule, it can be said that they act oppositely, that any organ receiving both will respond in opposite directions to stimulation by one or the other type. Thus the heart responds to vagal stimualtion by a decrease in rate, to sympathetic stimulation by an increase; blood vessels in general respond to sympathetic stimulation by constricting, to parasympathetic stimulation by dilating, etc.

Although carrying nerve impulses to the various organs, these nerves do not *create* such nerve impulses. The nervous system functions in the form of *reflex arcs*. Each reflex arc consists of three elements: a sensory fibre, whose tip is capable of being stimulated, a central connection either in the brain or spinal cord, and a motor fibre which passes to a muscle or some other organ and initiates a response. So, in studying the action of these nerves in regulating the heart rate, we must deal with reflexes.

The Bainbridge reflex

When exercise is undertaken, the blood-return to the heart is increased. With the heart beating at its resting rate, the blood is not removed quickly enough and tends to pile up in the atria and in the great veins near the heart. This stretches the walls of these vessels. Buried in their walls are sensory nerve endings which are stimulated by

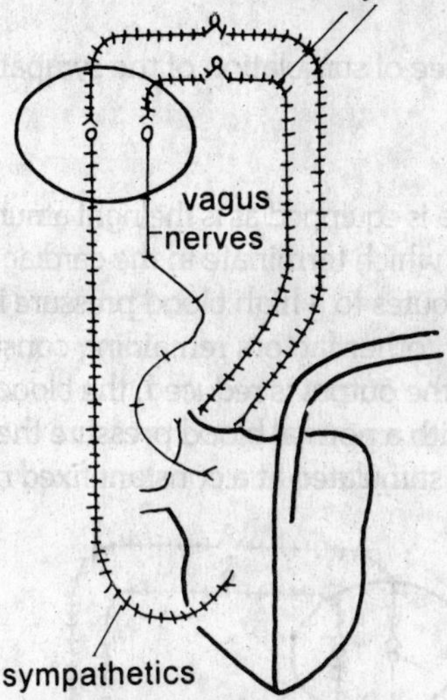

Fig. 7.7. Figure showing the Bainbridge Reflex.

the stretch of the walls. Impulses are carried over these nerve fibres to a region in the medulla of the brain called the *cardiac center.* It is so called because all its nerve cells are concerned with regulating the heart rate. Arriving there, two things can happen to the nerve impulses to bring about the desired increase in heart rate and relieve the accumulation of blood in the atria and the great veins. (1) The impulses may be transmitted to the portion of the center where the sympathetics originate, be sent down the sympathetics, and so accelerate the heart. Or (2) they may be applied to *inhibit* that portion of the center from which impulses are constantly being sent out over the vagus. Since impulses over the vagus slow the heart rate, inhibition of this region will accelerate the rate. In the accompanying sketch, the Bainbridge reflex is being accomplished by a combination of these two means of control.

The heart rate is normally controlled by a *balance* between impulses reaching it over the vagus and over the sympathetics. It is a preponderance of one kind of impulse over the other that produces an effect. Thus, inhibition of the vagus center produces a preponderance of sympathetic impulses, and thus speeds up the heart. By the same token, inhibition of the sympathetic center would produce a preponderance of vagal impulses and thus slow down the heart. It seems that in the human most of the effect is achieved by inhibition of the vagal center although there may be some degree of stimulation of the sympathetic region as well.

The aortic reflex

The arch of the aorta is equipped as is the right atrium, with sensory nerve endings in its wall which terminate in the cardiac center. One of the factors which contributes to a high blood pressure is the output of the heart. If this is large, other factors remaining constant, the blood pressure will be high; if the output is reduced, the blood pressure falls. We must assume that with a normal blood pressure the sensory nerve endings in the aorta are stimulated at a constant fixed rate.

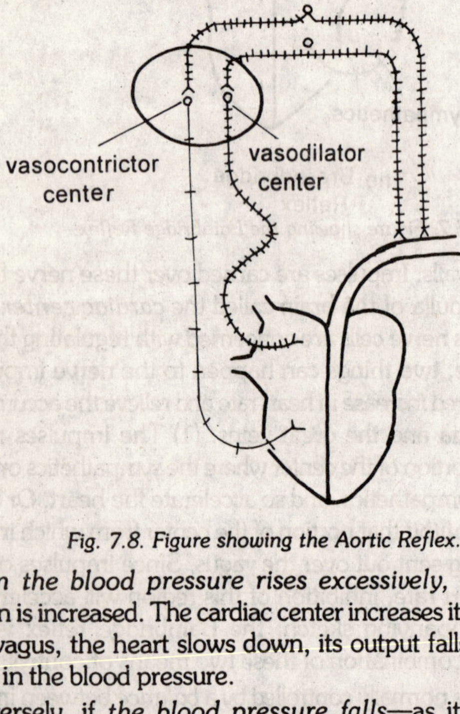

Fig. 7.8. Figure showing the Aortic Reflex.

When the blood pressure rises excessively, the intensity of stimulation is increased. The cardiac center increases its nervous output over the vagus, the heart slows down, its output falls, and there is a reduction in the blood pressure.

Conversely, if *the blood pressure falls*—as it does following hemorrhage, for example— the intensity of stimulation is decreased, the cardiac center sends fewer than the normal number of impulses out over the vagus, and the heart rate increases. The rapid heart rate following hemorrhage—frequently the tip-off that an internal hemorrhage has occurred—is due to the operation of this mechanism.

It is well known that intense emotion produces changes in the normal rhythm of the heart. One may literally be so scared that the heart stops, or almost; with passing of the fright, one becomes aware that his heart is racing. Here the stimulus to the cardiac center somes frum a region higher in the brain, the *hypothalamus*. (Evidence will be presented later that emotional experiences are centered in that region). From it, impulses descend to the cardiac center and from the center are relayed, either by way of the vagus or the sympathetics, and the heart rate is altered accordingly. The student may object to calling this a reflex, but it must be remembered that for the emotion to be experienced something must have happened, that is, sensory nerves must have been stimulated.

Chemical Control of Heart Rate

The two chief chemical substances involved in regulating heart rate are thyroxin and adrenalin. *Adrenalin* is a hormone secreted by the adrenal medulla. Its real name is "epinephrine". Adrenaline (reduced to "adrenalin" by common usage) is the trade name of a widely used preparation, but since "adrenalin" has become so familiar, I prefer to use it in this text. We will be running into it frequently, so that its action will now be described, as it operates in the emergency situation illustrated.

Imagine a furry little animal feeding peacefully in a bright, sundrenched meadow. Suddenly, from the brush, emerges this enormous, ferocious wolf. It is clear that only by an all-out efforts can the little animal escape. Its intense fright registers in its hypothalamus, which stimulates the sympathetic nervous system to great activity. Certain fibres of the sympathetics reach the adrenal medulla, stimulate it, and its secretion, adrenalin, floods the blood stream. Let us briefly list the *actions of adrenalin:*

1. It increases the rate and strength of heart beat, thus increasing the amount of blood circulating through the body, which, in the present emergency, is bound to be helpful.
2. It causes a rise in blood pressure. This means that increased blood supply is delivered to the muscles under a high lead of pressure. Also the increased pressure insures a generous flow of blood to the brain.
3. It causes the liberation of much glucose from the liver, where it is stored. Glucose is the fuel which the animal's muscles must burn in his effort to escape.
4. It causes the spleen to contract and thereby increases both the total volume of circulating blood and the red count as well, enabling it to transport more oxygen to the muscles.
5. It relaxes the bronchioles, making breathing easier. Adrenalin does not actually stimulate respiration; as we shall see later, this is done

by excessive production of carbon dioxide. But adrenalin does make it easier to get the larger volumes of air needed into and out of the lungs.

6. It dialates the pupils of the eyes. It is conceivable that the little animal might run into the jungle or down a hole, where the light was poor. Dialation of the pupil permits more light to the eye.
7. It causes the hair to stand on end. Each hair is equipped with a small muscle, called the *erector pylii,* contraction of which elevates the hair. This gives the animal a more ferocious appearance, and while it is not likely, in the case under discussion, that this would intimidate the wolf, it would, at least, make it more difficult for the wolf of bite him.
8. It causes the blood to clot more rapidly so that, if the wolf *does* bite him, he isn't so likely to bleed to death.

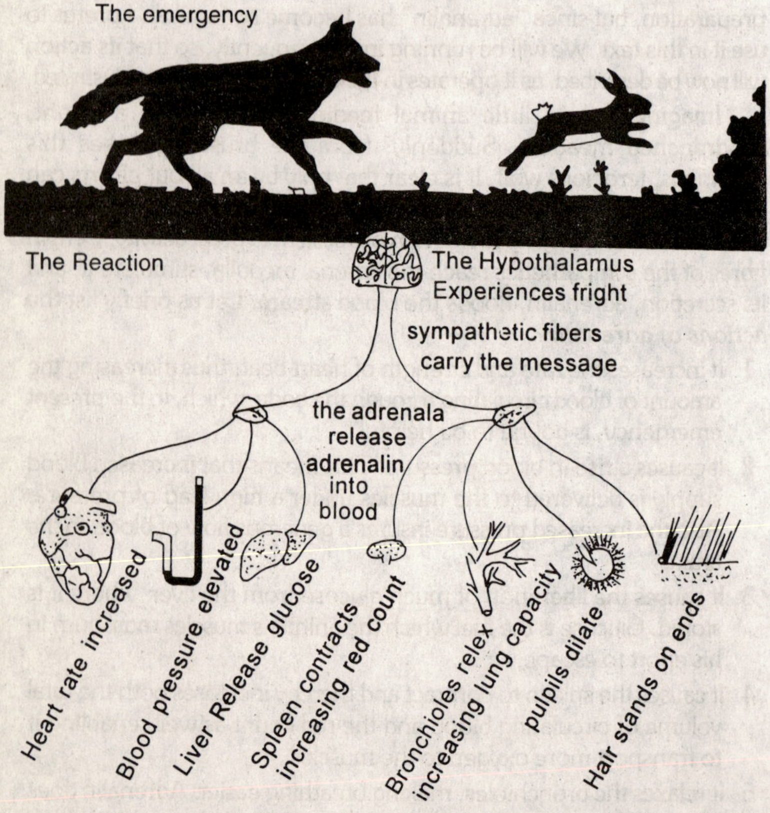

Fig. 7.9. Cannon's Emergency Theory.

All of these happenings, as you will have noted, increase the animal's chance of survival. This is expressed as the *Emergency Theory,* credited to the physiologist Walter B. Cannon. It simply states that the function of adrenalin in the body is to better enable an animal to survive an emergency.

These same things happen, of course, when the sympathetics are stimulated. It may look as though the body wanted to make doubly sure and had developed the adrenal mechanism to aid and abet the nervous system. The actual "why" of it—how it comes about that adrenalin duplicates the action of the sympathetic nervous system—will be discussed later. It is again a case of common embryonic origin.

Thyroxin, the second of the chemicals involved, has no such emergency function. It is being secreted continually and is constantly present in the blood. It is only in thyroid disease that its effect on the heart rate is shown. When the thyroid is overactive the heart rate is increased; when underactive the heart is slowed. It is believed that thyroxin, which is a general stimulant to the metabolism of food, acts on the heart as a whole, not only on the S-A node.

The Stroke Volume

The second factor affecting *cardiace output* is the stroke volume, that is, the amount of blood which the left ventricle ejects per beat. Within limits, the stroke volume depends upon the "venous return". The normal heart is capable of a considerable degree of enlargement; when the venous return is increased—as it is in exercise—the chambers of the heart are able, up to a point, to accommodate the additional blood. The walls of the right atrium and the great veins ae thin and stretch readily, so that the Binbridge reflex is evolved and the heart rate is increased.

The increased venous return is exercise is brought about in the following manner: (1) When muscles contract, they exert a "milking" effect on the blood vessels which they contain. With each contraction, blood is squeezed out into the veins—it cannot be squeezed back into the arteries because the arterial pressure is so high—and with each relaxation the blood vessels of the muscle again fill up with blood. (2) In exercise, breathing becomes deeper. The heart lies within the thorax; when the thorax expands, blood is "sucked into" the heart. These two factors, working together, cause the amount of venous blood returned to be greatly increased. First the right side of the heart, and within a few beats the left, are dilated and take bigger "bites" of blood. Thus the stroke volume is increased.

Increasing the stroke volume permanently

The extent to which this enlargement of the heart is possible depends, first of all, on the size of the heart. it would naturally be greater in a large man than in a small woman. In the second place, it depends on the previous training to which the heart has been subjected. We can add two new words to our physiological vocabulary in this connection: *Dilatation* simply means the stretching of the walls of the ventricles by greater than usual amounts of blood. *Hypertrophy* means actual enlargement, i.e., the thickening of the muscle walls due to increased thickening of the individual fibres of cardiac muscle. It is a property of cardiac muscle that dilatation, no matter how produced, is always followed by hypertrophy. There are disease conditions in which this occurs, but we are not talking about them now; we are talking of the physiological dilatation and hypertrophy of exercise. As an active child grows, he runs about, he plays games. As he grows older and takes part in high school and college athletics, the process continues; with each athletic effort, the venous return is increased, the ventricles are dilated, dilatation is followed by hypertrophy, and so the stroke volume is increased. In fact, it can be said that one of the chief purposes of athletic training is to produce such a heart.

We should here point out to what an extraordinary degree the heart is capable of increasing its output. At rest, the average heart beat is 72 beats per minute and the stroke volume is 70 cc., making a cardiac output of about 5 liters. (It will be remembered that the total volume of blood in the body is about the same amount, namely 5-6 liters, which means that a volume of blood equal to all that in the body passes through the left ventricle about once per minute). Under conditions of very violent exertion the heart of a well-trained athlete is capable, for short periods of time, of beating at the rate of 150 per minute, with a stroke volume of 200 cc., giving a cardiac output of 30 litres per minute. It is said to Paavo Nurmi, noted marathon runner, that his cardiac output was in the neighbourhood of 40 litres per minute. These are the extremes between rest and very violent exertion; any degree of activity is between will make intermediate demands. Thus simply moving from the horizontal to the erect position causes a momentary increase in the heart rate of several beats per second and a slight increase in the stroke volume.

What is the best way to increase cardiac output?

Since the heart can increase its output either by increasing its rate of by increasing its stroke volume, the question is in order. "Does either method have an advantage? Would it be better, when increased amounts of blood are necessary, to have the demand satisfied mainly by an increase in rate or by an increase in stroke volume?"

This question can be answered unequivocally that it is better done by increasing the stroke volume. This is true for two reasons: (1) A rapidly beating heart is an inefficient heart, because the time intervals between beats are so short that they do not permit adequate filling of the ventricles, and if time does not permit the ventricles to fill, their output will be small. (2) The stretched ventricular wall contracts with greater force and with less expenditure of energy than slack ventricular muscle. This discovery by Starling is illustrated schematically in the accompanying drawings. Taut muscle as represented in the right-hand drawing required less food and less oxygen for a given amount of work than did the slack muscle. To summarize, the more slowly beating hearts has time for the ventricles to be a wall filled; this stretches the walls; and the contraction that follows is of a high degree of efficiency.

8

EXCRETION

The word *metabolism* is used to designated the different chemical changes that take place in the tissues of the body upon which heat production, muscular energy, growth, and maintenance of vital life processes depend. The part played by the end products of digestion, after they passed into the circulatory system, in building up protoplasm and in providing materials to be used in essential body processes is but one phase of metabolism. The term anabolism is used to designate this positive, or building-up, phase. Catabolism is the expression which is used to signify the reactions involved in the breaking down of chemical substances in the body to simpler compounds by oxidation, with the release of energy in its various forms.

Some of the products of metabolism cannot be used as a source of energy. They are called the end products of metabolism, or *wastes*. The waste products formed within the cells pass slowly out of the cells into the tissue fluid and hence into the lymphatic system and bloodstream. Since they are of no further use to the body, they must be eliminated. Among the waste products of metabolism are carbon dioxide, urea, ammonia, uric acid, creatinine, various pigments, and inorganic salts. Carbon dioxide is eliminated for the most part, through the gills in aquatic forms, through the skin and lungs in amphibians, and through the lungs in terrestrial vertebrates. The remaining substances, most of which are products of the metabolism of proteins, are excreted almost entirely through what are called the *excretory* or *urinary organs*. It has been demonstrated, however, that in certain fishes nitrogenous and other wastes may be excreted in considerable quantity by way of the gills. In man, small amounts may be eliminated through the sweat glands of the skin. It should be recalled at this point that although feces may contain substances excreted through the bile or by the cells of the intestine, the greater bulk of it is made up not of true excreta but of indigestible material which has never entered the body proper.

Most excretory substances are in solution in water. Water itself is not considered to be a waste product, but any excess is eliminated along with substances dissolved in it. The final product is known as *urine*. The organs which eliminate the urine are called *kidneys*.

Although the excretory system, particularly in the male, is usually closely associated with the reproductive system, nevertheless so far as is

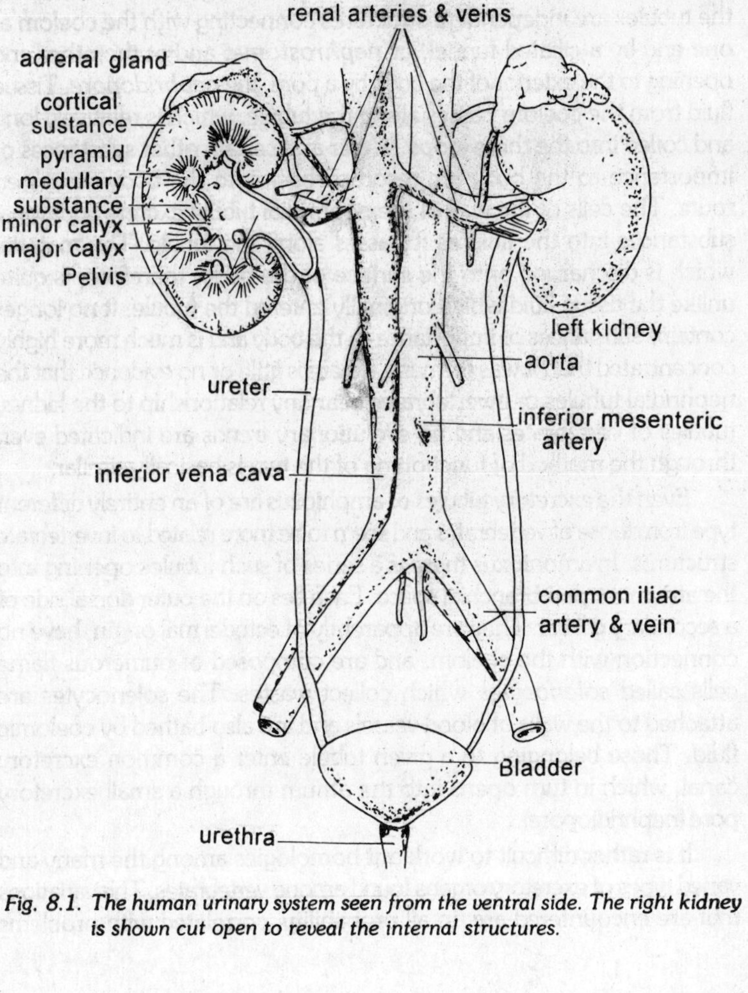

Fig. 8.1. The human urinary system seen from the ventral side. The right kidney is shown cut open to reveal the internal structures.

possible the two systems will be discussed separately. Frequently the excretory and reproductive systems are grouped together and spoken of as the parts *urogenital*, or *urinogenital system.* All parts of the excretory system is vertebrates, except for a few structures near the terminal duct openings, are mesodermal structures which are derived from the embryonic mesomere.

Types of Kidneys

In some of the invertebrates, excretory organs called *nephridia,* or *nephridial tubules,* are present. These are frequently segmentally arranged, with a pair in practically every segment. In the earthworm each nephridium occupies parts of two adjacent segments. In this animal the tubules are independent structures connecting with the coelom at one end by a ciliated funnel, or *nephrostome,* and at the other end opening to the exterior of the body by a pore, the *nephridopore.* Tissue fluid from the coelom passes along the tubule, which is relatively long and coiled into the three loops. Water and certain other substances of importance to the body are resorbed back into the body tissues en route. The cells of the wall of the nephridial tubule excrete additional substances into the fluid as it passes along the tubule. The material which is discharged onto the surface of the body, therefore, is quite unlike the tissue fluid which originally entered the tubule. It no longer contains substances of importance to the body and is much more highly concentrated than it was formerly. There is little or no evidence that the nephridial tubules or invertebrates bear any relationship to the kidney tubules of vertebrates and no evolutionary trends are indicated even through the method of functioning of the two is basically similar.

Even the excretory tubules of amphioxus are of an entirely different type from those of vertebrates and seem to be more related to invertebrate structures. In amphioxus there is a series of such tubules opening into the atrium, or peri-branchial space. Each lies on the outer dorsal side of a secondary gill bar. They are apparently of ectodermal origin, have no connection with the coelom, and are composed of numerous flame cells called *solenocytes* which collect wastes. The solenocytes are attached to the walls of blood vessels and are also bathed by coelomic fluid. Those belonging to a given tubule enter a common excretory canal, which in turn opens into the atrium through a small excretory pore (nephridiopore).

It is rather difficult to work out homologies among the many and varied types of excretory organs found among vertebrates. The variations that are encountered are, in all probability, correlated with problems with which vertebrates have had to cope in adapting themselves to the different environmental conditions under which they have lived. Whether animals remained aquatic or took to terrestrial existence posed important problems in connection with excretion. Whether aquatic forms lived in fresh water or salt water made a great difference in their adapting themselves to changes in osmotic pressures, variations in salt concentration, and the like. Comparative anatomists have long speculated concerning the evolution of the various types of vertebrate kidneys. Gradually a fairly logical sequence of events has been postulated. There has been, and still is, much confusion in this field, particularly in regard to terminology. Investigators who have studied kidney development in embryos of birds and mammals in the past have attempted to homologize these structures with the excretory organs of lower forms. It now appears that many of these speculations were erroneous. The student, therefore, should be wary in accepting certain interpretations of homologies that appear in the older literature on the subject. Often the ideas advanced have been correct, yet the terminology used has confused the tissue.

Archinephros

The belief is now generally held that the primitive vertebrate ancestor possessed an excretory organ which is referred to as an *archinephros,* or *holonephros.* This is supposed to have consisted of a pair of *archinephric ducts* located on the dorsal side of the body cavity and extending the length of the coelom. Each duct was joined by a series of segmentally arranged tubules, one pair of tubules to a segment. At its other end, the tubule opened into the coelom by a ciliated, funnel-shaped, peritoneal aperture called the *nephrostome.* Also formed in connection with each tubule was a so-called *external glomerulus,* a small knot or cluster of capillaries interposed within the course of an arteriole (a small knot or cluster of capillaries interposed within the course of an arteriole (a small branch of an artery) and located in close proximity to the nephrostome. A thin layer of peritoneal epithelium was reflected over the projecting surface of the external glomerulus. Tissue fluid, exuded at the glomeruli, passed into the coelom and thence through the nephrostomes into the kidney tubules and finally through the archinephric ducts to the outside. It is believed that the various kidneys of present-day forms may originally have been derived from a primitive type similar to the archinephros. It is interesting to note that even today the larval form of *Myxine,* the hagfish, and the larvae of some of the caecilians, or apodan amphibians, possess kidneys of the archinephric type.

The Anamniote Kidney

The anterior part of the archinephric kidney has persisted in only a few vertebrates, and even in the adult hagfish it has been somewhat modified. It appears in the embryos of most vertebrates as a transitory structure, usually referred to as the *pronephros*, which degenerates soon after it has formed. In the few anamniote vertebrates in which the pronephros persists in the adult stage it is called the *head kidney.* The remainder of the kidney, posterior to the pronephric region, is known as the *opisthonephros.*

There is much confusion as to the correct term to be applied to the kidney duct among vertebrates. It has been variously called the pronephric duct, mesonephric duct, Wolffian duct, and ureter. When appropriated by the reproductive system for the transport of spermatozoa, parts of this same duct are called the ductus epididymidis and the ductus deferens. It is well to keep these terms in mind since different authors use them all rather freely and students of anatomy and embryology may become rather confused by the variations in nomenclature. The term *ureter,* as applied to this duct, should be avoided, however, as it is most properly applied only to the kidney duct in reptiles, birds, and mammals (*Amniota*). We shall use the term *archinephric duct* in referring to the primitive kidney duct as it appears in cyclostomes, fishes, and amphibians (*Anamniota*).

Pronephros

The pronephros in anamniotes actually consists of a varying number of anteriorly located pronephric tubules, together with a pair of archinephric ducts. The tubules and ducts lie in the dorsolateral mesoderm on either side of the mesentery that supports the gut. The ducts extend posteriorly and usually open into the cloaca. One end of each of the segmentally arranged tubules connects with the archinephric duct near its anterior end; in fact the duct is actually formed by successive tubules bending posteriorly and fusing with adjacent tubules. The other end of the tubule opens into the coelom by means of a nephrostome. The nephrostome and the part of the tubule near the nephrostome are ciliated. In a few forms, rounded external glomeruli, arterial vessels of capillary dimensions covered with peritoneal epithelium, coming from segmental branches of the dorsal aorta, project into the coelom near the nephrostome. Most forms, however, possess *internal glomeruli.* These are small knots of interarterial capillaries, each surrounded by a double walled structure called *Bowman's capsule,* the two together being known as a *renal,* or *Malpighian*, *corpuscle.* The outer wall of

Bowman's capsule is the *parietal layer;* the inner wall, everywhere in contact with the blood vessels of the glomerulus, is the *visceral layer.* Blood is brought to the internal glomerulus by an *afferent arteriole* and leaves through an *efferent arteriole.* The latter breaks up into true capillaries along the course of a pronephric, tubule, and the blood is ultimately returned to the heart through one of the post-cardinal veins. Even though pronephric tubules have internal glomeruli associated with them, a connection with the coelom via the nephrostome is retained. Whether glomeruli are of the external or internal type is of no particular significance in their relation to the pronephros.

From the glomerulus a protein-free filtrate of blood plasma (tissue fluid) passes into the coelom or into the cavity of Bowman's capsule, depending upon whether the glomerulus is of the external or internal type. Cilia sweep this fluid into the tubule by way of the nephrostome or down the tubule from the capsule. Certain cells in the wall of the tubule probably excrete wastes into the lumen, where they are added to those in the fluid filtered from the glomerulus. A selective reabsorption of water and other constituents occurs as the fluid passes along the tubule so that a much smaller amount of fluid with a higher concentration of wastes in solution finally enters the archinephric duct. Electron microscopic observations on the functional pronephric tubules of *Rana pipiens* tadpoles indicate that their fine structure is similar to the tubules of adult kidneys with microvilli, pinocytotic vesicles, etc.

Sometimes several glomeruli unite to form a larger *glomus.* In some cases pronephric tubules expand so as to form *pronephric chambers,* or else several tubules may fuse to form one large chamber. The importance of the pronephros consists chiefly in the part it plays in forming the archinephric duct, which persists even though the pronephros itself disappears. In members of the class Chondrichthyes, the pronephros degenerates soon after it is formed and no vestiges remain. In some larval forms the pronephros functions in getting rid of wastes at a time when the opisthonephros is in the process of formation. In many fishes and larval amphibians it becomes modified and functions for a time as a true excretory organ. It is only in the hagfish and certain teleost fishes, such as *Zoarces, Fierasfer,* and *Lepadogaster,* that the head kidney with its coelomic connections persists in adult life. In myxinoids, the pronephros and opisthonephros become completely separated by degeneration of the portion between them. The nephrostomes of the pronephros connect with the pericardial cavity where a single large glomus is present. The fluid drained by the tubules atypically passes into a nearby vein instead of entering a duct.

Opisthonephros

Since the pronephros in most cases is but a transient structure, the opisthonephros is actually the more important part. It serves as the adult kidney in lampreys, most fishes, and amphibians.

In many accounts the term *mesonephros* is used in describing what we are here calling the opisthonephros. Biologists have come to realize that the opisthonephros of cyclostomes, fishes, and amphibians is not quite comparable to the mesonephros of embryonic amniotes even though the two are structurally similar in many ways. We shall, therefore, in the following account, use the term mesonephros in referring to the structure which appears only *during embryonic development* in reptiles, birds, and mammals.

The reason for the distinction lies in the fact that three types of kidneys which appear in the embryos of amniotes—pronephros, mesonephros, and metanephros— represent developments from different levels of the primitive archinephros which appear in succession in an anterior-posterior direction. The transition from one to the next is almost imperceptible. The opisthonephros of cyclostomes, fishes, and amphibians actually extends over a region which in amniotes will form the mesonephros and metanephros. In forms possessing an opisthonephros there is a general tendency toward a concentration of kidney tubules toward the posterior end of the organ. The anterior portion frequently loses its significance as an excretory organ and, in the male, may become part of the reproductive system.

The opisthonephros differ from the pronephros in several respects. As we shall see, the persistent archinephric duct may be taken over almost entirely in males by the reproductive system, in which case additional, or accessory, urinary ducts are formed to carry away waste materials. The chief difference between the pronephros and opisthonephros lies in the fact that the segmental arrangement of the kidney tubules no longer exists in the latter and numerous tubules may lie within the confines of a single segment, this being particularly evident in the posterior portion of the opisthonephros. Furthermore, the connection of the kidney tubules with the coelom is lost in most cases, the presence of renal corpuscles with internal glomeruli being typical.

The kidney tubules, of which the bulk of the opisthonephros is composed, develop from the intermediate cell mass (mesomere, nephrotome, intermediate mesoderm) which lies between the somite and the lateral-plate mesoderm of the embryo or larva. The archinephric duct, formed during development of the pronephros, is already present in this region. In some forms the anterior tubules of the opisthonephros appear in the same segments as the posterior pronephric tubules, so that there is some degree of overlapping.

At first the opisthonephric tubules are solid structures, but they soon develop lumina and begin to grow. One end of each tubule extends out to establish a connection with the archinephric duct. The tubule grows and becomes S-shaped. Its other end enlarges. This is soon invaginated by a tuft of tortuous interarterial capillaries formed by a small, segmental, renal artery coming from the dorsal aorta. The end of the tubule thus becomes a double walled Bowman's capsule with its contained internal glomerulus, the two together forming a renal (Malpighian) corpuscle. The tubules themselves vary somewhat in structure in different forms. Generally speaking, however, each tubule differentiates into a narrow neck at the end near the renal corpuscle, followed in turn by secretory and collecting portions. The collecting portions connects with the archinephric duct. The secretory part of the tubule forms two loops named the *proximal* and *distal convoluted segments,* or *tubules,* in relation to their proximity to the renal corpuscle.

At first the opisthonephric tubules have a metameric arrangement, but secondary, tertiary, etc., tubules soon arise so that the segmental character of the opisthonephros becomes obscured. Each tubule has its own renal corpuscle, but the collecting ends of several tubules may unite to form a ureter-like duct which either opens into the archinephric duct or establishes an independent opening into the cloaca. In some forms several such urinary ducts may be present.

An afferent arteriole brings blood to the glomerulus which, it will be recalled, is really a tuft of tortuous small vessels of capillary dimensions which intervene within the course of an arteriole. Such an arrangement is sometimes called a *rete mirabile.* The afferent arteriole divides into a few main branches, each supplying a subdivision or "lobule" of the glomerulus. Anastomoses between capillaries within a lobule are common, but no between those of separate lobules. Main, or preferred, channels within the glomerulus are indicated. In any event, the small vessels finally converge and leave the glomerulus through an efferent arteriole. As this small vessel leaves the glomerulus, it is distributed along the proximal part of the kidney tubule where it breaks up into true capillaries. The vessels that collect the blood enter the postcardinal veins, which also receive blood from the renal portal system supplying the greater part of the kidney tubules proper.

There is no complete agreement as to the method of functioning of opisthonephric tubules since there is considerable variation in their detailed structure among the various groups comprising the Anamniota. Animals inhabiting salt water have different problems with which to contend than do fresh-water and land-dwelling species. In some, the renal corpuscles are unusually large; in others, they may be altogether absent. In some, the renal corpuscles are unusually large; in others, they may be altogether absent. In some, portions of the kidney tubules may be ciliated; in others, cilia are lacking. In the frog, for example, long cilia project into the lumen from the cells of the neck of the tubule. Shorter cilia are present for a slight distance in the part of the tubule just beyond the neck. The beat of the cilia is directed toward the proximal convoluted tubule. Experiments on amphibians indicate that at the glomerulus a protein-free filtrate of blood plasma is eliminated, probably because of pressure differences. A higher pressure (both blood and osmotic) exists in the glomerular capillaries than on the other side of the visceral layer of Bowman's capsule which closely invests the glomerulus. Although the afferent arteriole leading to the glomerulus is about twice the diameter of the efferent arteriole which leaves it, their lumina are of approximately the same diameter. The wall of the afferent vessel is thicker, thus accounting for the discrepancy in size. If the efferent vessel were a vein, the diameter of its lumen would be larger than that of its afferent companion. At any rate, the fact that the lumina of the two vessels are similar in size accounts for the fact that there is some resistance to the passage of blood through the glomerular capillaries, resulting in a building up of pressure within the glomerulus. The cells of the secretory portion of the tubule excrete some urinary wastes into the lumen of the tubule. The secretory tubule, particularly in its distal portion, is known to reabsorb water, chlorides, bicarbonates, and reducing substances from the contents of its lumen. The rate of reabsorption of each of these substances differs. Thus the fluid (urine) which reaches the collecting tubule is in a more concentrated form. It then passes into the archinephric duct or into an accessory, ureterlike duct which leads to the cloaca. According to this concept, then, glomerular filtration, excretion by certain cells of the tubule, and selective reabsorption by the secretory tubules are the important factors in kidney function.

Only internal glomeruli are usually present in the opisthonephros. Peritoneal funnels connect the tubules with the coelom in certain forms, but this is unusual. In most species such connections are lost, and the coelom is no longer continuous with the outside through the excretory system.

The Amniote Kidney

In reptiles, birds, and mammals, three types of kidneys are usually recognized: pronephros, mesonephros, and metanephros. These appear in succession during embryonic development, but only one, the metanephros, persists to become the functional adult kidney. Mesonephros and metanephros actually represent different levels of the opisthonephros of the anamniota, the metanephros being the equivalent of the posterior portion. In all forms, an anteriorly located pronephros is present during very early stages of development, but it soon degenerates and the more posterior mesonephros then develops. The duct of the pronephros, however, persists to become the duct of the mesonephros. This is actually the same as the archinephric duct, to which considerable reference has been made. Since most accounts refer to this duct as the *mesonephric,* or *Wolffian duct,* to avoid further confusion on the part of the student, we shall use the term Wolffian duct here in referring to the archinephric duct as it appears in amniotes.

The mesonephros persists for a time and then degenerates. In the meantime the metanephros has begun to develop from the region posterior to the mesonephros. Portion of the mesonephros may persist to contribute to the reproductive system in the malefactor to remain as mere vestigial structures without any apparent function.

The concept that the various kinds of vertebrate kidneys have evolved in a craniocaudal sequence from an ancient archinephros gains some support from the fact that even in primitive forms the structure of the kidney shows a gradient of complexity from cranial to caudal ends. This may involve nothing more than better development of kidney tubules and glomeruli but differences are evident, nevertheless.

Pronephros

The pronephros in amniotes forms in a manner similar to that of anamniotes. Segmentally arrangement pronephric tubules appear in the intermediate cell mass in some of the anterior segments of the body. These are at first solid structures, but they soon hollow out, one end establishing a connection with the coelom. The tubules appear in succession in a craniocaudal direction. The distal end of each tubule bends backward and fuses with the adjacent tubule. A variable number of such tubules forms in different species. In the chick, for example, 10 or 11 pronephric tubules form on each side from the fifth to the fifteenth or sixteenth segment. The last tubule grows caudally and establishes a connection with the cloaca. Thus, a long pronephric duct is formed, the anterior end of which is connected to a series of tubules with coelomic connections. The tubules soon disappear; in fact the anterior tubules may degenerate before the posterior ones even form. External glomeruli may or may not form. If they do, the sequence of their appearance may not even coincide with the time when the tubules are forming. In mammals, pronephric tubules appear only as the merest of vestiges. Hence, the pronephric duct can scarcely be said to be formed as the result of fusion of tubules. Nevertheless, it appears in the nephrotome region, first as a solid cord which grows back to the cloaca, hollowing out to become a typical pronephric duct. Thus the amniote pronephros is little more than a transient structure which undoubtedly has no value at all as an excretory organ. To the comparative anatomist it is of interest because its presence indicates that the amniote kidney, like that of the anamniotes, has evolved from the ancestral archinephric, or holonephric, type. When the duct of the pronephros becomes the duct of the mesonephros it is called Wolffian duct.

Mesonephros

The mesonephros is sometimes called the *Wolffian body.* It extends over a greater number of segments than does the pronephros and develops after the pronephros has formed. Mesonephric tubules appear in the intermediate cell mass in the region where the pronephric duct is located, but posterior to the pronephric tubules. In some forms the anterior mesonephric tubules appear in the same segments as the posterior pronephric tubules, so that there is a slight degree of overlapping. The mesonephric tubules differentiate in the same manner as the opisthonephric tubules described earlier. Some of the anterior tubules may form peritoneal connections, but this is unusual. Contrary to the condition in the pronephros, several tubules may form in a single segment. In the human embryo from 2 to 9 mesonephric tubules may develop in this manner, making a total of between 30 to 40 for each mesonephros. In other forms the number may be even greater. As a result, the mesonephros, at least in some forms, as in the pig, becomes rather voluminous, occupying an extensive portion of the body cavity.

In reptiles, birds, and mammals the mesonephros exists only temporarily and is followed in development by the metanephros, which becomes the functional adult kidney. In the embryonic stages of the mouse, rat, and guinea pig, the mesonephros is so poorly developed and appears so early that it can scarcely be regarded as functional at any time. Certain embryonic membranes in these animals may possibly help to rid the body of excretory products in the interval before the metanephros develops. On the other hand, in reptiles and in such mammals as echindnas and certain marsupials, the mesonephros may persist for a time after birth. Well-developed peritoneal funnels, associated with the mesonephros, have been described for the monotremes.

As long as the mesonephros functions, the Wolffian duct is the urinary passage; but as soon as the metanephros becomes functional this dut degenerates in the female although it persists in the male. The remnants of the mesonephros, which are left after the main part has degenerated, include portions which becomes associated with the reproductive system. The male structures derived from mesonephric components include the *epididymis, ductus deferens, seminal vesicles, paradidymis,* and *ductus aberrans.* In the female such rudimentary structures as the *paraoophoron, epoophoron,* and *canal of Gartner* represent remains of the mesonephros.

Metanephros

The metanephros organizes from the caudal end of the nephrogenic mesoderm, which becomes displaced cephalad and lateral during development. This is the same mesoderm that gives rise to the caudal part of the mesonephros of fishes and amphibians. The number of tubules that form from this caudal section in amniotes is extremely large—up to an estimated 4.5 million. Differentiation of the metanephric kidney begins when a metanephric bud sprouts from the caudal end of the mesonephric duct. Surrounding the bud is nephrogenic mesoderm. The but grows cephalad, carrying the metanephric blastema along with it. Eventually it gives rise to the metanephric duct, or ureter, and the *pelvis of the kidney.* From the pelvis many fingerlike outgrowths invade the surrounding kidney blastema and become collecting tubules. Meanwhile, S-shaped tubules are organizing within the blastema. One end of each tubule grows toward and encapsulates a glomerulus to form a renal corpuscle; the other end grows toward, and finally empties into, a collecting tubule.

The tubules of mammalian kidneys have a long, thin, U-shaped *loop of Henle* interposed between proximal and distal convolutions. As the loops of the Henle elongate, they grow away from the surface of the kidney where the glomeruli are located and toward the renal pelvis. The kidney therefore has a *cortex*, containing renal corpuscles, and a *medulla*, consisting of hundreds of thousands of loops of Henle and common collecting tubules. The loops and collecting tubules are aggregated into a number of conical *pyramids*, depending on the species, and give the medulla a striated appearance in frontal section. Each pyramid—as many as 20 in some mammalian kidneys—tapers to a

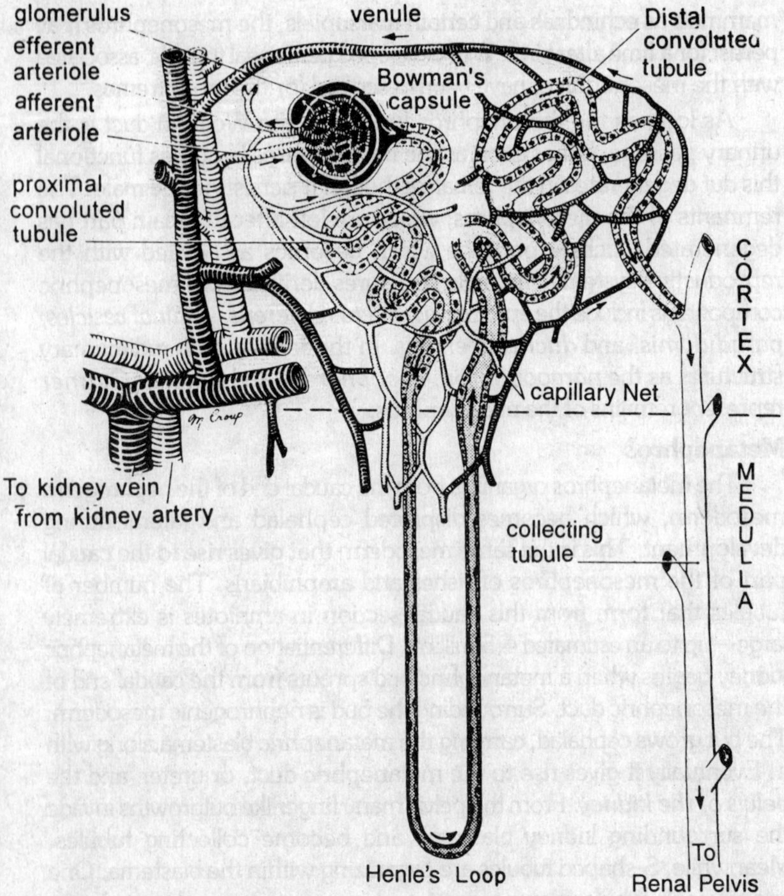

Fig. 8.2. Diagram of single kidney tubule and its blood vessels.

blunt tip, the *renal papilla*, that projects into a funnel-shaped out pocketing (*calyx*) of the pelvis. Each collecting tubule drains a small number of metanephric tubules (7 to 10 in humans), and empties into the pelvis at the top of a papilla.

Blood in the straight capillaries (*vasa recta*) that parallel the two arms of each loop of Henle flows in a direction opposite that of the glomerular filtrate, another instance of countercurrent flow and its functional ramifications. The filtrate enters the proximal convoluted tubule at a given concentration of ions, depending on the species and the physiological state of the organism. In the descending arm the solute concentration normally increases progressively toward the bottom of the loop as water moves osmotically from the filtrate into the epithelial cells of the tubule and then into ascending vasa recta. In the ascending arm, under nondehydration conditions, the epithelial cells of the tubule are impermeable to water and the filtrate in that arm becomes increasingly hypotonic to the tissue fluid as a result of *active transport* of sodium out of the filtrate under the influence of the sodium-regulating hormone aldosterone. Consequently, a dilute urine of relatively high volume (depending on the expendibility of water) is excreted by a terrestrial mammal. When dehydration threatens, antidiuretic hormone enters the circulation; the epithelial cells of the ascending arm, distal convoluted tubule, and collecting duct become highly permeable to water; water enters them from the glomerular filtrate as a result of a favourable osmotic gradient; and the glomerular filtrate beyond the base of the loop of Henle becomes increasingly hypertonic as a result of the extraction of water. Under these conditions the excreted urine is hypertonic, and the volume is small. The physics and chemistry of the process of urine formation from glomerular filtrate is a complex phenomenon the details of which are beyond the scope of this text.

Reptilian, avian, and some mammalian kidneys are lobulated, each lobe consisting of clusters of many tubules. The kidneys of snakes, legless lizards, and apodans are elongated, conforming to the slender body. Those of birds lie snugly against the contours of the sacrum and ilium. In most mammals the kidneys are smooth and bean shaped, and the arteries, veins, nerves, and ureters enter and leave at a median notch, the *hilum*.

Because of their embryonic origin as buds off the mesonephric ducts, the ureters at first terminate in these ducts, and they do so throughout life in male sphenodons and lizards. In other reptiles and in birds and monotremes they ultimately open into the cloaca. In placental mammals the ureters open into the urinary bladder because of differential growth of components of the embryonic cloaca, to be explained later.

External Salt Excretion in Vertebrates

Vertebrates that live in an environment laden with salt, or that inhibit arid environments and cannot afford much body water to carry off accumulated salts, have extrarenal structures for salt excretion. Marine fishes have chloride-secreting glands on the gills, and elasmobranchs have rectal glands that perform this function. (The rectal glands of bullsharks caught in fresh water are smaller than those caught in salt water, and show regressive changes). Marine reptiles and birds that scoop fish out of salt water have nasal glands that excrete salt. So do terrestrial lizards and snakes that live in arid habitats. These same lizards and snakes have atrophied glomeruli, which also conserves water.

The salt-secreting nasal glands of lizards are located outside the olfactory capsule and empty into the nasal canals via small ducts. Whitish incrustations of sodium chloride and potassium can be seen in the nasal canal or at the nostrils.

The nasal gland of marine birds is a large paired gland located above the orbit. It is drained by a long duct that opens close to a nostril. A groove extends from the opening to the tip of the beak. Within 15 minutes after these birds have drunk water containing sodium chloride and potassium, minute drops of fluid containing these salts trickle down the groove and drip or are shaken off the beak.

Sweat glands eliminate some salt in mammals, but salt loss by this route is incidental to secretion of water for its evaporative cooling effect. Whereas the excretion of electrolytes via most routes is regulated by hormones, their loss via sweat glands is unregulated. In fact, salt lost by this route usually must be replaced.

Urinary Bladders

Most tetrapods have a urinary bladder that serves as a storage site for urine before it is voided. Its chief survival value for terrestrial vertebrates seems to reside in the fact that it is a reservoir of water that may be needed later, hence should not be wasted. Infrequently, in one species or another, the fluid is put to another use. Antidiuretic hormone from the pituitary gland, the release of which is evoked when dehydration threatens, causes active water resorption from the bladder, after which the water is recycled. Most fishes have a small unpaired or bicornuate swelling at the end of the urinary pathway that is called a urinary bladder; however, these appear to be inconsequential as holding sites for urine. In some lower vertebrates the bladder may serve as a recovery site for certain essential ions that are scarce in the environment.

Fishes

The urinary bladder in most fishes, when present, is an insignificant terminal enlargement of the conjoined caudal ends of the urinary ducts. The urogenital sinus of male sharks and the urinary sinus of females, both of which lie within a urogenital or urinary papilla, are the closest structures to a urinary bladder in sharks, but the sinuses are so small as to be useless as a *reservoir* for urine. Homologous sinuses in Rajiformes, sometimes called urinary bladders, are only slightly larger, and the same thing may be said of the swelling at the conjoined ends of the urinary ducts of lower actinopterygians. The latter have been called *tubal bladder* in recognition of their location. They are frequently larger in females than in males. Dipnoans have a small diverticulum of the dorsal wall of the cloaca that is called a urinary bladder.

The most conspicuous structures bearing the name urinary bladder in fishes are vesicles that arise ontogenetically as evaginations from the caudal end of the conjoined embryonic mesonephric ducts, as in pikes, which are primitive teleosts. Urine must back into these, as it does into a amphibian bladders.

There is no *apparent* survival value in the presence of a urinary bladder in most fishes for water conservation, inasmuch as freshwater fishes are immersed in water and as has been pointed out, many marine fishes have the capacity to extract fresh water from sea water by drinking sea water and quickly excreting the salts. Cyclostomes have no structure that can be called a urinary bladder.

Tetrapods

The urinary bladders or amphibians through mammals arise during organogenesis as an evagination of the embryonic cloaca, and they empty into the cloaca in adults, except in placental mammals. They are lacking in crocodilians, snakes, some lizards, and in birds than ostriches. Their absence in birds reduces the energy requirements for flight.

In amniote embryos the cloacal evagination that gives rise to the bladder is prolonged beyond the ventral body wall as an extraembryonic membrane, the allantois. Only the base of the allantois—the part proximal to the cloaca—contributes to the adult bladder. After birth the part of allantois within the coelom and distal to the bladder remains in mammals as a *urachus* (*middle umbilical ligament*) connecting the tip of the bladder with the umbilicus. The urachus, which is more prominent in some species (primates, for instance) than in others (cats, for instance) lies in the anterior border of the ventral mesentery of the bladder along with the obliterated umbilical arteries.

Anurans, turtles, and some lizards unusually large bladders, and some freshwater turtles have *accessory urinary bladder.* The latter are used by females to carry water for softening and moistening the soil when a nest for the eggs is being prepared; if they have an additional function it has not been demonstrated. Because the kidney ducts of amphibians, reptiles, and monotremes empty into the cloaca, urine in these tetrapods collects in the cloaca and then backs up or is forced into the bladder while the sphincter at the vent is closed. In placental mammals the kidney ducts empty directly into the bladder, and the bladder is drained by the urethra. More will be said about the urethra shortly.

The smooth muscle in the wall of the tetrapod bladder is disposed in a reverse pattern to that of the muscular coat of the digestive tract, there being an inner longitudinal and outer circular layer. The size of the lumen at all times is only sufficient to accommodate the quantity of contained urine. As an adaptation, the muscle fibres are interspersed with a larger than usual amount of loose connective tissue, which facilitates adjustments in the thickness of the bladder wall as the bladder fills or empties. Also, the cells of the epithelial lining possess, to an unusual degree, the ability to change their position by sliding over each other, thinning out as the bladder fills, and heaping as it empties. As a result, there is generally little unfilled space in the bladder.

The Formation of Urine

The combination of the three processes of *filtration, reabsorption* and *ougmentation* or *tubular secretion* enables the kidney to remove wastes but conserve the useful components of the blood. Filtration occurs at the junction between the glomerular capillaries and the wall of Bowman's capsule. The blood is "filtered" as it passes through the capillaries so that water, salts, sugar, urea and all the substances in the blood, except the blood cells and large molecules such as the plasma proteins, pass into the cavity of Bowman's capsule to become the *glomerular filtrate*. The total blood flow through the kidneys is about 1200 ml. per minute, or one quarter of all the blood pumped by the heart! The plasma going through the glomerulus loses about 20 per cent of its volume to the glomerular filtrate; the rest leaves the glomerulus in the efferent arteriole. The mechanism underlying this efferent arteriole. The mechanism underlying this process is the purely physical one of pressure filtration, and depends on the fact that the small artery entering the glomerulus is larger than the vessel leaving it, the efferent arteriole. Consequently the blood pressure in the glomerular capillaries is relatively high, and a fraction of the plasma filters across into the capsule. By introducing a fine glass syringe into the Bowman's capsule of a frog's kidney and actually collecting and analyzing some of the glomerular filtrate, A. N. Richards of the University of Pennsylvania showed that it has the same concentration of urea, salts, glucose, and so forth, as the plasma, but lacks its proteins. The cells of Bowman's capsule are thin

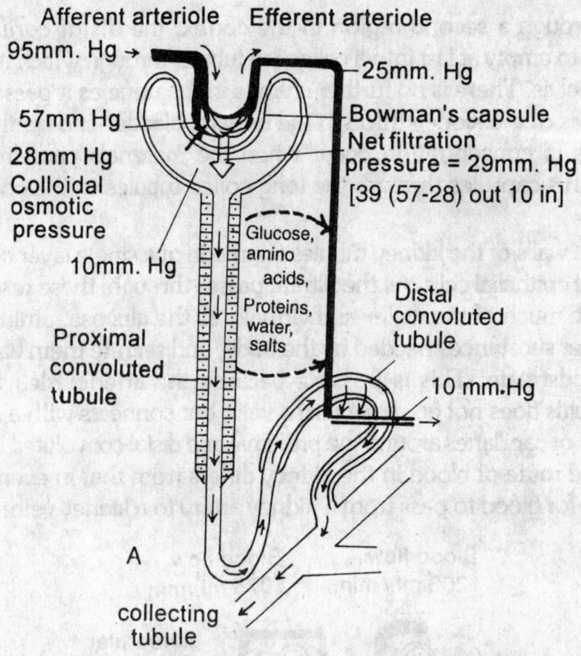

Fig. 8.3. Diagram of a kidney tubule showing the pressure gradients that move fluids from blood to glomerular filtrate (filtration pressure). Substances are reabsorbed, largely in the proximal convoluted tubules, by processes of active transport.

and unable to move materials from the capillaries; the work of pushing the filtrate from the plasma into the capsule is done by the heart. It can be shown experimentally that as the blood pressure, and consequently the filtration pressure, rise and fall, the quantity of glomerular filtrate varies accordingly. The amount filtered is also regulated by the constriction or dilation of the arterioles leading to and from the glomerulus. The amount filtered is increased by the constriction of the efferent arterioles and dilation of the afferent arterioles.

If the composition of the urine excreted were like that of the glomerular filtrate, excretion would be a wasteful process, and a great deal of water, glucose, amino acids and other useful substances would be lost. However, the kind and quantity of substances present in the urine are quite different from those in the plasma and glomerular filtrate. From each Bowman's capsule, located in the cortex, the filtrate passes first through a *proximal convoluted tubule* (also in the cortex), then through a long loop going into the medulla and back into the cortex, then through a second region in the cortex, the *distal convoluted tubule,* to empty at last into a collecting tubule, through which it passes to the pelvis. There is no further change in the urine as it passes from the pelvis of the kidney through the ureters, bladder and urethra; the changes in concentration occur when the materials pass from the Bowman's capsules through the long coiled tubules to the collecting tubule.

The walls of the kidney tubules are made of a single layer of flat or cuboidal epithelial cells. As the filtrate passes through, these remove or reabsorb much of the water and virtually all the glucose, amino acids and other substances needed by the body, and secrete them back into the bloodstream. This is possible because the arteriole leaving the glomerulus does not go directly to a vein, but connects with a second network of capillaries around the proximal and distal convoluted tubules. Thus the route of blood in the kidney differs from that in every other organ—for blood to pass from a kidney artery to a kidney vein, it must

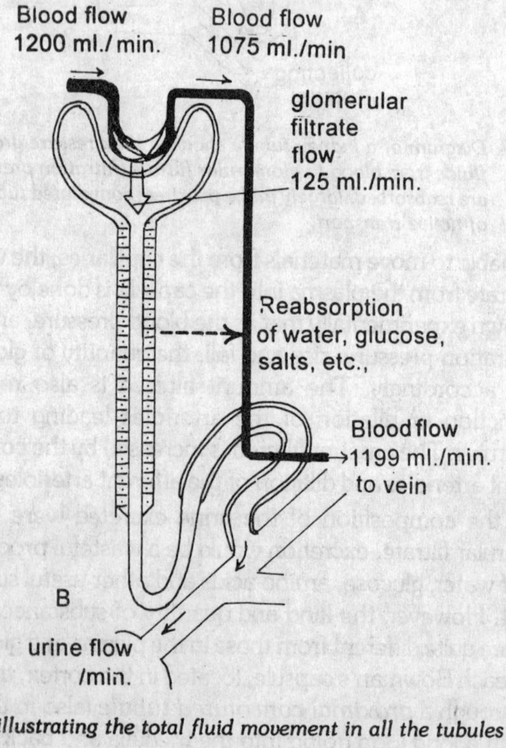

Fig. 8.4. Diagram illustrating the total fluid movement in all the tubules of the kidneys.

pass through *two sets* of capillaries. The ability of the kidney to excrete urine and regulate the composition of the blood depends upon this fact. Substances are reabsorbed into the bloodstream selectively, according to the momentary needs of the body; if there is too much glucose in the blood, as in a patient suffering from diabetes, glucose is not reabsorbed, but passes on in the urine. The cells lining the tubules must expend energy (utilize ATP) and do work to secrete these substances by the process of "active transport" back into the bloodstream, usually against a diffusion gradient. In fact, a given amount of kidney tissue consumes a larger amount of oxygen per hour than an equivalent weight of heart muscle, indicating that the kidneys work harder than the heart. They obtain the energy for this work from biological oxidations within their cells; when the kidney is deprived of oxygen, reabsorption (though not filtration) ceases.

The human kidney produces about 125 liters of filtrate for every liter of urine formed; the other 124 liters of water are reabsorbed. In this way the waste products, such as urea, are greatly concentrated as the filtrate passes down the tubules. The concentration of urea in the urine is about sixty-five times that in the glomerular filtrate and would be even higher but for the fact that a small amount of urea is reabsorbed by the tubules. The quantity of water reabsorbed depends also on the body's current needs for it and is controlled by the antidiuretic hormone (ADH) secreted by the posterior lobe of the pituitary. If a large quantity of water or beer is drunk, less water is reabsorbed and a copious, dilute urine is excreted. If water intake is restricted, a maximum amount of water is reabsorbed by the cells of the tubules, conserving water for the body, and a scanty but concentrated urine is excreted.

The cells of the kidney tubules not only remove substances from the filtrate and put them back into the bloodstream, but also excrete additional waste materials from the bloodstream into the filtrate by an active transport mechanism. This process, called *augmentation* or tubular secretion, probably plays only a minor role in human kidney function, but in animals like the toadfish, whose kidneys lack glomeruli and Bowman's capsules, excretion by the tubules is the only method used. When the blood pressure, and consequently the filtration pressure, drop below a certain point, filtration ceases in man, although urine is still formed by tubular excretion. Dyes injected into experimental animals can be seen to pass from the bloodstream into the urine through the cells lining the tubules. Experiments have revealed that drugs such as penicillin and Atabrine are removed from the blood and excreted by the process of augmentation. There is no doubt, then, that augmentation can occur in man and other animals, but just how large a role it normally plays in the process of excretion is unknown.

When the fluid reaches the end of the distal convoluted tubule, and some substances have been reabsorbed and others added, the glomerular filtrate has become urine.

The Regulatory Functions of the Kidney

By the selective excretion of certain substances and the reabsorption of others, the kidney plays an extremely important role in regulating the composition and volume of the blood and other body fluids. Any excess of acid or base released during metabolism is excreted by the kidneys, thereby maintaining the proper pH of the blood. The kidneys regulate the osmotic pressure of the body fluids bathing the cells by regulating the concentration of salts in the blood. This is important, because if the concentration of salts in the body fluids exceeds that within the cells, water passes out of the cells and they shrivel and die. Or if the concentration of salts in the body fluids falls below that within the cells, water passes into the cells, causing them to swell and burst.

Although there is glucose in the glomerular filtrate, there normally is none in the urine because it is reabsorbed by the cells of the tubules. But if there is so much glucose present in the blood, and consequently in the glomerular filtrate (as there is in diabetics) that not all of it can be reabsorbed in the time the filtrate passes through the tubules, some glucose will appear in the urine. When the concentration of glucose in the blood reaches the "kidney threshold," which, for glucose, is about 150 mg. per 100 ml. of blood. There are comparable "kidney thresholds" for many other substances but the concentration at which the substance begins to appear in the urine is different for them.

The kidney regulates the total volume of blood as well as the concentration of the substances dissolved in it. When the total amount of blood is decreased after a hemorrhage, the blood pressure is lessened. Since the filtration pressure depends on the blood pressure, it is decreased correspondingly and less liquid is filtered through the glomeruli into the Bowman's capsules. The kidney produces a smaller volume of urine, and body fluids are conserved. When a great deal of liquid is taken in, the blood volume, blood pressure and filtration pressure are raised, a larger volume of urine is formed, and blood volume is restored to normal.

The quantity of urine excreted depends not only on how much liquid is consumed, but also on the amount of salts and other solids to be excreted from the blood. When food is unusually salty, the kidney must excrete a proportionately larger amount of salt to maintain the proper osmotic pressure of the blood, and the urine volume is increased. Since the solids excreted are dissolved in the urine, and increased amount requires more water to pass them. That is why *diabetes,* of which one of the chief symptoms is the presence of sugar in the urine, is also characterized by a copious flow of urine. An increased amount of dissolved solids in the glomerular filtrate increases its osmotic pressure and consequently decreases the rate of reabsorption of water, resulting in the increase in urine volume. This mechanism enables the kidney to respond to increases in the concentration of urea and other wastes in the blood, since the higher concentration itself stimulates the flow of urine and the excretion of wastes.

Another factor controlling urine volume is the antidiuretic hormone secreted by the posterior lobe of the pituitary which controls the rate at which water is reabsorbed by the kidney tubules. In a comparatively rare disease, *diabetes insipidus* (insipid indicating a lack of sugar), due to a deficiency of this hormone, the output of urine may reach 30 or 40 liters per day instead of the normal 1.2 to 1.5 liters, and the patient suffers from an insatiable thirst.

Certain of the adrenal cortical hormones—aldosterone, deoxycorticosterone—play a role in regulating the reabsorption of sodium and potassium in the proximal and distal convoluted tubules.

Substance Present in the Urine

Since the kidneys excrete, in addition to the normal products of metabolism, most of the abnormal substances that may enter or be formed in the body, analyzing the urine is a valuable method for determining the general metabolic condition of the body. Normal urine is about 96 per cent water, 1.5 per cent salts and 2.5 per cent organic wastes, chiefly urea; its pH ranges from 4.8 to 8.0. The salts in the urine are the same as those in the blood and tissues, chiefly sodium chloride, with small amounts of potassium, calcium, magnesium and ammonium sulfate, phosphate and bicarbonate.

The 1200 to 1500 ml. of urine excreted daily contain about 60 gm. of solids; half of this is *urea*, synthesized primarily in the liver from carbon dioxide and from ammonia derived from the deamination of amino acids. Other normal constituents of urine are *uric acid*, a breakdown product of the purines present in nucleic acids; *creatinine*, which is derived from creatine phosphate, present in muscle and serving as a reservoir of ~P; and *ammonia*, derived from the metabolism of amino acids.

The yellow colour of urine is not due to any of the substances mentioned, but results from the presence of a pigment called *urochrome*, a breakdown product of hemoglobin related to the bile pigments.

9

MUSCULAR CONTRACTION

The body's physiological actions and responses to stimuli are actually performed by its muscles, glands, and other *effectors*. Movement being the major such action of animals, this chapter examines the means by which muscular contraction is accomplished. Brief consideration is also given to two other actions: ciliary movement and the production of currents by the electric organs of certain fishes. Although the principal subject of this chapter is the morphology and function of vertebrate skeletal muscle, the basic principles presented are generally applicable to other types of vertebrate muscle and to the wide variety of muscle cell types found in invertebrates.

MUSCLE STRUCTURE AND ARRANGEMENT

Vertebrates have three major types of muscle: cardiac muscle, of the heart, smooth muscle, of the viscera, and striated muscle, of the skeletal system.

Cardiac Muscle

Cardiac muscle cells are unusual in that they are all laterally fused to, or anastomosed with, each other. Such fusion is thought to provide more efficient conduction of the action potentials generated by the heart's pacemaker than if there were no fusion of cells. Also unique to the heart muscle, or *myocardium*, are numerous *intercalated discs* of tough connective tissue that join the muscle cells to each other at their ends. These discs give great tensile strength to the entire myocardium.

Smooth Muscle

The cells of smooth muscle, also known as involuntary or nonstriated muscle, are spindle-shaped, being tapered at both ends. Each fiber, as a

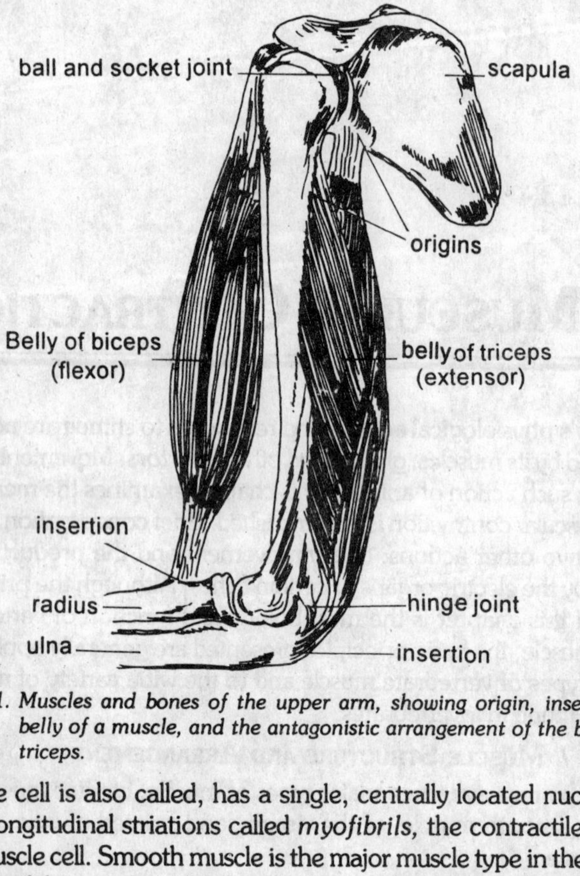

Fig. 9.1. Muscles and bones of the upper arm, showing origin, insection and belly of a muscle, and the antagonistic arrangement of the biceps and triceps.

muscle cell is also called, has a single, centrally located nucleus and faint longitudinal striations called *myofibrils,* the contractile units of the muscle cell. Smooth muscle is the major muscle type in the internal organs of the vertebrate body; it makes up the muscular component of artery walls, gland-ducts, and digestive tract and is controlled by the autonomic nervous system. Smooth muscle constitutes but a small fraction of the vertebrate body's musculature. In many invertebrates, such as the common earthworm, smooth muscle is the only muscle tissue in their bodies.

Striated Muscle

By far the greatest mass of the vertebrate body is composed of striated muscle. Striated muscle is so named because its fibers are prominently striped with alternate dark and light bands. These bands are readily visible under the light microscope, even without staining. Like smooth muscle, striated muscle is made up of many individual muscle fibers. Striated muscle is under conscious control and therefore is also known as skeletal or voluntary muscle.

Striated muscle fibers are multinucleate, with the nuclei lying along the cell periphery, just inside the plasma membrane, which is known as the *sarcolemma*. Mitochondria in striated muscle cells are large and numerous, as is appropriate for cells that expend great amounts of energy. Each muscle fiber contains many myofibrils. Striated muscle cells are found throughout the animal kingdom, from sponges to insects to mammals. Even some protozoans have striated contractile fibrils. Striated muscles are most common, however, in arthropods and vertebrates.

Groups of striated muscle fibers are organized into muscle bundles, or *fasciculi*. A muscle is composed of several to many faciculi. The bundles, as well as the entire muscle are enclosed in a sheath of tough connective tissue called *perimysium* or *fascia*. Individual muscle fibers are also ensheathed in connective tissue. In many skeletal muscles, the connective tissue sheath at one end is continuous with a *tendon*, a glistening white cord of tough connective tissue that attaches the muscle to a movable part of the skeleton, as in lower leg. The end of the muscle attached by tendon to the more movable skeletal part is termed its *insertion*. The other end of such a muscle, called its *origin*, is attached by short connective tissue *ligaments* to a less movable part of the skeleton. For example, the insertions of the lower leg muscles, which move the foot, are on the bones of the foot; their origins are on the upper and lower leg.

The Arrangement of Muscles within the Body

Most, but not all, muscles are arranged in antagonistic sets, in which the action of one or more muscles is opposed by that of others. For example, the muscles that lower the jaw are antagonized in their action by those that raise the jaw. The only movement possible for a muscle is *contraction*, that is, shortening. For a structure to be moved in any direction, it must be pulled in that direction by one or more muscles. Thus, all moving body parts have at least two sets of muscles. One of the clearest examples of the antagonism of two striated muscles is provided by the actions of the *biceps femoris* and the *triceps femoris* of the upper arm. The biceps flexes the forearm, whereas the triceps extends it. The circular and longitudinal smooth muscles of the intestine and other tubular organs are also arranged in antagonistic sets. Contraction of the longitudinal fibers of the intestine shortens and thickens it at the same time, stretching the circular muscles. When the circular muscles contract, the intestinal diameter is reduced, forcing it to lengthen—thereby stretching the longitudinal muscles.

Physiology of Striated Muscle

Muscle function has long been investigated at two levels, one being the classical physiology and mechanics of entire muscles, the other being the biochemistry and microscopic structure of the muscle cell. Classical muscle physiology studies the bases for muscular fatigue, physical factors in contraction, and the varying responses of muscles to stimuli of different strengths and frequencies. Although the principles discussed in this and the following section on microscopy and biochemistry of muscle contraction refer to striated muscle, most of them apply to cardiac and smooth muscle as well.

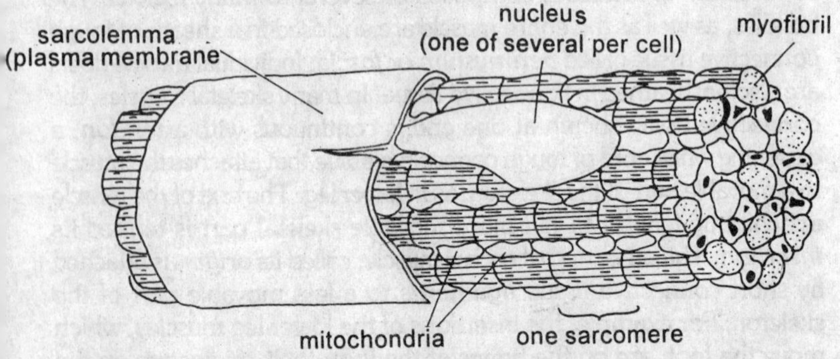

Fig. 9.2. Sketch of a portion of a single striated muscle cell, or fiber, showing details visible in electron microscopy.

The Motor Unit

When a particular muscle, such as the calf muscle of the leg, contracts, seldom do all of its fibers respond at the same time. Instead, the fibers work in groups, and, as fatigue occurs in one group, another group takes over; the various groups usually take turns in the execution of a given task. It was explained that each motor neuron, with its cell body in the spinal cord, sends out an axon to a skeletal muscle. At the muscle, the axon gives off many branches, each branch innervating a separate muscle fiber through its *motor end plate*.

The muscle fibers served by the branches of a single motor neuron are all stimulated whenever that neuron generates a train of impulses. When an action potential reaches a branch point of an axon—always at a node of Ranvier—impulses normally go down both of the branches.

As a result, all the muscle fibers served by the branches of one axon are stimulated and contract together. A motor neuron, its branches, and all the muscle fibers they serve from a *motor unit.* For example, a motor unit of a leg muscle may contain as many as 150 muscle fibers. The muscles that control such delicate actions as the movements of the eye, on the other hand, have an average ratio of only three muscle fibers to each motor neuron serving them, providing a much finer degree of control.

Fatigue

When a muscle performs a task for a long time, such as lifting and holding a heavy weight, a large number of motor units is involved. As time goes on, motor units continually take turns, gaining a period of rest when not contracting. Eventually, depending on the physical condition of the individual, the muscle is unable to do any further work; at this point fatigue occurs. One factor in fatigue is the adequacy of the muscle's blood supply, of importance because blood supplies the oxygen and glucose needed for energy, besides removing the waste products of muscle metabolism. One benefit of a warm-up before an athletic contest is an improvement in blood supply to the muscles. Engaging in a program of regular exercise strengthens the muscles and improves their circulation on a more lasting basis. As a result, fatigue does not occur as readily in people who exercise regularly.

The physiological basis of fatigue at the level of the motor unit is of two types. In one type, continual direct stimulation of a nerve attached to a particular muscle results in the depletion of the acetylcholine-containing synaptic vesicles at its motor end plates. The muscle then fails to respond until more acetylcholine is synthesized. That the muscle itself is not fatigued and is still capable of contracting can be shown by stimulating it directly. It contracts, showing that in such a case, fatigue has occurred at the synapse. Another cause of fatigue occurs in an excised muscle deprived of a blood supply. Continued stimulation will deplete the glycogen it contains and will result in an ultimate failure to respond to any stimulation, direct or not. Although both these types of fatigue are readily demonstrated in the laboratory, they are not thought to occur under normal circumstances.

Elastic Components and the Smoothness of Muscular Action

Although isolated muscle fibers contract in a rapid, jerky fashion when stimulated, intact muscles in the body contract very smoothly. One major reason for this is the *elasticity* of muscles, that is, their

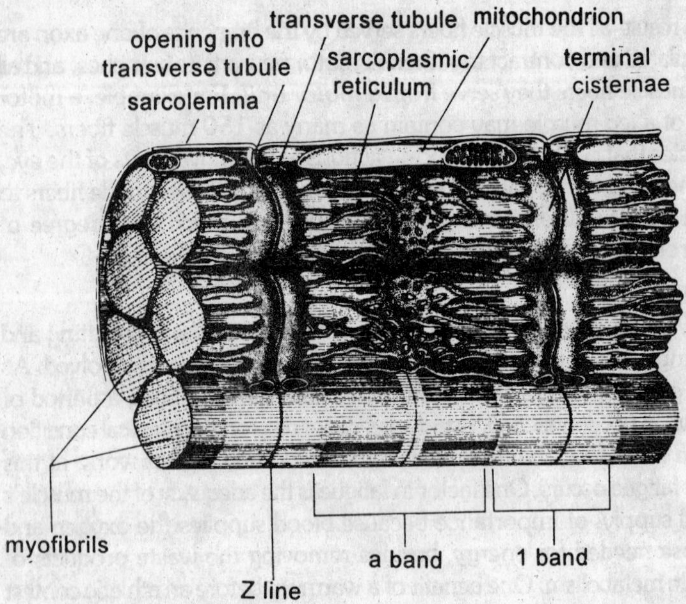

Fig. 9.3. Control of filament sliding in muscle fibers. A drawing showing the relationships of myofibrils, sarcoplasmic reticulum, and transverse tubules (T tubules) in a muscle fiber.

ability to be stretched. When a given muscle contracts, neighbouring muscles that are not contracting, especially ones that are antagonistic to the contracting muscle, resist and slow its contraction. Another important contribution to muscle elasticity is made by the connective tissues associated with a muscle. Their natural eleaticity also slows and thereby imparts smoothness of muscular contraction. The combined elasticity of muscles and associated connective tissues causes the tension, or stretching force, of contraction to be shared by these other tissues. Such internal friction in other tissues slows not only the rate of contraction but of relaxation as well. Although additional factors are involved, the sharing of tension by other tissues is a major reason for the fact that muscles remain contracted for a time after the stimulus that caused their contraction is discontinued. Even an isolated muscle fiber undergoing a single contraction lasting but a few milliseconds may take a full second—1,000 times as long—to relax again. Without the elasticity of muscle and connective tissue, our movements would be quite jerky. This tissue elasticity also keeps contracting muscles from literally tearing themselves apart.

Types of Muscular Contraction

The contraction of muscles can be classified in various ways. Descriptions of some types of muscular contraction follow.

1. In *isometric contraction,* the muscle is not permitted to shorten. Examples include the muscular activity involved in maintaining posture or that involved in attempting to pick up an object that is bolted to the floor. In such cases, the muscle actually does shorten somewhat but is not perceived to do so because of the elasticity of the attached tendons and other noncontractile connective tissue.
2. In *isotonic contraction*, the degree of tension within the muscle remains the same while the muscle contracts. Theoretically, a steady pull of the arm muscles during a slow swimming stroke could constitute an isotonic contraction. In reality, during ordinary movements of the body's muscles, the load of tension on a muscle continually changes. True isotonic contractions therefore rarely occur normally. Because such contractions can be induced experimentally, they have been a favorite subject of study by physiologists.
3. In *tonic contraction*, a muscle engages in sustained, partial contraction. This is actually a type of isometric contraction maintained over a long period of time. The term is applied to the normal but slight amount of contraction or tension exhibited by all healthy muscles, also referred to as muscle tone. An individual exhibits less muscle tone when deeply anesthetized than when awake.
4. In a *twitch,* a skeletal muscle responds to a single volley of nerve impulses by briefly contracting all of its fibers. A twitch can be produced isometrically or isotonically. The time and total duration of a twitch varies in different animals and in different types of muscles in the same animal. For example, the gastrocnemius (calf) muscle of a domestic cat contracts in 0.039 second and relaxes in 0.04 second, whereas the adjacent soleus muscle contracts in 0.077 second and relaxes in 0.079 second. Like some other types of contraction, a single twitch rarely occurs in a normally functioning animal.
5. In *tetanus,* the muscle produces sustained, maximal contraction. This type of contraction is discussed more fully in the following section.

The Effects of Repeated Maximal Stimulus

A maximal stimulus is one that achieves the fullest possible contraction; even if the strength of the maximal stimulus is doubled, no greater shortening of the muscle occurs. In a twitch, the fibers of a muscle contract following a single maximal stimulus. However, if a second maximal stimulus is applied before the first contraction cycle is completed, the second contraction is added to the residue of the first and a greater shortening of the muscle results. The two contractions are said to have summed, and the process is referred to as *summation*. The additional shortening results from the recoil of the noncontracting but elastic connective tissue components of the muscle. Even as the muscle fibers begin to relax, these elastic elements contribute to the twitch by "pulling back," much as a stretched rubber band resists being stretched. When a second stimulus is applied while these elastic components are still recoiling, the second contraction is added to this shortening.

If a third maximal stimulus is applied before the end of the second cycle, it is added to the first two, and an even greater shortening of the muscle results. Because the effect, as recorded by a measuring device, appears as a tracing resembling a series of steps on a staircase, it was referred to by early muscle physiologists as treppe (German for "stairs"), or the staircase effect.

What happens if a long series of maximal stimuli is applied to a muscle at short intervals? At first, the muscle continues to shorten with each new stimulus. However, the muscle shortens less and less with each new stimulus until it remains contracted at a constant length, even if stimuli are continued at the same rate. The muscle is then in complete tetanus, or fusion. For most muscles, the tension generated by a tentanic contraction is about four times that of a twitch. In normal functioning, most muscular contractions reach incomplete tetanus.

The Structure and Biochemistry of Striated Muscle Contraction

Muscle fiber structure and function have been intensively investigated by light and electron microscopy and at the molecular level. Our present understanding of muscular contraction reflects the combined results of observations made at all three of these levels of investigation.

Microscopic Appearance of Striated Muscle

Birefringent, or doubly refractive, materials split a light beam into two beams as they refract, or bend, it. In contrast to this reaction, a beam of light entering an isotropic substance such as glass, water, or air at an oblique angle from another medium is refracted to a new direction but still appears as a single beam. When examined by light microscopy in polarized light—light vibrating in only one plane—the dark cross striations of skeletal muscle fibers are found to be anisotropic, or birefringent, while the lighter bands are isotropic. This observation led to the designation of the dark striations as *A bands* and the lighter striations as *I bands*. Close examination under the light microscope or

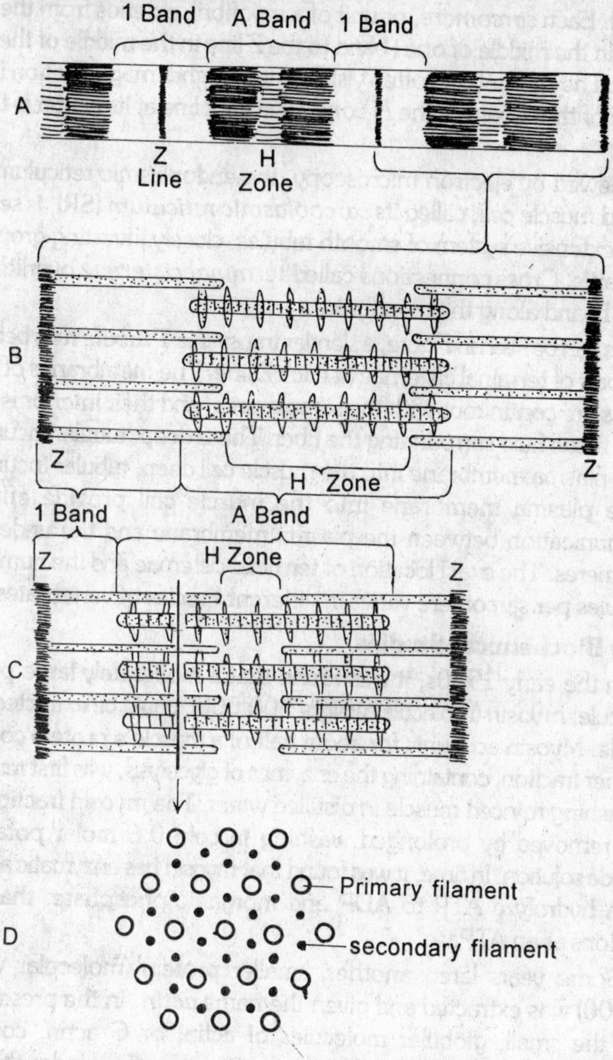

Fig. 9.4. Diagram illustrating the sliding filament hypothesis of the mechanism of muscle contraction.

by electron microscopy reveals that the banded appearance of each fiber is due to the A and I bandings of the many fine, longitudinal myofibrils of which the fiber is principally composed. Because the myofibrils are closely aligned with each other, the entire fiber appears striped. Each *sarcomere,* or unit of a myofibril, extends from the dark Z *line* in the middle of one I band to the Z line in the middle of the next I band. The lighter band often visible under higher magnification in the middle of the A band is the *H zone.* The dark line at its center is the *M line.*

Viewed by electron microscopy, the endoplasmic reticulum of a striated muscle cell, called its *sarcoplasmic reticulum* (SR), is seen to be an extensive system of smooth tubules, closely investing groups of myofibrils. Cross connections called *terminal cisternae* overlie most of the I band along the myofibrils.

As can be seen in Figure, a slender transverse *T tubule* runs between each pair of terminal cisternae at each Z line. The membranes of the T tubules are continuous with the sarcolemma, and their interior is open to the tissue fluid surrounding the fiber. These deep, tubular incursions of the plasma membrane into the muscle cell deep, tubular incursions of the plasma membrane into the muscle cell provide efficient communication between the plasma membrane and the underlying sarcomeres. The exact location of terminal cisternae and the number of T tubules per sarcomere varies in different species of vertebrates.

Early Biochemical Studies

In the early 1930s, it was found that a moderately large protein molecule, *myosin* (molecular weight 500,000), could be extracted from muscle. Myosin accounts for about half of a muscle's protein content. Another fraction, containing the enzymes of glycolysis, was first removed by washing minced muscle in distilled water. The myosin fraction was then removed by prolonged washing in cold 0.6 molar potassium chloride solution. In time, it was found that myosin has enzymatic activity: it can hydrolyze ATP to ADP and inorganic phosphate, that is, it functions as an ATPase.

Some years later, another, smaller protein (molecular weight 42,000) was extracted and given the name *actin*. In the presence of ATP the small, globular molecules of actin, or G actin, could be polymerized to form a long fibrous protein, F actin. Under the right conditions, myosin could also be made to form a long filament in the presence of actin. This substances was called *actomyosin*.

Actomyosin fibers—prepared by squirting the substance through a fine pipet—contracted when placed in a solution of Ca^{2+} and ATP. This landmark achievement of the 1940s excited the scientific world. For the first time, the principal biochemical components of the contractile mechanism were clearly identified; they had been separately removed from the muscle cell, reassembled, and then made to perform their essential role of contraction.

In the years since the identification of actin and myosin as the principal molecules of vertebrate muscle, contractile elements composed of actin and myosin molecules have been discovered not only in all animals but in many protists and some plants as well. Actin and myosin appear to have evolved very early in the history of life on earth.

Changes in Sarcomere Structure During Stretching and Contraction

Several important observations were made possible when comparisons were made of muscles prepared for examination by light and electron microscopy while (1) being stretched (2) undergoing contraction, and (3) relaxed. Both light and electron microscopy of these preparations showed that A bands remained the same length under all three of these conditions. I bands, on the other hand, lengthened when the fiber was stretched and shortened during its contraction. Similar findings had been reported since the late 1800s but were not generally accepted until the 1950s.

The Fine Structure and Chemistry of Myofibrils: Two Types of Filaments

Myofibrils that were sectioned transversely and viewed by electron microscopy were found to be composed of filaments of two thicknesses. Extracting myosin from the preparations resulted in the disappearance of the thicker of the two filaments; removed of actin caused the disappearance of most of the thin filament structure.

In many cross sections of myofibrils, each thick filament (myosin) was surrounded by several fine filaments (actin). Exceptions were cross sections that passed through the center of the A band or through most regions of the I band. Sections through the center region of the A band usually revealed only the thick myosin filaments. Sections through the I band contained only in the thin filaments of actin.

The Constant Length of the A Band

Whether relaxed or contracted, the A band remains the same length. When a myofibril in a relaxed (resting) condition is viewed by electron microscopy, the myosin and actin filaments are arranged. In the darkest regions of the A band, filaments of myosin and actin overlap. The light (H zone) area within A band in rest-length myofibrils lacks actin filaments. The dark (M) line at the center of the H zone is due to cross connections between myosin filaments.

The Changing Length of the I Band

Only actin filaments are found in the I band regions. If a fibril contracts, the I band shortens—as does the H zone—because of the further insertion of the actin filaments among the myosin filaments. Note that the Z lines, denoting the ends of the sarcomere, are much farther apart at rest than when the myofibril is contracted.

In the fully contracted state—usually achieved only if the myofibril has been freed of its connective tissue—the H band completely disappears. In full contraction, the actin filaments meet and may even overlap at the center of the sarcomere. The I band almost completely disappears as well, and the Z line approach the edge of the A band

The Sliding Filament Hypothesis

From these observations of the changes in sarcomere structure during stretching and contraction, largely the work of British biologists Andrew F. and Hugh E. Huxely, it become clear that the contraction of muscles is not due to a shortening of any of its components molecules—a view once widely held. Instead, myosin and actin filaments shorten the myofibril by actively sliding past each other. The myosin filaments can be regarded as retaining their position while the actin filaments move in reference to them. These discoveries changed existing theories of contractility and provided the molecular basis for what is known as the sliding filament theory or muscle contraction.

How can chemical energy be transduced into the mechanical energy needed to move the filaments past each other? What forces could cause such movement? What connections, if any, do the parallel thick and thin filaments have with each other? Some of these questions will be considered next.

Cross Bridges Between Myosin and Actin

High-resolution electron micrographs of myofibrils show the existence of fine cross bridges between the myosin and actin filaments. The bridges project at right angles from the surface of the myosin filaments at regular intervals except in a sizable central region in the H zone from which they are absent. They do not project in only one plane

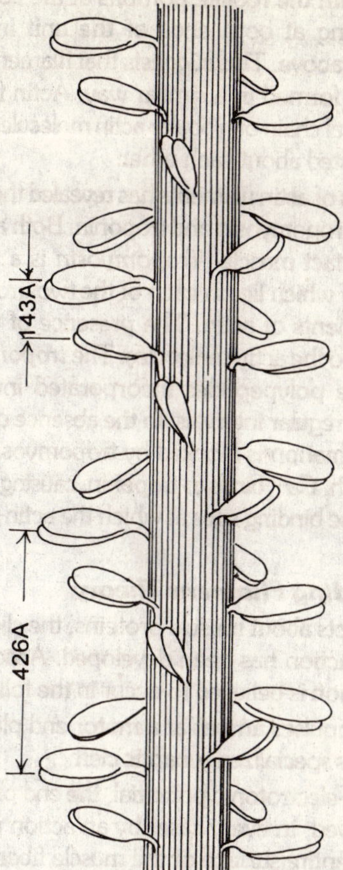

Fig. 9.5. Diagram of cross bridges on a thick (myosin) filament of a striated muscle myofibril.

but form a helical (coiled) arrangement, with six cross bridges for each full turn of the helix. Each of the six thin (actin) filaments surrounding any one thick (myosin) filament is touched by that thick filament as well as by the thick filament on its other sides. Each of the six cross bridges in one full turn of the helix is bounded to the nearest of the six actin filaments surrounding it.

The Proteins of Muscle

The myosin molecule as seen in an electron micrograph is rodlike, with a globular head at one end. Each complex, made up of a globular head and a small part of the shaft, termed *heavy meromyosin*, has ATPase activity. Most of the rodlike part of the myosin molecule is called *light meromyosin*. Units of several myosin molecules become assembled in solution with the rodlike portions at the center and the globular heads projecting at both ends of the unit in the helical arrangement referred to above. This suggests that filaments of myosin in the intact muscle are formed in a similar way. Actin filaments are made up of a double helical chain of globular actin molecules, resembling two strings of pearls twisted about each other.

Biochemical analysis of actin filaments has revealed the presence of two additional proteins: tropomyosin and troponin. Both are present in small amounts in the intact muscle. *Tropomyosin* is a long, fibrous protein, one molecule of which lies in each of the two grooves formed by the two twisted filaments of actin. The presence of tropomyosin apparently gives rigidity to the actin molecules. The *troponin* molecules are complexes of three polypeptides incorporated into the actin-tropomyosin filament at regular intervals. In the absence of Ca^{2+}, actin-myosin cross-bridge formation is blocked by tropomyosin. But when Ca^{2+} concentration is high, Ca^{2+} binds to troponin, causing tropomyosin to move aside and expose binding sites at which the actin-myosin cross bridges can form.

Summary of the Sliding Filament Theory

From the known facts about muscle proteins, the sliding filament theory of muscle contraction has been developed. According to this theory, muscle contraction is believed to occur in the following way:

1. A series of nerve impulses arrives at a motor end plate, releasing acetylcholine into its specialized synaptic cleft.
2. A local depolarizing electrotonic potential, the end plate potential, is generated, followed, in every case, by an action potential that sweeps across the entire surface of the muscle fiber, including its invaginations, the T tubules.
3. The depolarization of the T tubule membrane affects the adjacent terminal cisterna of the sarcoplasmic reticulum by increasing its permeability to Ca^{2+}. Because Ca^{2+} is 500 times more concentrated within the SR than in the sarcoplasm, the increase in the membrane's permeability to Ca^{2+} results in a rapid diffusion of Ca^{2+} into the sarcomere.
4. The flood of Ca^{2+} binds calcium to troponin, which displaces tropomyosin, permitting the globular heads of the myosin molecules to extend outward to form temporary bridges with adjacent actin filaments.

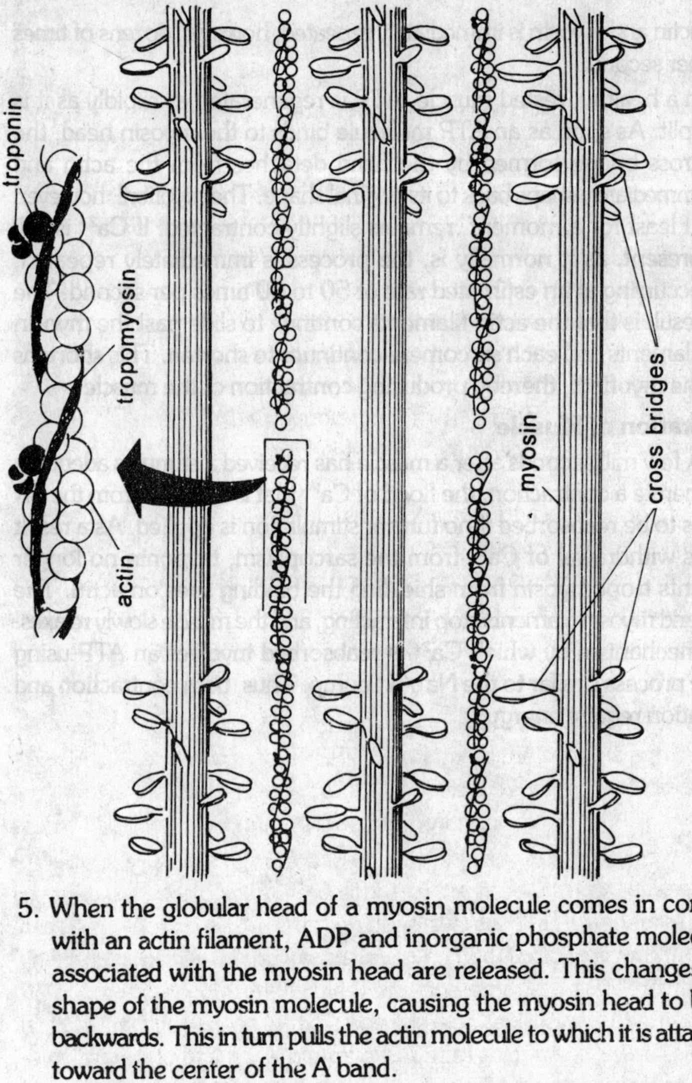

Fig. 9.6. Probable relationship of actin and myosin filaments to each other during contraction. Note the two connecting cross bridges in the lowest figure.

5. When the globular head of a myosin molecule comes in contact with an actin filament, ADP and inorganic phosphate molecules associated with the myosin head are released. This changes the shape of the myosin molecule, causing the myosin head to bend backwards. This in turn pulls the actin molecule to which it is attached toward the center of the A band.
6. As all the cross bridges exert this pull at essentially the same time throughout the myofibril—and, indeed, throughout the entire muscle fiber—the actin filaments are forced to slide past the myosin filaments. The result is that each sarcomere contracts very slightly, no more than the length of a cross bridge. This interaction between

actin and myosin is immediately repeated, however, dozens of times per second.

7. In a healthy, rested muscle, ATP is regenerated as rapidly as it is split. As soon as an ATP molecule binds to the myosin head, the cross bridge formed by the head detaches from the actin and immediately snaps back to its original shape. The myofibril, however, at least for a moment, remains slightly contracted. If Ca^{2+} is still present, as it normally is, the process is immediately repeated, occurring at an estimated rate of 50 to 60 times per second. The result is that the actin filaments continue to slide past the myosin filaments and each sarcomere continues to shorten. This shortens the myofibril, thereby producing contraction of the muscle.

Relaxation of Muscle

A few milliseconds after a muscle has received a stimulus adequate to generate a contraction, the flood of Ca^{2+} that is released from the SR begins to be reabsorbed if no further stimulation is applied. As a result of this withdrawal of Ca^{2+} from the sarcoplasm, troponin no longer prevents tropomyosin from shielding the binding sites on actin. The actin and myosin filaments stop interacting, and the muscle slowly relaxes. The mechanism by which Ca^{2+} is reabsorbed involves an ATP-using active process similar to the Na^+/K^+ pump. Thus, both contraction and relaxation require energy.

10

Transmission of Nerve Impulses

The twentieth century has brought an explosion of knowldege in neurobiology, due in large part to the perfection of microchemical methods—the analysis of mintue quantities of material—electronic technology, such as recording voltages from within small cells, and electron microscopy. As in other areas of cell biology, refined techniques have been used to analyze mintue details of cellular composition and function. In neurobiology, the cell most often studied in the neuron.

The Electrical Basis of Nerve Impulse Conduction

Just as it would be impossible to understand how a telegraph or telephone work without some knowledge of electric currents and magnetism, an understanding of resting potential is needed for any appreciation of how neurons carry information. The *resting potential* is the difference in electrical charge between the inside and the outside of a neuron's plasma membrane. In effect, the neuron is a tiny, weak battery, that is, a device that produces electrical energy by chemical action. Comparing the inside of the neuron to the outside, the potential difference, or voltage, maintained by most neurons is about –70 to –90 millivolts (mV). Most cells have the ability to maintain a potential difference between the two sides of their plasma membrane. However, *excitable cells,* such as neurons and muscle fibers, can use this potential difference to convey information from one end of the cell to the other and even, in some cases, to communicate information from one cell to the next.

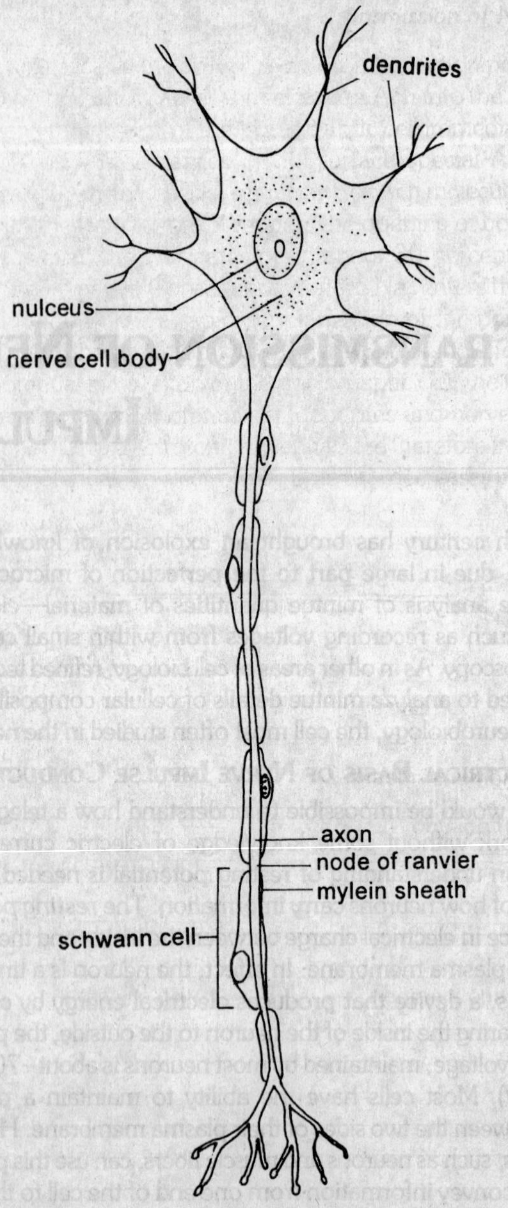

Fig. 10.1. Structure of a vertebrate motor neuron.

Ion Distribution

To understand the special situation that exists in a neuron, consider these principles that apply to all cells:

1. The total number of all kinds of diffusible particles, such as molecules or inorgaic ions, on one side of a plasma membrane tends to be balanced by an equal number of such particles on the other side of that membrane. However, because of the differing permeability of plasma membranes to various solutes, few, if any, particles of the same kind are ever evenly distributed on the two sides of the same membrane.
2. The positively charged particles (cations) on either side of the plasma membrane also tend to be balanced by an equal number of negatively charged particles (anions) on the same side. However, whenever one type of ion, such as charged proteins, cannot diffuse through a membrane, the smaller ions that are free to cross it will become unequally distributed on the two sides of the membrane.

Potassium, Sodium, and other Important Ions

In neurons, as in cells generally, the cytoplasm contains a number of dissolved amino acids and proteins, most of which are anions. These large, negatively charged molecules cannot diffuse out of the cell because they cannot penetrate the plasma membrane. However, because the plasma membrane of the neuron is quite permeable to potassium ions (K^+), these small ions are free to enter the cell, where, to some extent, they neutralize the cell's organic anions. In addition, the plasma membrane has a mechanism called a pump that actively transports K^+ into the cell despite an opposing concentration gradient. As a result, potassium ions are about 25 times more concentrated inside the cell than in the tissue fluid surrounding it. The distribution of ions on the two sides of the plasma membrane of a giant axon of the squid is given in table.

Table 10.1. Ionic Concentrations on Both Sides of the Plasma Membrane of a Squid Giant Axon

	Outside (mM/liter H_2O)	*Inside (mM/liter H_2O)*	*Ratio (o/i)*
K^+	15	400	$^1/_{25}$
Na^+	450	50	$^{10}/_1$
Cl^-	545	50-100	$^7/_1$
Ca^+	10	0.4	$^{25}/_1$
Organic anions	0	355	

Because the plasma membrane of a neuron that is not conducting a nerve impulse—the so-called resting neuron—is more permeable to K^+ than to any other ion, the neutralization of the electrical charge of the proteins and other nondiffusible anions inside the cell is almost entirely achieved by K^+. If K^+ were the only cation to enter, the resting potential would be about –90 mV, a value known as the equilibrium potential for K^+. However, because other ions, principally Na^+, also enter to some extent, the resting potential achieved in most cells is somewhat less than –90 mV. (Two other ions present in or about the neuron in significant amounts are Cl^- and Ca^{2+}. Both are more concentrated outside the cell than inside).

The distribution of Na^+ on the two sides of the membrane is strikingly different from that of K^+. Na^+ is about 10 times more concentrated outside the cell than within it, resulting in large part from the action of the same ion pump that transports K^+ into the cell. Another reason for this sodium distribution is that, in the resting state, the neuron's plasma membrane maintains an almost complete impermeability to Na^+. In other words, the sodium "gates" in the sodium channels within the membrane are almost completely closed.

Some inward leakage of Na^+ and some outward loss of K^+ down their respective concentration gradients occurs continulally, as has been demonstrated by radioactive tracer studies. It could be said that the neuron "battery" continually leaks some of its charge. This leakage is fully compensated for by the sodium-postassium pump, however.

The Sodium-Potassium Pump

As long as the plasma membrane of a neuron is intact and has an energy supply in the form of ATP—both being absolute requirements for active transport—the neuron continually discharges as much Na^+ as enters the neuron. That the process is truly an active one, requiring energy, can be shown experimentally. For example, when a giant axon of a squid is perfused with dinitrophenol (DNP), a poison that blocks the formation of ATP, the active transport of Na^+ out of the neuron stops. When the axon is washed free of DNP, normal active transport of Na^+ resumes.

When the mechanism that actively transports Na^+ was first discovered, it was called the sodium pump. Eventually, it was found that K^+ is also actively pumped into the cell by the same transport mechanism. Accordingly, the mechanism was named the sodium-potassium pump or the ion pump.

The pump works only if both Na^+ and K^+ are present. And it works only in one direction for each ion: Na^+ out and K^+ in. The ratio of exchange is three Na^+ ions out for every two K^+ ions in, for every ATP molecule utilized. The Na^+-K^+ pump thus contributes somewhat to the maintenance of the resting potential.

In summary, the distribution of ions in the resting neuron is as follows:

1. High concentrations of Na^+, Ca^{2+}, and Cl^- exist outside the cell.
2. High concentrations of K^+ and organic anions exist inside the cell.
3. The plasma membrane is highly impermeable to Na^+ and Cl^-, less so to Ca^{2+}, somewhat permeable to K^+, and altogether impermeable to organic ions.
4. The Na^+-K^+ pump actively maintains the Na^+ and K^+ concentrations at a steady "resting" level.
5. Following its concentration gradient, K^+ tends to leak out of the cell. However, K^+ ions are held within the cell and to the outside of the membrane by attraction to the organic anions that cannot follow them through the membrane. The result is that the outside follow them through the membrane. The result is that the outside of the membrane is electrically positive, and the inside negative.

These conditions provide the physical basis for the resting potential. The electrical signals by which parts of the nervous system communicate with each other can be regarded as disturbances produced in the resting potential. Such signals are of two types: electronic potentials and action potentials, or nerve impulses.

Electrotonic Potentials

If the electric charge inside a neuron is slightly lowered by introducing cations into the cell, a partial *depolarization* of the membrane occurs; that is, the inside becomes less negative in relation to the outside. The degree of depolarization depends on how many cations enter the neuron. If an electrode carrying a direct electric current is used to induce depolarization, the amount of depolarization is proportional to the strength, or voltage, of the current. If the administered current flows in the opposite direction, however, the membrane becomes *hyperpolarized;* that is, the inside becomes more, not less, negative and is less likely to generate a nerve impulse.

However, no matter whether the stimulus is depolarizing or hyperpolarizing, the effect on the neuron is to produce a disturbance known as an *electrotonic potential*, also referred to as a local, graded response.

Such a disturbance weakens with both time and distance as it spreads from the point where the stimulus was given, along the inside and outside of the membrane, before fading away entirely after a short distance. Because of this decrease in intensity with time and distance, electrotonic potentials are useful only for local signaling.

The Neuron as Capacitor

Because the neuron's plasma membrane is thin and relatively impermeable to various ions, the neuron has the properties of an electrical *capacitor*, or condenser, a device that accumulates and stores electric charges. In the neuron's role as capacitor, the plasma membrane serves as a thn layer of insulation separating the two oppositely charged elements. The positive charges on the outside of the membrane accumulate there because they are attracted to and held to the outer surface of the membrane by the negative charges just inside the membrane. The negative charges are provided by the organic anions that cannot diffuse out of the cell, whereas the positive charges are provided mostly by Na^+ and K^+ ions that are held to the outer surface of the membrane by electrostatic attraction to the organic anions on its inner surface. If the negative charge inside the membrane is reduced, by a depolarizing current, for example, the positively charged ions on the outside of the cell are no longer held to the membrane's surface. These cations promptly disperse into the surrounding tissue fluid. This flowing away of the membrane's outer charge is the first effect of a reduction of the negative charge inside a neuron. Together, the reduction of the inner charge and the resulting loss of outer charge comprise the depolarization.

Sodium Ionic Current

Any depolarization of the resting neuron's plasma membrane has a special effect on the membrane: it immediately increases its permeability to Na^+. Depolarization causes the gates in the sodium channels to open. Because sodium is in higher concentration outside the cell than within it, the increased permeability to sodium immediately allows a current of sodium ions to flow into the cell in the depolarized region. The amount of initial depolarization and the strength of the resulting sodium current are directly proportional to the strength of the depolarizing stimulus. A weak stimulus is followed by the opening of only a few sodium gates and the entry of only a small amount of Na^+. But even the entry of a weak cationic current further neutralizes the negative charge inside the cell, that is, it further depolarizes the membrane, including the adjacent regions as well. This additional depolarization increase the membrane's permeability to Na^+ and, for a time, produces a *positive feedback effect,* in which the entry of Na^+ as a result of depolarization thus itself becomes a depolarizing stimulus. However, there is a check upon this process. After the membrane has undergone depolarization for a time in a given region, the sodium gates in that region close. They remain closed until that part of the membrane has once again become polarized by returning to the resting potential.

The Potassium Channels and Repolarization of the Membrane

When depolarization begins and the sodium gates begin to open, the gates in the potassium channels also start to open, but much more slowly. Accordingly, K^+ begins to flow ot of the cell at the point of stimulation. The potassium current attains its peak outflow about half a millisecond after the sodium gates have closed. (Weak electrical currents of as brief duration as a millisecond or less can be amplified and displayed as tracing by the device known as a *cathode ray oscilloscope.*) The closing of the Na^+ gates, the opening of the K^+ gates, and the resulting outflow of K^+ repolarize the membrane in the affected region, restoring it to the resting potential condition.

If the stimulus given to a neuron is weak and does not persist very long, only a few sodium channels are affected and few sodium ions enter the cell. The effect on the neighbouring regions of the neuron is therefore minimal, and the electrotonic potential generated is limited to the immediate region and soon dies out. However, if a strong stimulus is given, opening may sodium gates, a considerable amount of Na^+ will enter, causing depolarization not only at the point of stimulation but of the immediately adjacent regions as well. The slowly reacting potassium gates are not able to repolarize the membrane before the strong sodium current has begun to flow through the cytoplasm of the neuron, completely depolarizing adjacent regions of the membrane. The process renews itself again and again, spreading rapidly down the fiber. An *action potential,* or *nerve impulse,* has been generated.

Action Potentials

In sharp contrast to the strength of the electrotonic potential, the strength of the action potential does not decrease with distance or time. Once triggered, action potentials pass along the entire length of the longest axon. Action potentials are thus well suited to long-distance communication.

The CRO record of an action potential is a spike-shaped curve. The upslope of the action potential curve displayed by a cathode ray

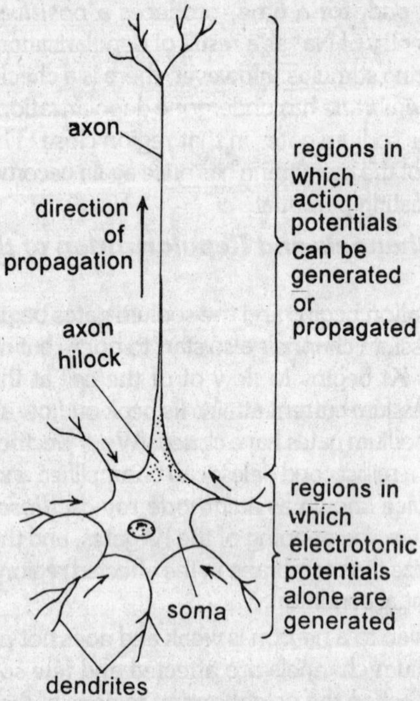

Fig. 10.2. Action potentials normally arise only at the axon hillock and propagate down the axon.

oscilloscope results from increased membrane permeability to Na^+; the downslope results from increased permeability to K^+ with sodium channel inactivation. In the initial phase of an action potential, the sodium gates open wide and Na^+ is free to enter the cell. As a result, the process does not end at complete depolarization, or zero membrane potential. Instead, Na^+ continues to enter and the electrical change continues beyond, or overshoots, the zero point. The inside becomes about 30 to 50 mV positive in relation to the outside, a value near the point at which the sodium concentrations inside and outside, a value near the point at which the sodium concentrations inside and outside the membrane are at equilibrium. The sodium current has now run its course and the countercurrent of K^+ exerts its full effect by repolarizing the membrane.

As the K^+ gates open wide, the Na^+ gates close and cannot open again for 1 or more milliseconds, no matter how strong a depolarizing stimulus is given. As a result, this region of the neuron is temporarily inexcitable to further stimulation; it has entered what is termed the *absolute refractory period*. The rapid exit of K^+ that then occurs quickly returns the membrane to near the resting potential level. The entire action potential takes only a millisecond or so. The shape of the record of an action potential on the CRO screen suggested the term *spike potential* or simply spike.

The Threshold-Level Stimulus

As we have seen, an electrotonic potential or other stimulus strong enough to depolarize a resting potential to about –40mV causes an extensive opening of sodium gates followed by a strong inrush of Na^+. Instead of the disturbance fading and the membrane returning to its resting potential, a complete depolarization occurs and spreads to neighbouring regions of the neuron. The minimum stimulus strength needed for an action potential to occur is said to be of *threshold-level intensity*. As with squeezing the trigger of a loaded gun, once the process reaches a certain point, it continues to completion even if the triggering stimulus is discontinued.

As noted earlier, one way to cause a depolarization of threshold-level intensity is to introduce cations into the cytoplasm of a neuron. An action potential can also be generated in any part of a neuron by the arrival of a potential from another part of the same neuron. Because the action potential depolarizes the neuron, it produces a strong electrotonic potential in the region of the membrane immediately ahead of it, one great enough to generate another action potential. It is this process that occurs over and over again down the entire axon.

The generation and passage of an action potential is not altogether unlike the way a fuse, such as one attached to a stick of a dynamite, burns after being ignited. Once ignited, it continues to burn down to it end. The action potential is a classic example of an "all-or-none" phenomenon, one that, like the burning of a fuse or the firing of a gun, either occurs entirely or not at all.

Of the various part of a neuron, only the axon normally initiates action potentials. They arise at the region called the *axon hillock*, where the axon joins the soma. Regions of the neuron such as the dendrites and cell bodies normally initiate only electrotonic potentials. Although these areas can also generate spikes, they seldom do because of their high thresholds. Thus, when a depolarizing stimulus is received from another neuron by a dendrite or soma, only an electrotonic potential is generated. It spreads passively and with decreasing strength across the cell surface. The stronger the stimulus, the logner the resulting electrotonic potential lasts, and the farther it spreads before it dies out.

Only if its remaining strength is sufficient to depolarize the membrane of the axon hillock when it arrives there will a spike potential be generated in the neuron. Action potentials arising in an axon hillock, however, are propagated in both directions—down the axon and back to the soma and dendrites where they die out uneventfully.

Axon Diameter and Conduction Velocity

Several invertebrate phyla have evolved the same general method for increasing the conduction velocity of action potentials in key axons: by increasing their diameter. The giant axons of earthworms, crustaceans, insects, and squids are good examples. Probably the best studied of these is the giant axon of the common squid. As in all neurons, a flow of ionic current in the axoplasm (cytoplasm of the axon) precedes the generation of each action potential. Just as a large-diameter pipe permits a more rapid flow of water than does a pipe of small diameter, ions flow more freely and thus more rapidly in a giant fiber, increasing the velocity of nerve impulse propagation.

Myelination and Salatory Conduction

Vertebrates have evolved a different approach to increasing conduction velocity of nerve impulses. Surrounding many neurons are layers of a whitish, fatty substance called myelin, which acts much like the insulation around an electrical wire. Each *myelin* layer is actually layer upon layer, up to 100 layers, of plasma membranes of the adjacent neuroglial cell. Such cells associated with peripheral nerves are known as *Schwann cells*. Because myelin is an effective insulator, it prevents leakage of ionic currents through the plasma membrane. Moreover, Na^+ and K^+ are virtually absent in the myelinated regions of the axon. As a result, no spike potentials can be generated in any region of the axon covered by a myelin sheath. The myelin sheath, however, is interrupted at regular intervals by bare regions called *nodes of Ranvier*. At each node, the axon plasma membrane is exposed on all sides to the surrounding tissue fluid. At the nodes, the axon's plasma membrane has a very high concentration of Na^+ and K^+ channels.

Whenever an action potential is generated at the unmyelinated axon hillock, it is propagated along the neuron until it reaches the myelinated region of the axon. There the ionic currents of Na^+ and K^+ generated by the spike are unable to flow through the myelin-covered membrane. Ionic currents can flow through the axoplasm, however, as well as through the tissue fluid surrounding the axon. When an action potential arises at the axon hillock, the potential difference (voltage difference) between the hillock and the first node is great enough to trigger the opening of

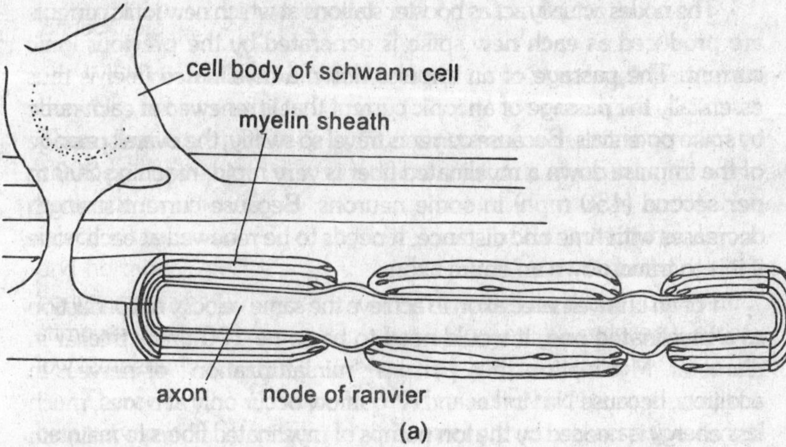

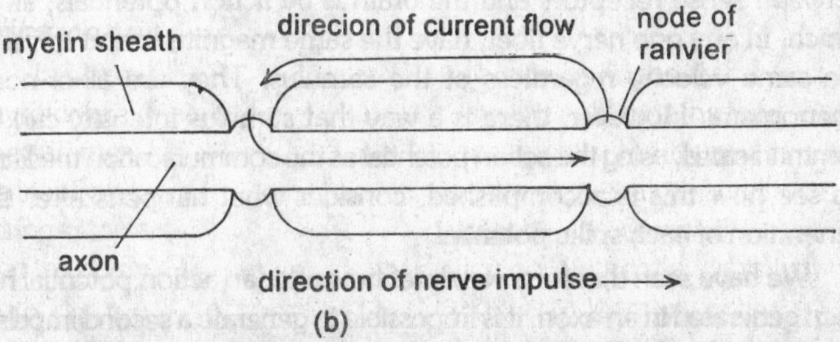

Fig. 10.3. Saltatory conduction. (a) Relationship of a neurogial cell to the myelin sheath in a myelinated fiber in a cat's spinal cord. (b) Conduction of the nerve impulse between nodes in a myelinated nerve is accomplished by a flow of current. Current flow is represented by heavy arrows. Action potentials are generated at the bare nodes, which in turn generate another current, which produces another action potential at the next node.

the Na^+ gates at the node, producing an action potential there. This immediately generates an ionic current, which travels between the first and second node. Another spike is thereupon produced at the second node and so on at each node down the entire myelinated axon. Because the action potential thus appears to leap from one node to the next, this type of propagation has been called *saltatory conduction* (from the Latin, *saltare*, "to leap").

The nodes actually act as booster stations at which new ionic currents are produced as each new spike is generated by the previous ionic current. The passage of an impulse down a myelinated fiber is thus essentially the passage of an ionic current that is renewed at each node by spike potentials. Because currents travel so swiftly, the overall passage of the impulse down a myelinated fiber is very rapid, reaching 200 m per second (450 mph) in some neurons. Because current strength decreases with time and distance, it needs to be renewed at each node if it is to travel down an entire axon.

For an unmyelinated axon to achieve the same velocity of conduction as a myelinated one, it would need to be some 100 times thicker in diameter. Myelination thus permits "miniaturization" of nerves. In addition, because Na^+ influx and K^+ outflow occur only at nodes, much less energy is needed by the ion pumps of myelinated fibers to maintain their resting potentials.

Communication by Means of Nerve Impulses

It might be wondered what, if anything, can be communicated by nerve impulses. Recall that the principal means of communication between sense receptors and the brain is by action potentials, all of which, in any one nerve fiber, have the same magnitude and travel at the same velocity regardless of the stimulus. They are all-or-none phenomena. However, there is a way that stimulus intensity can be communicated, using the action potential as the communication medium. To see how this is accomplished, consider what happens after the generation of each spike potential.

We have seen that for a very brief time after an action potential has been generated in an axon, it is impossible to generate a second impulse. This interval is the absolute refractory period. If a neuron is continually stimulated by a threshold-level stimulus, it will generate and conduct impulses at a certain constant frequency. The time between the generation of one impulse and the next is the total refractory period for that neuron.

At the end of a neuron's absolute refractory period it is possible to generate another nerve impulse only if the stimulus intensity is above the normal threshold level. Viewed in another way, following the absolute refractory period, the threshold level for a neuron is higher than normal; this time period is known as the *relative refractory period*. Thus, if a stimulus that is more intese than the threshold level is continuously applied, the total refractory period is shorter than it is with continuous application of a stimulus that is barely at threshold level. Impulses are

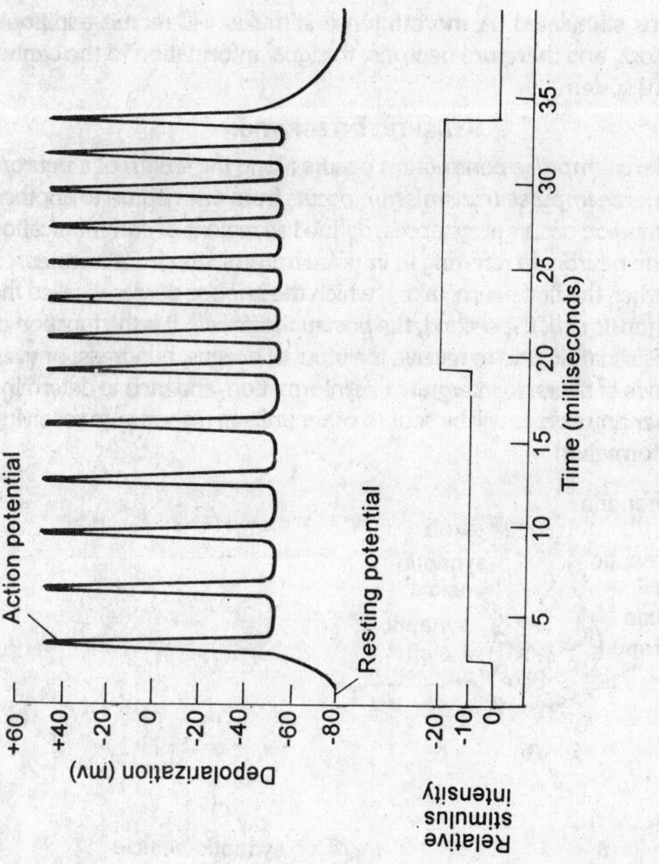

Fig. 10.4. A CRO record illustrating that stimulus intensity is coded as, or translated into equivalents of, nerve impulse frequency.

thus generated more frequently when the stimulus is appreciably above threshold intensity. If the stimulus intensity is increased even more, the refractory period will shorten further, generating impulses even more frequently. But there is a limit to how much a neuron's refractory period can be shortened by increasing the stimulus intensity. Once that limit is reached, no matter how much the stimulus intensity is increased, no further increases in the frequency of impulses result.

This capability of responding differently to stimuli of different intensities means that within a certain range of intensities, neurons translate stimuli into impulses whose frequencies correspond to the intensity of the stimulus. Therefore, the rate at which impulses are received by the brain is directly related to the intensity of the stimulus. Another index of the intensity of a stimulus is the total number of neurons stimulated. A more intense stimulus will recruit additional receptors, and therefore neurons, to signal information to the central nervous system.

SYNAPTIC INTEGRATION

Nerve impulse conduction occurs along the length of a neuron, while *nerve impulse transmission* occurs from one neuron to another. Transmission occurs at synapses, defined as regions of communication between neuron. In referring to impulse transmission from one neuron to another, the first neuron along which the impulse passes is called the *presynaptic cell*, the second, the *postsynaptic cell.* It is the function of the postsynatpic cells to receive the input of dozens, hundreds, or even thouands of fibers, to intergrate this information, and then to determine whether any signals will be sent to other units in response to receiving this information.

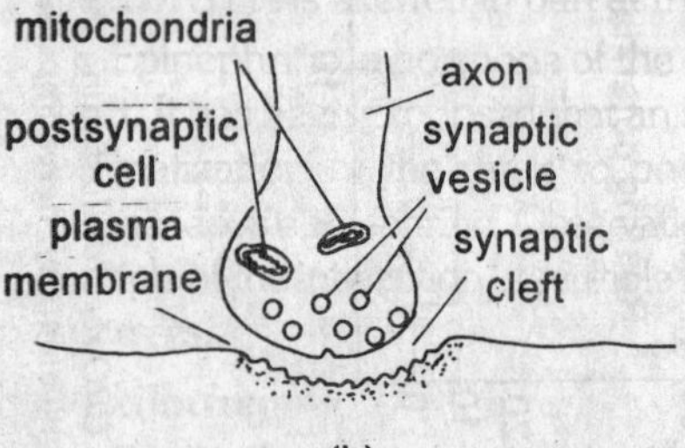

(b)

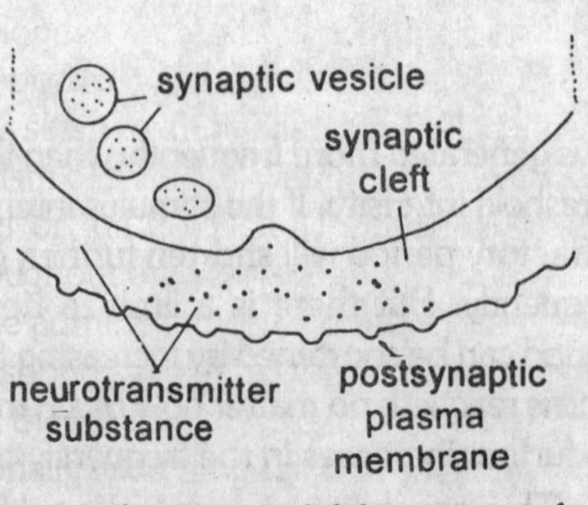

Fig. 10.5. Release of neurotransmitter into the synaptic cleft by exocytosis from synaptic vesicles.

Basic Mechanisms of Synaptic Transmission

Synapses are not unlike executive centers in a large organization. By a combination of chemical and electrical signals, incoming data are evaluated in computerlike fashion by the postsynaptic cell, which immediately determines what information is to be sent on to other parts of the body.

Transmission by Neurotransmitters

With nearly all synapses, action potentials arriving at the axon tips, called *synaptic knobs* of a presynaptic cell result in the release of precise amounts of a *neurotransmitter,* a special chemical substance contained in minute synaptic vesicles. The released neurotransmitter enters the synaptic clefts, the spaces between the axon tips and the postsynaptic cell's plasma membrane. The neurotransmitter acts on receptors on the plasma membrane of the postsynaptic cell to generate electrotonic potentials. The entire process, including the diffusion of the transmitter across the cleft, usually takes about a half a millisecond but may be slightly longer. It was this "synatpic delay" that first led physiologists to suspect that a chemical transmitter was involved and that a minute space separated the two cells.

In one common type of synapse, the arrival of action potentials at synaptic knobs produces a depolarizing *excitatory postsynaptic potential* (EPSP). If the EPSP or the sum of multiple EPSPs spreads as far as the axon hillock of the postsynaptic cell and results in a threshold depolarization of the hillock, one or more action potentials are generated in that cell. The best known of the neurotransmitters involved in generating EPSPs is acetylcholine (Ach).

Another type of postsynaptic potential hyperpolarizes the postsynaptic cell, making it more negative and therefore less likely to generate an action potential. This type is accordingly termed an *inhibitory postsynaptic potential* (IPSP).

At most synapses at which EPSPs are generated, a small amount of neurotransmitter continually leaks into the synaptic cleft. Although it is not enough to induce an action potential in the postsynaptic cell, the small postsynaptic potential thus created makes it easier for a spike to be generated when action potentials do arrive at the synapse.

Calcium ions apparently play a role in synaptic transmission. The arrival of the action potential is thought to open the Ca^{2+} gates in the plasma membrane of the presynaptic cell. Because Ca^{2+} is over 100 times as concentrated outside the cell as inside, a rapid influx of Ca^{2+} into the presynaptic cell occurs. Ca^{2+} plays an essential part in causing the membranes of the synaptic vesicles to fuse with the plasma membrane of the presynaptic cell, thereby releasing neurotransmitter into the synaptic cleft.

In a typical case, the arrival of one or more action potentials at the tips of axon branches causes the release of Ach into the synaptic cleft. The plasma membrane of the postsynaptic cell immediately opposite the points of release carries on its surface special Ach receptors, molecules with an enzymelike specificity for Ach molecules. When Ach unites with these receptors, it triggers the opening of both Na^+ and K^+ gates. However, the Na^+ gates open about 30 percent more widely than do the K^+ gates. Consequently, more Na^+ enters the cell than K^+ leaves it. The result is a rapid depolarization of the postsynaptic cell membrane. Although this might exceed the threshold for inducing an action potential in the axon of the postsynaptic cell, the threshold level for generating action potentials in dendrites and nerve cell bodies is much higher. Only electrotonic potentials are therefore induced. Unlike an action potential, an electrotonic potential decreases in intensity as it

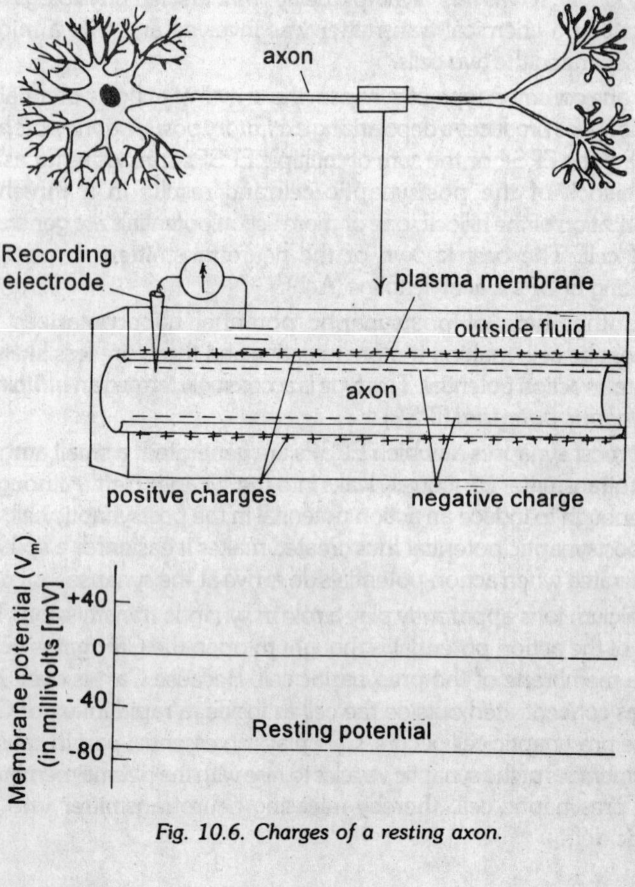

Fig. 10.6. Charges of a resting axon.

passes along the dendrites and nerve cell body. Only if it is still of threshold intensity by the time it reaches the axn hillock will it generate an action potential in the axon of the postsynaptic cell.

Destruction of Acetylcholine

It might be wondered what would prevent acetylcholine, once released into a synaptic cleft, from continuing to stimulate the postsynaptic cell. Actually, nothing, if it were not for the presence of an enzyme: acetylcholinesterase. In the presence of this enzyme, acetylcholine is hydrolyzed to choline and acetic acid. Once this is accomplished, the material is no longer capable of stimulating the receptors in the plasma membrane of the postsynaptic cell. The effect of the enzyme is not so rapid, however, as to prevent the neurotransmitter from acting. The enzyme merely ensures that acetylcholine does not persist and continue to stimulate the postsynaptic cell. Similarly, Ca^{2+} that has caused the release of Ach from a presynaptic cell promptly becomes sequestered by calcium binding proteins, mitochondria, or both. Consequently, for each action potential that reaches a synaptic knob, only a brief burst of Ach release occurs.

Postsynaptic Inhibition

As noted, release of a transmitter substance by the tips of axons does not always produce EPSPs. Sometimes the result is an IPSP, in which case the postsynaptic cell is temporarily rendered less capable of firing, that is, of generating an action potential.

Because of the minute size of most interneuronal synapses, they are very difficult to study, and some of the most intriguing questions about inhibitory synapses remain unanswered. This is understandable when we consider the problems involved in determining variations in the nature of receptor molecules form one synapse to another among the hundreds present on a single postsynaptic cell. Despite this, evidence is accumulating that transmitters other than acetylcholine are released at many synapses. Some of these are inhibitory. One example is *gamma-aminobutyric acid* (GABA). The effect of GABA on a postsynaptic cell is to open the Cl^- channels; the Na^+ and K^+ channels remains closed. Because Cl^- ions are in higher concentration outside the cell than within, a rapid influx of Cl^- ions occurs. This hyperpolarizes the membrane, thereby producing an IPSP. In the peripheral nervous system, however, the inhibitory quality of a particular synapse is often due to a difference in the nature of the receptor in the plasma membrane of the postsynaptic cell. The same neurotransmitter could thus generate either an EPSP or an IPSP, depending on the kind of receptor molecule affected. These differences in response depend on which channels are opened by the particular receptor-transmitter complex. When Na^+ channels are opened, an EPSP results; when anion channels are opened, the result is an IPSP.

At any given moment, the potential generated across the entire surface of a postsynaptic cell is composite of the effects—both excitatory and inhibitory—achieved at the several snyapses on its surface.

Kinds of Chemical Synapses

Electron microscopy has shown that there is an enormous variety of synaptic types, providing a wealth of opportunities for integration. Most vertebrate synapses are chemical synapses, characterized by the presence of synpatic vesicles and mitochondria in the cytoplasm of the synaptic knobs of presynaptic cells. These organelles are absent from the synaptic region of the postsynaptic cell. Although any one chemical synapse usually possesses but one kind of vesicle, there are at least four kinds of vesicles and over 30 known or suspected transmitters in CNS.

Electrical Synapses

Some unusual synapses lack vesicles, and the space between pre- and postsynaptic cells is only 20 to 50Å wide, rather than the usual 200Å or greater. Such cells are connected by intercellular channels that provide a low resistance to the passage of ions. Such a synapse is termed a *gap junction,* or nexus (Latin for "a fastening"). Because an action potential can pass without delay and without a chemical mediator between such cells, they are also known as electrical synapses. In effect, they unite the two cells into one long cell that can usually conduct impulses in either direction. Chemical synapses, on the other hand, are always polarized, that is, they transmit in one direction only, from the tips of axons, which contain synaptic vesicles, to dendrites or nerve cell bodies.

It is important to distinguish gap junctions, or electrical synapses, from so-called tight junctions. The latter are found between a variety of cells, including neurons. But they are not synapses. They do not have intercellular channels. Instead, they form high-resistance seals between cells and prevent electrical conduction between cells. They insulate rather than communicate.

The Integrative Capabilities of the Interneuronal Synapse

By their wide diversity, interneuronal synapses provide great opportunity for integration. For example, in some parts of the brain, hundreds or even thousands of synaptic knobs may form synapses on the dendrites and soma of a single neuron. While many of these may be the branches of a single axon, many represent input from multiple neurons. Moreover, while many presynaptic neurons are excitatory, many, if not most, are inhibitory. The postsynaptic cell integrates the input from these various sources and generates action potential accordingly. A few of the many types of postsynaptic integration are described in the following sections.

Summation of EPSPs

A single action potential arriving at an excitatory synapse almost never releases enough neurotransmitter to generate an action potential in the postsynaptic cell. Instead, a large number of EPSPs usually must be produced within a short time—usually at many synapses impinging on the same postsynaptic neuron—for the resulting depolarization to be sufficient to produce a spike potential at its axon hillock. The properties of the axon hillock determine whether action potentials are generated in the postsynaptic cell. The process by which EPSPs are added to each other to generate action potentials is termed *summation;* it has both temporal and spatial dimensions.

Temporal Summation

If the time interval between impulses arriving at an excitatory synapse is sufficiently brief, the EPSP produced by the first impulse will not have died away completely before a second is generated and added to the residue of the first. As a result, a greater EPSP is produced than from a single impulse. If a third impulse arrives before the summed potentials induced by the first two have decayed, the third will be added to them, resulting in an even greater EPSP. Such addition of successive potentials to the remnants of previous ones is termed termporal summation.

Spatial Summation

In temporal summation, EPSPs produced at close intervals of time by the same synapse are added to each other. In spatial summation, EPSPs produced at nearly the same time by neighbouring synapses are added to each other. The effect is similar: a stronger EPSP is produced in either case. The principles of temporal and spatial summation apply to the production of IPSPs as well.

Facilitation and Antifacilitation

Facilitation, another type of integrative activity occurring at synapses, is easily confused with temporal summation. As in temporal summation, facilitation depends on the arrival of two or more impulses at the synaptic surface or surfaces of the same postsynatpic cell within a certain space of time. But the result in this case is that the second EPSP generated is greater than the first, the third is greater than the second, and so on. In each case, moreover, the previous EPSP has died away completely before the next develops; so this is not an instance of adding to the residue of the previous potential, this is, it is not a case of temporal summation. Somehow, in faciliatation, the first impulse to arrive at a synapse increases the sensitivity of the mechanism by which neurotransmitter is mobilized and released; the arrival of the second further enhances this sensitivity, and the effect recurs at the arrival of each successive impulse to a point at which no further enhancement results. The length of time that facilitation lasts may vary greatly from one type of synapse to another. In some, the effect persists only for milliseconds, in others for several minutes, and in some for a day or longer.

In sharp contrast to what happens in facilitation, in some postsynaptic cell junctions, the arrival of several impulses at short intervals results in decreased sensitivity to further impulses. This phenomenon is appropriately termed antifacilitation.

Both facilitation and antifacilitation can be useful integrative activities under certain circumstances. Facilitation probably plays a part in perfecting an animal's response to specific stimuli. As such, it would probably be one of the benefits provided by a "warm-up practice" just before an athletic contest. One elegant refinement of facilitative integration is seen in certain types of synapses in which the junction can be temporarily changed from one that does not exhibit facilitation to one that does. In such cases, the postsynaptic cell becomes more sensitive to the arrival of impulses from a certain type of presynaptic cell if impulses have recently arrived at neighbouring synapses formed by presynaptic cells of another type. Understandably, the phenomenon is difficult to distinguish from—and sometimes has been confused with—spatial summation. That it is not simply spatial summation is shown by the fact that the phenomenon does not occur when the two different types of presynaptic cells are stimulated in reverse order. Antifacilitation has a conservative function. It is one of te reasons we can become insensitive after a time to many recurrent or continuous stimuli— sounds, odors, even pain.

A quite different basis for decrease in sensitivity exists in *sensory adaptation,* a phenomenon seen in sensory receptors that become insensitive and fail to generate impulses when stimulated continuously or repeatedly at short intervals. Antifacilitation occurs at synapses, whereas sensory adaptation occurs at receptors. Such receptors, together with synaptic junctions capable of antifacilitation, spare animals the need to process much useless information.

Most of the mechanisms by which the various types of facilitation and antifacilitation are achieved are not yet understood.

11

Reproductive Organs

Male Reproductive Organs

The male reproductive system is associated with the urinary system from which it is partially derived. It consists of the following: (1) the primary sex organs which produce the sperm, i.e., the testes; (2) accessory sex glands which produce fluids that combine with and serve as a medium of transport for sperm during emission, i.e., the prostate, vesicular glands, bulbo-urethal glands, and lesser structures; (3) a series of ducts through which sperm and fluid are transported to the outside, i.e., the epididymis, ductus deferens, and urethra; and (4) the external genitalia, i.e., the penis and, in some forms, a scrotum. Embryonic development, anatomy, histology, and physiology of these structures will be discussed and species differences, will be illustrated.

Embryology of the Reproductive Organs

The genital system arises from mesodermal mesenchyme to form bilateral *genital ridges* between the *mesonephros* and dorsal aorta. The genital ridges are covered with *germinal epithelium,* which is continuous with the mesothelium of the *coelomi* cavity. Into the germinal epithelium move primordial germ cells from either extraembryonic endoderm or form the yolk *sac.* As the genital ridge thickens and moves away from the mesonephros, it develops its own mesentery, the *mesorchium.* Germinal epithelium, including *primordial germ cells,* invaginates into the germinal ridge and forms *primary sex cords* which become continuous with *rete cords* derived from *mesenchyme.*

The structure, which has bisexual potencies, may then be termed a gonad. It consists of an outer cortes composed of mesenchyme; and a medulla composed of sex cords, rete cords, and mesenchyme. In the male, the gonad is a *testis* in which the primary sex cords and rete cords become hollow and form *seminiferous tubules* and *rete tubules*, respectively. The germinal epithelium regresses and becomes identical to other coelomic epithelium. The cortex differentiates into a connective tissue sheath, the *tunica albuginea*. Other mesenchymal cells differentiate into *interstitial cells*. Still others from a connective tissue framework of septula, which are continuous with the peripheral albuginea and the medial *mediastinum testis* in which the rete testis is located.

The septula divide the testis into *lobules*, each of which contains a number of seminiferous tubules.

As the metanephros develops and the mesonephros degenerates, certain tubules of the mesonephros undergo reduction; they remain functional as a connection between the rete tubules and the mesonephric portion of the Wolffian duct, and are known as the efferent ductules of the *epididymis*. The mesonephric duct loses its urinary function; the cranial portion undergoes extreme convolution and becomes the *ductus epididymis* while the caudal portion remains relatively straight and

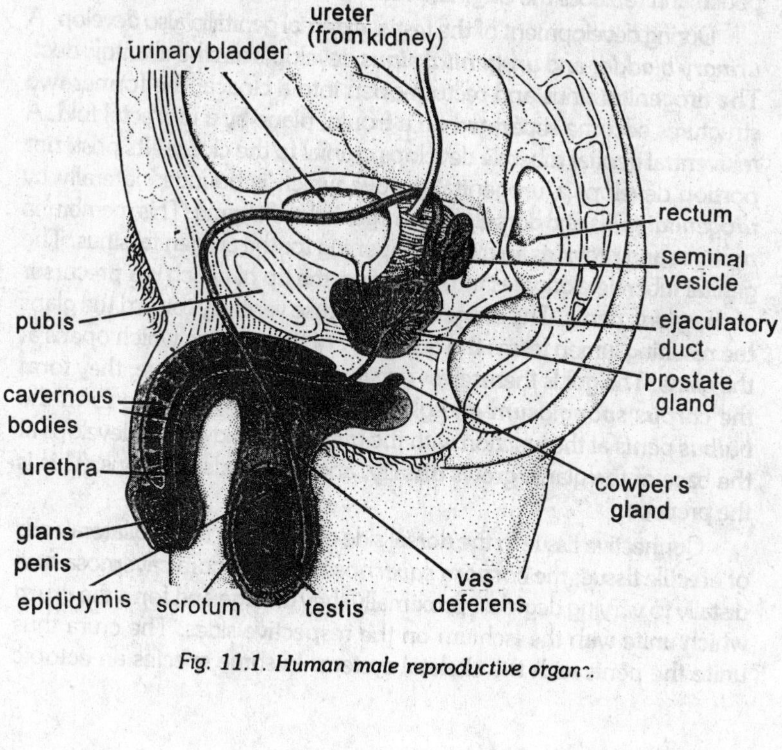

Fig. 11.1. Human male reproductive organs.

becomes the *ductus deferens*. The latter may expand into the *ampullary gland* cranial to its junction with the urethra; it may also develop an evagination just caudal to the ampullary gland, the vesicular gland.

A cyst lined with dolumnar epithelium, the *appendix epididymis*, is connected with the cranial end of the epididymis. A *paradidymis*, consisting of a group of blind tubules, is located in the spermatic cord just proximal to the cranial end of the epididymis. Both these structures are nonfunctional remnants of the mesonephros.

A groove parallel and lateral to the mesonephric duct develops in the urogenital ridge. The sides of the groove fold together and fuse to form the *Mullerian ducts*. Posteriorly, the genital ridges elevate, fold together, and fuse to form the median genital cord. In the process,, the *Mullerian ducts* are brought into apposition and fuse. They end in a middorsal projection into the urogenital sinus, *Muller's tubercle*. The tubercle forms at the entrance of the Wolffian ducts and will become the adult *colliculus seminalis*. In the male, the anterior *Mullerian ducts* degenerate; the posterior fused position remains as a midventral blind pouch, the *prostatic ulricle*. Epithelium from the urethra invades the pouch and replaces the original *Mullerian duct* epithelium.

During development of the testis, external genitalia also develop. A *urinary bladder* and *urogenital sinus* develop from the allantoic duct. The urogenital sinus and rectum open into a cloaca; the former two structures become separated on a frontal plane by a urorectal fold. A midventral genital tubercle develops cranial to the cloaca. Its posterior portion develops a urogenital groove which is bounded laterally by *urogenital folds* and dorsally by a urogenital membrane. This membrane ruptures and provides an external opening for the urogenital sinus. The genital tubercle elongates into a phallus, the tip of which is a precursor of the *glans penis*. The urogenital folds fuse ventrally toward the glans the resulting lumen forms the quali *part* of the *urethra* which opens at the glans. The golds themselves develop into erectile tissue; they form the *corpus spongiosum* and glans distally, and expand to form the *bulbus penis* at the junction with the urethra. A fold of skin develops at the base of the glans, grows distally, and surrounds the glans. This is the prepuce.

Connective tissue in the dorsal side of the penis forms bilateral rods of erectile tissue, the *corpora cavernosa*. The corpora cavernosa fuse distally to varying degrees; proximally they diverge and form the *crura* which unite with the ischium on the respective sides. The crura thus unite the penis with the skeletal system. In some species an ectopic bone, the *baculum* or *os penis*, develops between the two separate corpora or within the single corpus cavernosum.

Genital swellings develop lateral to the urogenital folds. In the male, these are destined to become the *scrotal swellings* from which will develop two *scrolal pouches*. The pouches are separated internally by a partition, the septum scorti, its external position is marked by the *scrotal raphe*.

In many species the testes descend into the scrotum regression of the mesonephros. The mesonephros joins with the peritoneum of the scrotal swelling and mesenchymal cells to form a connective tissue cord, the *gubernaculum testis,* between the testis and the serotal swelling. As soon as the gubernaculum forms, the peritoneum surrounding it evaginates into the body wall and thereby forms a lining for the scrotal pouches, the *scrotal sac*. Since formation of the gubernaculum precedes muscle formation, an *inguinal canal* is formed at the site of the evagination. The body lengthens, but the gubernaculum does not. As a result, the testis, with its associated tubules, blood vessels, and nerves, is drawn lateral to the ureter and into the scrotal sacs. The gubernaculum shortens, but this apparently is an effect, not a cause of testis descent.

As the testis descends into the scrotum, it becomes partly covered by a reflected fold of peritoneum, the *tunica vaginalis*, in species in which the testes are seasonally withdrawn into the body, the inguinal canal remains open. If the testes remain permanentely in the scrotum, the inguinal canal closes and the tunica vaginalis forms a closed sac with a *parietal layer* lining the scrotum and a visceral layer adjacent to the tunica albuginea of the testis. The two layers join at the peritoneal fold, and it is here that blood vessels and nerves enter the testis.

The Testes

The primary sex organs, the *testes,* consist principally of an oval or pear-shaped mass of highly coiled tubules. They enlarge at sexual maturity, and in some species, maintain a practically constant size thereafter; in such species the male is capable of breeding at any time. In other species, the testes undergo alternate regression and enlargement and the male is capable of breeding only at certain periods.

Position of the testes is highly variable. In omnivores such as the opossum, carnivores and primates, they descend into the scrotum and remain there permanently. In bats, rats, and rabbits, the testes are carried within the abdominal cavity part of the year. During the breeding season the scrotum of bats and squirrels enlarges and the testes descend through

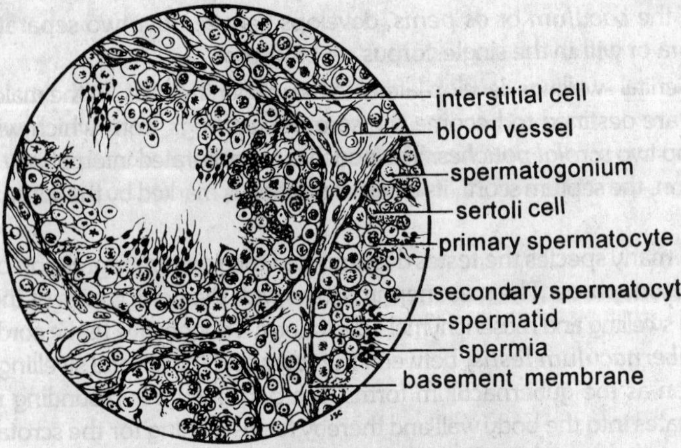

Fig. 11.2. Section of a human testis showing parts of several seminiferous tubules containing cells in various stages of spermatogenesis. Between the seminiferous tubules are evident the interstitial cells that synthesize and secrete testosterone.

the inguinal canal and enlarge. In other rodents and in rabbits, no true scrotum is formed and the testes lie in an inguinal pouch from which they may be withdrawn into the abdominal cavity even when they are enlarged.

A. Histology

Each testis is surrounded by a capsule of collagenous fibres, the *tunica albuginea,* from which there is an extension, the *mediastinum testis,* into the central portion of the testis. Membranes of more or less continuous connective tissue, the *septula, extend* from the tunica to the mediastinum and divide the testis into triangular lobules.

Within each lobule are several *seminiferous tubules* which are formed from branching of the sex cords. The highly coiled portion of the tubule near the tunica is the *tubuli contorli,* and the relatively straight portion near the mediastinum is the *tubuli recti.*

Each seminiferous tubule is surrounded by a basement membrane of collagenous fibers and lined with stratified epithelium. The epithelium consists of *spermatogenic cells* and *Sertoli's* or *sustentacular cells.* There are five morphologically distinguishable types of *spermatogenic cells* and *Sertoli's* or *sustentacular cells.* There are five morphologically distinguishable types of spermatogenic cells : spermatogonia, primary spermatocytes secondary, spermatocytes, spermatids and spermatozoa. They represent successive stages in the differentiation of mature male sex cells, and occur, in the order named, from the basement membrane, to which the spermatogonia are attached, to the lumen of the tubule (Plate 1). Sertoli's cells are attached to the basement membrane in spaces between spermatogonia, and therefore, are of irregular shape. They provide mechanical support and serve in the nutrition of spermatogenic cells.

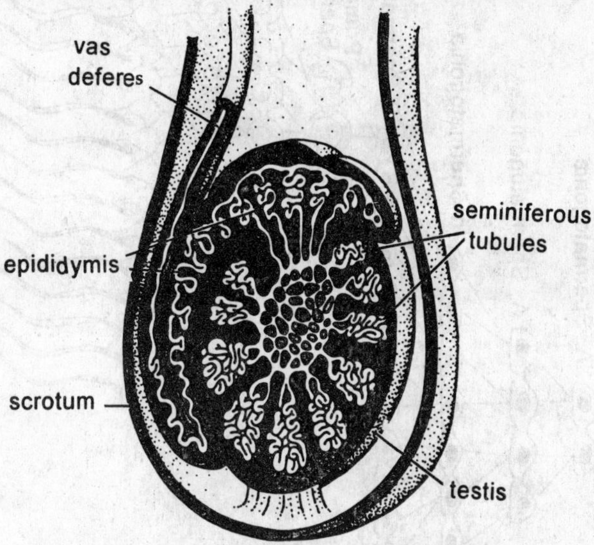

Fig. 11.3. Internal stucture of testis.

Tubuli recti open into the rete testis. Two types of rate testes have been recognized, based upon their position within the testis: (1) a superficial type, present in man, rat, and mouse, and (2) an axial type found in the rabbit, guinea pig, cat, dog, and cattle.

In the rat, the rete testis may be divided into three parts: (1) the intertesticular, the largest part, is a single flattened tubule, with flattened pillars near the midline, which lies on the seminiferous tubules near the cranial pole of the testis; (2) the intratunical rete, a transitional part of the intratesticular, which consists of a series of irregular passage ways running through the tunica albuginea; and (3) the extratesticular rete, an irregular subdivided cavity which is located on the eranial end of the testis, and which is joined by the ductuli efferentes near its most eranial extension. The intratesticular rete originates from the gonadal blastema, the extratesticular rete from the rete blastema which is derived from the mesonephros.

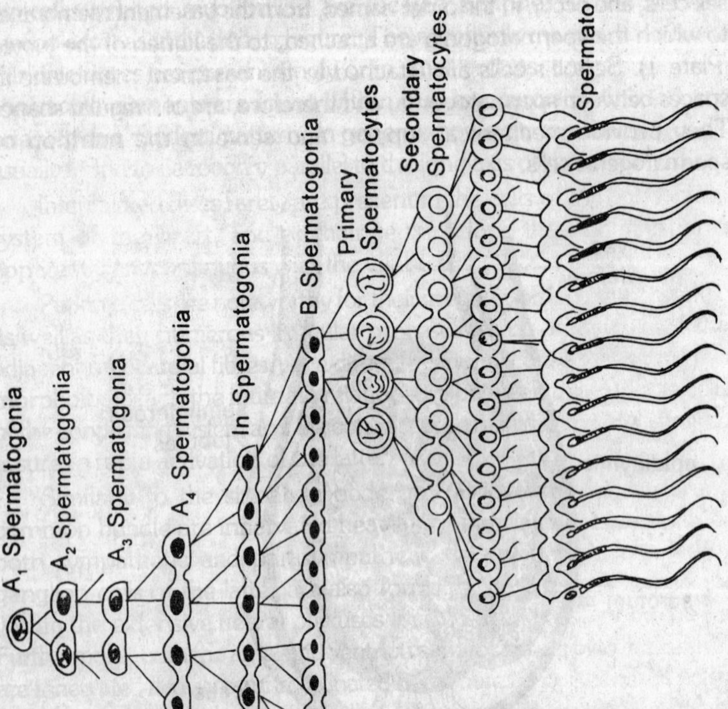

Fig. 11.4. The retention of intercellular bridges in spermatogenesis in several generations of cell division.

As found in the cat, the axial type of rete testis consists of a group of relatively straight tubules surrounded by connective tissue of the mediastinum and located approximately in the long axis of the testis. Tubules pass from the body of the testis at its proximal end and join ductuli efferentes which, in convoluted form, constitute part of the caput epididymis.

Rete testis tubules are lined with simple epithelium which rests upon a basement membrane that is adjacent to tissue of the mediastinum testis. Epithelial cells are squamous in the main cavity, but cuboidal or low columnar near the tubuli recti and ductuli efferentes. The luminal surfaces have numerous microvill and a few cells have a single cilium.

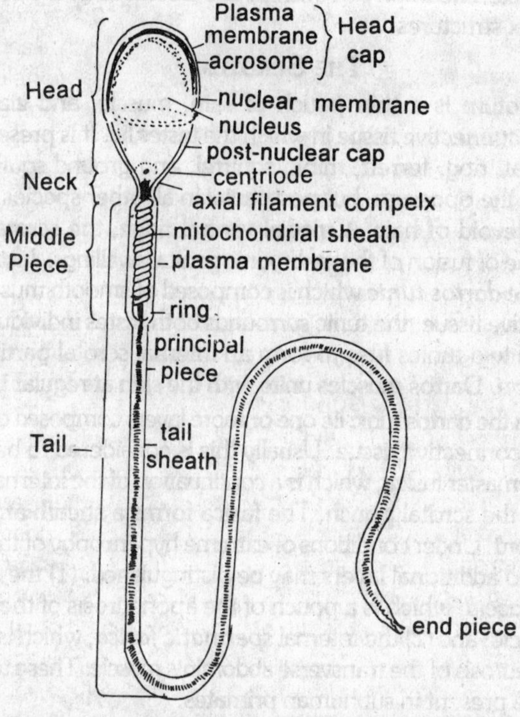

Fig. 11.5. Diagrammatic representation of a mammalian spermatozoa.

Interstitial cells of Leydig occur in clusters among the seminiferous tubules or in rows along blood vessels. They arise from cells which are indistingusihable from fibroblasts. In species with a cyclic reproductive function such as squirrels, the interstitial cells undergo atrophy and new cells differentiate the following breeding season. Pigment cells are present in some species, and, with atrophy of normal interstitial cells during sexual quiescence, become relatively more abundant and impart a dark brown colour to the testis.

B. Function

The primary function of the testes is production of male sex cells. The function of the seminiferous tubules is controlled by intestitial-cell stimulating hormone from the adenohypophysis. The interstitial cells are the primary source of testosterone which, supplemented by testosterone produced by Sertoli's controls the size and secretory activity of accessory sex glands and the development of secondary sexual characteristics. The interrelationships are discussed in the section on accessory sex structures.

The Scrotum

The scrotum is a thin pouch of skin, muscle, and elastic and collagenous connective tissue in which the testes lie. It is present in the opossum, cat, dog, ferrett, mink, squirrel, and ground squirrel. It is prepenial in the opossum, but postpenial in all other species. Usually the skin is devoid of hair. A medial ventral ridge, the scrotal raphe, marks the line of fusion of the embryonic genital swellings. Just beneath the skin is the *dartos tunic* which is composed of smooth muscle fibres and connective tissue; the tunic surrounds both testes individually, and medially the two tunics fuse to form an internal scrotal partition, the *septum scroti.* Dartos muscles unite with the skin at regular intervals.

Beneath the dartos tunic lie one or more layers composed of striated muscle and connective tissue. Usually this is considered to be a single layer, the cremaster fasica, which is a continuation of the internal oblique muscle into the scrotal pouch. The fasica forms a sheath around the spermatic cord. Under conditions of extreme hypertrophy of the human scrotum, two additional layers may be distinguished: (1) the *external spermalic fascia*, which is a pouch of the aponeurosis of the external oblique muscle; and (2) the internal *spermatic fasica*, which is a pouch of the aponeurosis of the transverse abdominis muscle. These two layers probably are present in subhuman primates.

The scrotum is lined with the parietal layer of the tunica vaginalis. This layer may be much attenuated, but is complete. Both the mesothelial cells, which constitute the tunica, and the marcophages lying within the tunica have phagocytic properties. In the opossum, this layer contains heavy deposits of melanin. The serotum of squirrels and ground squirrels also is pigmented.

Rabbits, bats and members of the Cricetidae and Muridae have an *inguinal pouch* rather then a scrotum. The pouch is formed by muscles and fascia of the body wall. In some species most of the pouch is partially devoid of hair; in others, only that part occupied by the epididymis is appreciably devoid of hair. In rats and mice the posterior portion is pigmented to varying degrees.

Function

Maintenance of the testes at the temperature which is optimum for sperm production is the function of the serotum. In most mammals, spermatogenesis takes place at less than body temperature. Contraction and relaxation of the dartos muscles raise and lower the testes with respect to the body, and their temperature is thereby regulated. In addition, the *pampiniform plexus*, which is formed by a network of veins as they leave the posterior border of the testes and pass up the spermatic cord, surrounds the convoluted testicular artery and serves as a thermoregulatory mechanism. Failure of a testis to descend into the scrotum results in a condition known as *cryptorchidism*. A bilaterally cryptorchid animal is sterile.

ACCESSORY TUBULES

A. Epididymis

The epididymis is a heavy cordlike structure, formed from highly coiled ducts, which closely adheres to the surface of the testis. It consists of three parts: (1) the *caput epididymis,* which lies at the end of the testis to which the spermatic cord is attached; (2) the *corpus epididymis* or central body; and (3) the *cauda epididymis,* which lies at the opposite pole of the testis from that occupied by the caput epididymis, and which joins the ductus deferens.

The *caput epididymis* consists in part of small lobes or cones formed from the highly convoluted *ducluti efferentes*. The apex of each cone is directed toward the mediastinum testis where the ductuli join tubules of the rete testis. Ductuli coalesce to form the single *ductus epididymidis,* which is highly convoluted duct that forms both corpus and caput epididymis.

Epithelium of the ductuli efferentes of a rodent, the nutria (*Moyocastor coypus*), has been described as composed of groups of tall cells alternating with short groups all being pseudostratified, ciliated columnar cells. Blebs of secretory material project from the free surface. This is quite similar to the description usually given for ductuli efferentes epithelium in man. In the cat, the epithelium is simple cuboidal with stereocilia on the free surface. Stereocilia are nonmotile, have neither basal bodies nor internal filaments, and are through to be very long microvill. In the opossum, ductuli are lined with simple columnar cells, some of which are ciliated and some nonciliated. These cells rest upon a thick basement membrane which is underlain with variable amounts of smooth muscle and connective tissue. Ductuli of the rat and rabbit are lined with simple columnar cells, at least part of which have stereocilia.

The *ductus epididymidis* maintains a relatively constant diameter as it traverses the caput and corpus epididymis, but its diameter approximately doubles in the caudal region. It lies embedded in collagenous connective tissue throughout its entire length. In sexually active animals the caudal tubules show distinctly through the connective tissue sheath.

Two cells types, holocrine and principal cells, are present in the rat epididymis. Basal apical, and clear cells represent different stages of the holocrine cell cycle as it accumulates secretory products and degenerates. Principal and holocrine cells differ in their histochemical reaction.

Ductus epithelial cells have long stereocilia projecting from the surface (Plates 1 & 2). Cell height and length of stereocilia diminish in the corpus epididymis. The epithelium of the greatly distended cauda epididymis may be simple lower columnar or low cuboidal; the cells may be without stereocilia, but with blebs of secretory material extruding from the free surface. But epithelium, in contact to basal nuclei of other species, is characterized by centrally located nuclei throughout the length of the ductus epididymis.

Function

Ductuli efferentes retain the absorptive function of the mesonephric tubules from which they were derived. A large volume of fluid is transferred across the wall to the lumen of the seminiferous tubules; this fluid serves to move the sperm along the tubule to the ductus deferens. Excess fluid which is not required for propulsion of the sperm in the ductus is resorbed by the ductuli efferentes, some of the cells of which have aggregations of canaliculi in the plasmalemma of the free surface. Movement of sperm is therefore more rapid in the ductuli efferentes than in the ductus epididymidis. The time required for movement of sperm through the latter structure varies from 25 days in the guinea pig to 22 days in the rat under normal conditions, 7 to 10 days in freely mating rats, 10 days in the hamster, and 8 to 10 days in the rabbit.

The ductus epididymidis serves as a site for maturation and storage of spermatozoa as well as a vehicle for their transport. Changes in maturation of sperm as they pass through the ductus epididymidis have been thought to be temporal ones which were regulated by intrinsic factors, but Paufler and Foote demonstrated that extrinsic factors from the ductus epididymidis were required for development of functional spermatozoa. Spermatozoa which were isolated in the caput epididymis underwent morphological changes typical of normal maturation, but they had a low degree of fertility. Such spermatozoa never attained the motility of those from the cauda epididymis, and were completely nonmotile after 4 weeks. Spermatozoa in the cauda epididymis, the main reservoir, remained motile for 8 weeks and were as fertile as ejaculated spermatozoa. Thus, substances from the corpora and cauda epididymis are necessary for production and nourishment of fertile spermatozoa.

The epididymis produces enzymes. Localization of different enzymes in different parts of the epididymis and in certain epithelial cells of a given part, as well as counteraction of substances in the stroma and epithelium of a given part, have been demonstrated by Allen and Slater. PAS positive granules are produced by both principal and holocrine cells in the rat, hamster, and guinea pig. Activity of the epithelial cell is controlled by testicular hormones.

B. Ductus Deferens

The ductus deferens continues the canal of the epididymis as a highly convoluted tube parallel to the corpora epididymis; it gradually straightens as it leave the epididymis, passes through the inguinal canal, loops around the ureter, and finally opens into the urethra at the colliculus seminalis. From the testes to where it loops around the ureter, the ductus deferens, the testicular and epididymal blood and lymphatic vessels, and nerves constitute and spermatic cord. Near its junction with the urethra, *ampullary glands* may form as either tubes or crypts in the wall of the ductus or as distinct lobes. Beyond the ampullary gland an evagination, the *seminal vesicle*, forms in some species. A short, straight section of the ductus just distal to its junction with the seminal vesicle is the *ejeculatory duct.*

Epithelium of the ductus deferens is ciliated or has stereocilia in most species and is highly variable: stratified squamous in the eat; simple squamous near the epididymis to columnar with a layer in basal cells in the area near the urethra in the guineing, high columnar with an incomplete layer of basal cells in the bat and opossum. The cells have centrally located nuclei in the bat, but greatly elongated basal ones in the opossum. Epithelium of the rabbit ductus is simple columnar; it may be folded to form crypts or tubules. The amount and proportion of elastic and collagenous tissue and smooth muscle varias from species to species. Most smooth muscle fibers are circularly arranged.

Accessory Sex Glands

A high degree of development of the accessory sex glands is characteristic of the Mammalia. Some glands are derived from each of the germ layers and, on the basis of their origin, they may be classified as: (1) those derived from the lower ductus deferens and therefore, of mesodermal origin (ampullary glands and vesicular glands); (2) those derived from the endodermal part of the urogenital sinus (prostatic, membranous, and penile urethra); (3) thoe derived from the proximal ends of the Mullerian ducts (prostatic utricle); (4) those derived from the ectoderm ofthe preputial or inguinal region (preputial glands and inguinal glands).

Products of these glands are combined with spermatozoa in the urethra to form the semen. There is much variation among species with respect to presence and degree of development of various glands and the chemical properties of their secretions.

A. Ampullary Glands

Ampullary glands (*ampullae ductus deferentis*) are highly variable structures formed from the ductus deferens near its junction with the urethra. The glands consists of tubules which are enclosed by a capsule of connective tissue and smooth muscle. Extensions of connective tissue from the capsule from septa between tubules. Smooth muscles are present in the anterior septa, but disappear as the ductus approaches the urethra. In shrews and bats, smooth muscle is absent in both the ampullary capsule and the septa. Some tubules may coalesce, and folds of epithelium may project into the lumen of others. In some species there is no distinct wall for the lumen of the ductus deferens within the ampullary gland; rather, it is formed by the beginning of the tubular evagination.

In shrews and bate the lumen of the ductus deferens is distinct and passes centrally through the ampullarium. In members of the families Cricetidae and Muridae, ampullary glands form distinct lobes from each ductus deferens. The lobes consists of ducts which pass through separetely and are therefore embedded in the muscularis of the ductus deferens for a distance. Eventually portions of the duct walls form the lumen of the ductus deferens just anterior to the ejaculatory duct.

Although a few sperm may be found in tubules where they open into the ductus deferens, there is no indication that ampullary glands are a storage depot for sperm. They are true glands.

B. Vesicular Glands

Vesicular glands (*vesiculae seminales*) arise as evaginations of the ductus deferens. Each consists of an elongated simple or branched tubule which is distended with fluid during the breeding season. The tubule may join with the ductus deferens to form the ejaculatory duct, or it may open independently into the urethra. The tubule wall consists of an inner mucous membrane, a medial layer of smooth muscle, and an external collagenous and elastic connective tissue layer.

In sexually potent animals, the tubule has a large central cavity filled with secretion into which folds of epithelium supported by delicate strands of connective tissue project. Union of the folds may produce small lateral erypts or subtubules. Compact glands are surrounded and divided into lobules by areolar connective tissue. Straight and curved glands are surrounded by variable proportions of smooth muscles and collagenous connective tissue.

Vesicular gland secretion varies from a viscous to gelationous, usually clear or slightly yellowish, fluid; free epithelial cells may be present. It serves as a medium for transport of sperm. In combination with anterior prostate or, in some species, bulbo-urethral secretion, it coagulates. In some species the coagulant forms a *copulatory* or *vaginal plug* and prevents loss of the semen. The plug subsequently breaks down and the sperm are released into the vagina in a steady stream. The copulatory plug is necessary for fertilization in some species of rodents.

C. Prostate Glands

Prostate glands (*glandulae prostaticae*) are present in all members of Eutheria. They occur either as a mass of tissue, which may be lobulated to varying degrees, located at the ductus deferens-urethral junction; or they may be of the disseminate type, which extend along the entire prostatic and most of the membranous urethra. Glands of the former type are encapsulated; they penetrate the urethral muscularis and open into the urethra by means of one pair to many ducts. They may form a single mass which is subdivided by septa or they may form two or three pairs of distinct lobes. Designation of the lobes in quardrupeds has followed two systems, of nomenclature. The first, modified from the used for man, designated anterior (ventral), middle, and posterior (dorsal). The second, designated anterior (cranial) middle, and posterior (caudal). A combination of the two systems has been used.

In some species, notably members of the family Cricetidae and Muridae, the anterior lobe of the prostate is elongated and closely appressed to the inner curvature of the vesicular glands; it is known s the coagulating gland. The "vasicular gland" of the European rabbit, which develops from a pair of dorsal evaginations of the urethra, is an homologous structure.

The prostate is a tubulo-alveolar or saccular gland. Saccular prostates have epithelial folds extending far into the lumen. Spherical globules of secretory products are released from the free surface and, during periods of high activity, the free surface is indistinguishable because of the presence of secretory globules. Three types of cells are present in the

Table 11.1. Comparative Anatomy of the Male Reproductive Structures

Species	Testis	Ampullary Glands	Seminal Vesicles	Prostate Gland	Paraprostate Glands	Bulbo-urethral Glands	Baculum	Urethrat Glands	Bulbar Gland	Preputial Glands	Inguinal Glands
Bat	3 x 2	4 x 15	A	5 x 3	A	1 x 1	F	F	A	A	A
Cat	14 x 8	A	A	5 x 2	A	4 x 3	F	A	A	A	A
Dog	40 x 30	F	A	25 x 16	A	A	F	A	A	A	A
Gerbil	14 x 9	6 x 5	20 x 12	9 x 7	A	6 x 4	F	F	A	A	A
Guinea pig	25 x 15	A	115 x 7	15 x 8	A	8 x 5	F	A	A	PD	A
Hamster	14 x 11	3 x 3	11 x 6	8 x 6	A	4 x 3	F	F	A	3 x 1	A
Mink	11 x 8	F	A	10 x 5	A	A	F	A	A	A	A
Mouse	6 x 4	2 x 1	13 x 4	4 x 4	A	3 x 2	F	F	A	6 x 5	A
Opossum	15 x 11	A	A	40 x 9	A	19 x 15	A	A	A	A	A
Rabbit-European	35 x 15	15 x 4	19 x 7	19 x 6	6 x 2	6 x 3	A	A	A	A	15 x 7
Rabbit-cottontail	35 x 17	F	A	25 x 6	A	6 x 2	A	A	A	A	10 x 4
Rat	20 x 12	4 x 4	20 x 10	13 x 10	A	5 x 3	F	F	A	16 x 4	A
Shrew	3 x 2	3 x 1	A	3 x 1.5	A	1 x 1	A	A	A	Diffuse	A
Squirrel-tree	30 x 12	F	7 x 4	28 x 8	A	13 x 13	F	A	10 x 7	A	A
Squirrel-ground	14 x 8	F	8 x 7	18 x 18	A	8 x 6	F	A	10 x 8	F	A
Subhuman-primate	10 x 7	A	14 x 7	6 x 5	A	A	F	A	A	Diffuse	A
Ferret	F	F	A	F	A	A	F	A	A	A	A

F-indicates that a functional gland is present, but was not measured; A-indicates that the structure does not occur; PD indicates that the gland is poorly developed.

anterior prostate or the coagulating gland of the rabbit: (a) light cells, (b) dark cells, and (c) true basal cells. Only one cell type is present in the posterior lobe. Differences in the staining reactions of the epithelium of the two lobes reflect differences in their secretions. In standard histological sections the lining of the anterior prostate of the rabbit and rat consists of cuboidal cells with oval hyperbasal nuclei, while the posterior prostate epithelium consists of columnar or high columnar cells with a round basal nucleus.

Table 11.2. Histology and Secretory Activity of Male Accessory Sex Glands and Urethra of Laboratory Animals

Organs	*Epithelium*	*Properties of Secretion*
Ampullary glands	Simple, cuboidal, or columnar	Yellowish liquid containing erogothianeine, fructose, sialic acid, and phosphorus
Vesicular glands	Simple or pseudo-stratified columnar	White, yellowish or bluish fluid or gelatinous substance containing fructose, protein, citric acid, and ascorbic acid
Prostate gland	Simple or pseudo-stratified columnar	Colourless liquid containing fibrinolysin, fibrinogenase, diastase, carbonic anhydrase, amylase, vesiculase, fructose, citric acid, free amino acids, and zinc
Paraprostate glands	Pseudostratified cuboidal or low columnar	No data
Bulbo-urethral glands	Simple columnar	Viscous fluid containing sialoprotein
Bulbar gland	Simple high columnar	Liquid which facilitates passage of bulbourethral secretion.
Urethral glands	Simple cuboidal	Clear, watery fluid containing mucoprotein
Preputial glands	Stratified squamous	Contain 7-dehydro-cholesterol
Prostatic urethra	Transitional	None
Membranous and proximal spongy urethra	Stratified or pseudo-stratified columnar	None
Distal spongy urethra	Stratified squamous	None

Table 11.3. Anatomical Characteristics of Male Accessory Sex Glands of Laboratory Animals

Organs 1	*Anatomical characteristics* 2	*Species* 3
Ampullary glands	Spindle-shaped enlargement composed of branchial tubular which open by ducts on entire circumference of the ductus deferens	Dog Mink Rabbit
	A distinct, abrupt enlargement composed of branched, tubular diverticulae, the proximal ends of which form the lumen of the ductus deferens	Bats Shrews
	Distinct lobes projecting from the ductus deferens	Mice Rat
	Shallow diverticulate in wall of ductus deferens	Guinea pig
Vesicular glands	Simple or branchial tubule which coils to form compact glands	Squirrels
	Simple, straight or curved tubules	Guinea pig
	Straight or curved saccules single and semibilobed bilateral structures	European rabbit Mice Rat
Prostate gland	Disseminate	Opossum
	semi-disseminate bilobed	Cat
	non-disseminate, bilobed	Bat
	mas subdivided by septa	Dog Mink Some primates
	Two or three pairs of distinct lobes	Mice Rats
Bulbo-urethral glands	Three lobed	Opossum
	Single lobed	
	ducts opening into distal spongy urethra via the bulbar gland and penile duct	Squirrels
	ducts opening into the proximal spongy urethra.	All others

Nondisseminate prostates are surrounded by a capsule of collagenous and elastic fibers and smooth muscle. Tissue identical to the capsule forms the septa which subdivide the lobes into lobules. Trabeculae of connective tissue partially subdivide the lobules.

Glands of the disseminate type of prostate do not penetrate the urethral muscularis and the collecting ducts common to several glands radiate from the entire urethral circumference. Three regions are distinguishable in the disseminate prostate of the opossum, each with a characteristic epithelial lining the glands and a characteristic amount of connective tissue separating the glands.

Function

Prostate secretion as a medium for transport for sperm to the female. It stimulates motility in the sperm and also provides nourishment. In the rabbit the anterior lobe secretion is high in fructose and citric acid, while the posterior lobe secretion is mainly a protein with a large amount of fructose and some citric acid. Production of prostatic secretion is primarily controlled by the testosterone level, and secondarily, by nervous stimulation. Production may be relatively constant, or three may be periodic high peaks of production.

The coagulating fraction of anterior prostate secretion is known as *vesiculase*. Coagulation of vesicular gland fluid is a double reaction in which procoagulase from the vesicular gland combines with vesiculase to form the enzyme coagulase. Coagulase then combines with a protein fraction of vesicular gland secretion, coagulinogen, to form coagulated protein.

D. Bulbo-urethral Glands

Bulbo-urethral or Cowper's glands (*glandula bulbo-urethrales*) are a pair lying dorsal to the bulb to the penis and are partially or completely embedded in the bulbocavernosus muscle. Ducts may be short, as in the rabbit and cat, and the glands may appear as a projection on the dorsal side other urethra; or they may be long, as in the gerbil, squirrel, and guinea pig, with the glands located near the base of the tail.

The compound tubulo-alveolar glands resemble mucous glands and are surrounded by a capsule of connective tissue and straited muscle; the connective tissue forms septa and trabeculae which subdivide the glands. In squirrels (*Sciurus*) neither intrinsic nor extrinsic muscle completely surround the gland; intrinsic muscles occur as short interwoven bundles on the portion nearest the duct. Ducts may be lined with pseudostratified or stratified cuboidal epithelium.

The bulbo-urethral secretion is a milky white, mucus-like material; it is not a true mucus since it does not precipitate when treated with acetic acid. Long through to be simply a lubricant, the secretion recently has been shown to contain a glucoprotein which undergoes a nonenzymatic, ionic reaction with a protein fraction of the vesicular glands to produce a coagulant in the rat, mouse, and hamster, but not in the guinea pig. The vesicular substrate upon which the bulbo-urethral secretion acts is different from that acted upon by the enzyme visiculase of the coagulating glands. Furthermore, the bulbo-urethral secretion reacts at an optimum at pH 5.9 in contrast to pH 7.4 for the coagulating gland enzyme. The foregoing reaction is probably responsible for formation of the vaginal plug in members of the genera *Sciurus and Citellus.*

E. The Bulbar Gland

The bulbar gland (*glanduli bulbi*) is confined to the Sciuridae and, in a modified form, in the Aplondontidae. The following description is based principally upon the foregoing publication. The gland is formed by fusion of the ducts of the bulbo-urethral glands; these ducts form the lateral lobes of the bulbar gland. Ducts of the median lobe and the lateral lobes coaleasee to form the penile duct which opens into the urethra at the ventral flexure of the penis. The glandular character of the bulbar epithelium gradually diminishes along the penile duct as it approaches its junction with the urethra.

The bulbar gland is located within the bulb of the corpus spongiosum. The gland is immediately surrounded by fibrous crectile tissue which constitutes part of the wall of the bulbo-urethral ducts and separates the two lateral lobes from the media lobe. The gland consists of complex folds. Secretion of the gland is thought to be liquid and to facilitate passage of the bulbo-urethral secretion through the penile duct.

F. Paraprostate Glands

Paraprostate, glands (*glandulac urehtrales paraprostalicus*) develop as evaginations of the urethra of the European rabbit. There is dication of them in the North American cottontail. These glands have been described in detail by Berii and Krichesky (1943). The following description is modified from their work.

The glands lie lateral to the urethra and ventral to the ampullae. In contrast to urethral glands, they lie outside the mucularis of the urethra. There are 3 to 8 glands on each side and each gland open lateral to the colliculus seminalis by an ejaculatory duct. The glands are tubulo-alveolar and resemble bulbo-urethral glands, but they lack striated muscle and have a greater amount of smooth muscle than the latter. Some paraprostates are indistinguishable from prostate glands.

G. The Prostatic Utricle

The prostatic utricle (*utriculus prostalicus*) is a small pouch commonly regarded as a persisting vaginal primordium which represents the fused posterior ends of the Mullerian ducts. It lies on the dorsal wall of prostatic urethra and may be imbedded in the prostate. There is a single middorsal opening on the colliculus seminalis. The pouch is lined with epithelium in a manner similar to that the prostate.

H. Urethral Glands

These (*glandulae urethrals*) are numerous small glands which lie within an areolar connective tissue layer between the epithelium and the muscularis of the prostatic and membranous urethra of certain bats. In *Corynorhinus,* they are simple tubulo-alveolar glands which open separately into the urethra; they are identical in staining reaction to bulbo-alveolar glands which open separately into the urethra; they are identical in staining reaction to bulbo-urethral glands.

In the hamster, compound tubulo-alveolar glands occur along the membranous and proximal penile urethra, where they partially penetrate the muscularis. Their staining reaction is much more basophilic than that of the bulbo-urethral glands.

I. Preputial Glands

Preputial or Tyson's glands (*glandulac preputiales*) are small, branched, and tubulo-alveolar. In rodents a pair of these glands lies on either side of the distal part of the penis and open into the preputial cavity near its orifice. In primates the glands are more diffuse and open at the base of the glans. The glands arise from evagination of preputial epithelium.

In the rat the glands are composed of two types of cells: (a) the squamous cells, which form a single layered wall and which later become stratified to form duct epithelium, and (b) the acinar cells, which accumulate fatty substance within their cytoplasm, fuse together, and disintegrate. In addition to the fatty substance, protein granules are also produced and released into the excretory duct when the cells degenerate. The holocrine secretion is called *smegma.*

Acinar cells do not regenerate and the glandular parenchyma gradually decreases with age. Connective tissue replaces the glandular parenchyma, so the gland does not decrease in size.

J. Inguinal Glands

These (*glandulae inguinales*) are present in rabbits as a pair of small compound tubulo-alveolar glands which lie between the skin and abdominal wall on either side of the penis. Each gland opens by a single duct into skin folds near the end of the prepuce. They do not contribute to the seminal fluid, but are sebaceous, odor-producing glands which are associated with sexual attraction. Socially dominant European rabbits have larger and more active glands than subordinate individuals. Castration during early life inhibits growth of the glands.

The Urethra

The common duct of the excretory and reproductive system, the urethra, serves as a vehicle for transport of spermatozoa and associated fluids to the end of the penis. It consists of three parts: the *pars prostatica, pars membranacea* and *pars spongion* or *pars cavernosa*. The portion from the bladder to the junction of the ductus deferens is formed from the primitive urethra which connected the embryonic bladder and the urogenital sinus. This portion is homologous to the entire female urethra.

The ductus deferens enters the urethra at a swelling on its dorsal wall, the *colliculus seminalis*. A single middorsal pouch, the *utriculus prostaticus,* may open on the colliculus. In some species, ducts of both the prostate and seminal vesicles also open on the colliculus.

That portion of the urethra from the colliculus seminalis to the base of the penis represents the persistent urogenital sinus, it is homologous to the female vestibule. The primitive urethra plus the most anterior part of the urogenital sinus constitute the adult membranous or muscular urethra.

The most distal part, the spongy or penile urethra, lies with the penis and is derived from the groove between the urogenital folds; it has no homologue in the female. Its opening at the end of the penis is the meatus or *external urethral orifice*. The bulbo-urethral glands and the urethral glands; the terminal part of the ejaculatory ducts of these glands has an epithelium similar to that portion of the urethra into which they open.

The Penis and Accessory Structures

A. The Penis

All mammals have a single intromittent or copulatory organs, the penis, for transfer of semen from male to female. It is located anterior to the scrotum in all Eutheria, but posterior to the scrotum in the opossum (Marsupialia). The penis consists principally of three cavernous

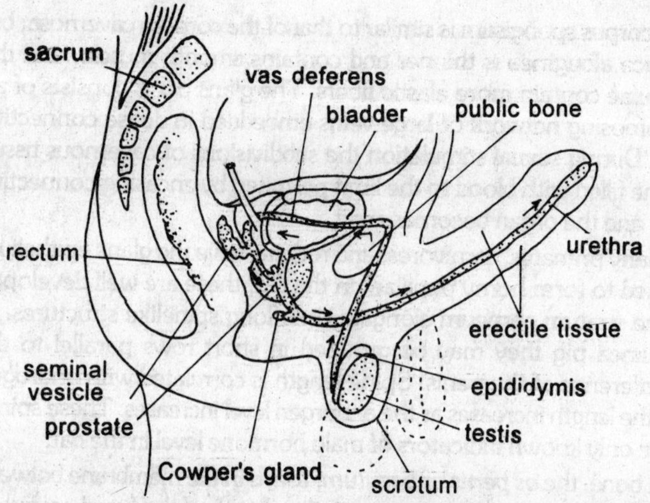

Fig. 11.6. The male reproductive system showing the location of the penis and scrotum (dashed lines) and during erection. Arrow indicate the pathway of sperm transport.

(erectile) tissue bodies; they are as follows: (a) the right and left *corpora cavernosa penis,* which are located on the ventral side and which undergo varying degrees of fusion, and (b) the *corpus spongiosum* penis, which lies dorsal to, and in a groove of the corpora cavernosa.

Proximally the corpora cavernosa diverge as the *crura* and each crus becomes attached to the posterior ischium on its own side. Each corpus cavernosum is surrounded by a thick membrane, the *tunica albuginea*, which consists principally of collagenous bundles with an elastic fiber network. The cavernous spaces are subdivided by trabeculae which are formed from collagenous bundles, elastic fibers, and smooth muscles. The subdivisions are lined with endothelium which is continuous with the lining of associated blood vessels.

Before birth and during the first week of postnatal life of the rat, tissue of the corpora cavernosa is vascular, undifferentiated mesenchyme. Complete differentiation does not occur until the animal is sexually mature at about 60 days of age. Previous researchers had implied that differentiation took place early in embryonic development.

The corpus spongiosum is traversed by the urethra and, in most species, it is expanded at its distal end into a caplike structure, the *glans penis.* Its proximal end is enlarged into the *bulbus penis* which lies between the crura; it is here that the urethra enters the penis. Framework of the corpus spongisum is similar to that of the corpora cavernosa; but its tunica albuginea is thinner and contains smooth muscle, and the trabeculae contain more elastic fibers. The glans penis consists of an anastomosing network of large veins embedded in dense connective tissue. During sexual stimulation the subdivisions of cavernous tissue become filled with blood to the limit permited by encasing connective tissue, and the organ becomes erect.

Many primates, carnivores, and rodents have the glans epithelium modified to form horny papillae. In the cat, these are well developed and the stratum corneum elongates into long spinelike structures. In the guinea pig they may be grouped in short rows parallel to the circumference of the penis. Spine length is correlated with androgen level; the length increases as the andorgen level increases. These spines are the only known indicators of male hormone level in the cat.

A bone, the *os penis* or *baculum,* forms in the membrane between the unfused corpora cavernosa or in the sheath of the fused corpus of bats, dogs, mustelids, squirrels, members of the Cricetidae and Muridae and many subhuman primates. This bone adds rigidity to the penis. It enlarges and may change shape as the animal increases in age, and has been used as an indicator of age in studies of some wild species. The baculum of each species has a characteristic shape, and it and the glans penis both serve as diagnostic taxonomic structures in certain groups.

B. The Prepuce

In most species the penis normally is enclosed in a fold of skin, the prepuce. In some primates the prepuce covers only the distal glans; in bats, the prepuce is free of the body wall; in the dog it lies in a fold of skin below the body wall; and, in other species, it lies immediately ventral to the body wall. The preputial cavity is lined with stratified squamous epithelium.

Female Reproductive Organs

The female reproductive organs are the ovaries, oviducts, uterus, vagina, and external genitalia. The internal organs are supported by the broad ligament, including a mesovarium for the ovary, a mesosalpinx for the oviduct, and a mesometrium for the uterus. The development of the reproductive organs is greatly enhanced by estrogens. Ovariectomy, for example retards their normal growth, whereas estrogens induce rapid growth of the oviduct, uterus, and vagina in immature or ovariectomized animals. Such growth involves increases in cell division and protein synthesis

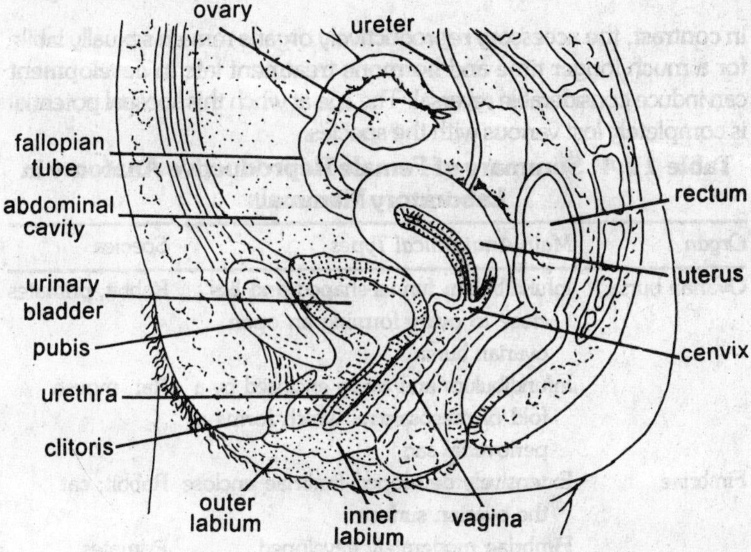

Fig. 11.7. Female reproductive organs.

The following discussion deals with the embryology, comparative anatomy, histology, physiology, and biochemistry of the female reproductive tract in laboratory mammals. The comparative anatomy is summarized in Table 11.4.

EMBRYOLOGY

The rudimentary reproductive system of the mammal consists of 2 sexually undifferentiated gonads, 2 pairs of ducts, a urogenital sinus, a genital tuberele, and vestibular folds. It arises primarily form 2 germinal ridges on the dorsal side of the abdominal cavity and can potentially differentiate into a male or female system (this is referred to as *embryonic bisexuality*). The developmental fate of the sexual rudiments in the male and female fetus is shown in table 11.5.

The sex of an animal depends on 3 distinct but related things: (a) inherited genes, (b) gonadogenesis, and (c) the formation and maturation of accessory reproductive organs. The sex of an embryo is basically determined by its genes, but the expression of the genetic sex is a developmental process depending on the function of the fetal gonads, and occasionally, the adrenal cortex.

Estrogen and androgen cause sex reversal in male and female embryos, respectively, during a brief period early in sexual differentiation.

In contrast, the accessory reproductively organs remain sexually labile for a much longer time and hormone treatment late in development can induce considerable reversal. The age at whch this libosual potential is completely lost various with the species.

Table 11.4. Summary of Female Reproductive Anatomy in Laboratory Mammals

Organ	*Main Anatomical Types*	*Species*
Ovarian bursa	Infundibulum funnel-shaped and lies close to ovary forming an open ovarian bursa	Rabbit, primates
	Infundibulum and ovary enclosed by a fold of mesosalpinx which forms a periovarial sac	Rat, mouse
Fimbriae	Extensively developed fimbriae enclose the ovarian surface	Rabbit, cat
	Fimbriae moderately developed	Primates
	Poorly developed fimbriae make very limited contact with ovarian surface	Rat, mouse
Uterotubal junction	Very complex; glandular mucosal lip; projecting papilla	Guinea pig
	Rosetter type projections; no intramural portion	Rabbit
	Single papilla; short intramural portion of oviduct	Rat
	Single mound; short intramural portion of oviduct	Dog
	Simple fold which projects into a pocket in uterine wall; no mound	Mink
	No projections; extended intramural portion	Primates
Uterus, cervix and vagina	Two uteri; two cervices; two vaginas	Opossum, kangaroo
	Two uteri; two cervices; one vagina	Rabbit, rat (Norwegian), chinchilla
	Two well developed horns, small uterine body; one cervix; one vagina	Cat, dog
	One uterine body without compartments; one cervix; one vagina	Primates

Gonads

The gonads form from a group of large granulated yolk sac cells which invade the germinal ridges. Two invasions occur in the female. The initial one is abortive, but the second results in the formation of sex cords which later break up into primordial germ cells (oogonia). The sex cords of the female are called *medullary cords;* those of the male are the *seminiferous tubules.*

The testis develops predominantly from the medulla of the sexually undifferentiated gonad, whereas the ovary arises primarily from its cortex. The primordial germ cells congregate in the developing gonad and proliferate the oogonia of the female. They and the secondary sex cords which are forming concurrently move into the cortex while the primary sex cords and the medulla regress in size.

Table 11.5. Some Anatomical Characteristics of the Female Reproductive Organs in Laboratory Mammals

Species	*Some Analomical Characteristics*	*Number of Mammary Glands*
1	2	3
Cat	Ovary is ovoid and lies with a peritoneal fold	8
	Cervix is remarkably short; lacks true internal os, and is directly continuous with dorsal wall of vagina; external os is V-shaped	(all or some may function at one time)
Chinchilla	Mesosalpinx tends to enclose ovary	6
	Accessory corpora lutea during pregnancy	(2 inguinal and 4 lateral thoracic)
	Vaginal closure membrane	
Dog	Ovary is flattened and completely enclosed in a roomy peritoneal pouch	10 (arranged in two ventrolateral series)
	Slender uterine horns are long and straight	
	Cervix is a short thick-walled segment	
	Vagina is wider above (cranially), that below	
Guinea pig	Two internal cervical openings, but only one common external os	2 (inguinal)

	Intestinal and urinary tracts open into a groove (the "*fossa anovaginoure-thralis*") Lower end of the vagina is closed by an epithelial membrane, but opens periodically at estrus and during parturition.	
Hamster	Ovary is compact and encapsulated; oviducts and uterus similar to those of the mouse Two cervical canals remain separate for about two-thirds of the length of the cervix, but then fuse Vagina has a mucified type of epithelium; its wall contains urethral glands similar to those in the female prostate	12 or 14 (thoracic and abdominal)
Mink	Ovary has abundant interstitial tissue; slightly developed Uterine glands are sparse External os of the cervix is a transverse uterine slit Vagina is long and has a transverse fold across its dorsal wall	6 or 8 (30 percent (nonfunctional)
Monkey	Uterus is divided into an upper segment, consisting of the body and fundus, and a lower segment formed by the thick-walled cervix; the fundus is irregularly pear-shaped and has a shallow depression Sexual skin does not usually swell, but undergoes variations in colour which are not always clearly related to the menstural cycle	2 (poorly developed) unless pregnant; supernumerary nipples sometimes present below normal nipples)
Mouse	Ovaries lie just below (ventral) the kidneys within transparent ovarian capsules A narrow, tunnel-like passage connects the periovarial space with the peritoneal cavity	10 (6 thoracic and 4 abdominoingenial)

Rabbit	Complete duplication of the uterine segment; two long uterine horns and two entirely separate cervical canals, each of which has an internal and external os; endometrium arranged in numerous transverse and longitudinal folds which are particularly prominent along the mesometrial borders; cervical canals have a narrower lumen and a more extensively folded mucous membrane than the uterine horns Vaginal portions of the cervical segments are surrounded by a complete ring of fornices	8 (arranged in ventrolateral series)
Rat	Ovary lies within ovarian bursa Periovarial space opens into the peritoneal cavity through a slit on the antimesometrial side of the bursa at the tip of each uterine horn	12 (two ventrolateral) series along thoracic and inguinal regions)

Table 11.6. Developmental Fate of the Sexual Rudiments in the Male and Female Mammalian Fetus

Sexual Rudiment	*Male*	*Female*
Gonad		
Cortex	Regresses	Ovary
Medulla	Testis	Regresses
Mullerian ducts	Vestiges	Uterus, oviducts parts of vagina
Wolffian duct	Epididymis, vas deferens	Vestiges
Urogenital sinus	Urethra, prostate Bulbourethral glands	Part of vagina, urethra
Genital tubercle (phallus)	Penis	Clitoris
Vestibular folds	Scrotum	Labia

Reproductive Ducts

Wolffian and Mullerian ducts are both present in the sexually undifferentiated embryo. In the female, the Mullerian ducts develop into a gonaduct system and the Wolffian ducts atrophy. The opposite is true for the male. The female Mullerian ducts fuse caudally to form a uterus a cervix and the anterior part of a vagina. The oviduct becomes coiled and acquires differentiated epithelia and fimbriae just before birth. Signs of epithelial differentiation include increased mucosal height, pseudostratification, the presence of peg cens, and evidence of secretory activity. The onset of these development events varies with the species.

In the male fetus, testicular androgen plays a role in the persistence and development of the Wolffian ducts and the atrophy of the Mullerian ducts. However, the growth of the female Mullerian ducts beyond the ambisexual stage is apparently hormonally independent and the duct is capable of considerable autonomous growth, coiling, and epithelial differentiation. Adrenal hormones are thought to stimulate oviductal growth in the guinea pig.

Urogenital Sinus

The urogenital sinus gives rise to the vestibule. The folds of skin which border the sinus form the lips of the vulva. The female phallus or clitoris, homologous to the male penis, grows little in size.

OVARY

The ovaries are paired organs suspended from the body wall by a *mesovarium*. Unlike the testes, they remain in the abdominal cavity. They are both endocrine (estrogenprogesterone, and androgen-secreting) and exocrine (egg-producing) glands.

The rabbit ovary is located in an open ovarian bursa, whereas that of the rat, mouse, and gerbil lies in a closed capsule called the *periovarial sac*. If the latter ruptures, at least in the rat, the reproductive capability is reduced because the eggs can then escape into the body cavity at ovulation.

The ovary is enveloped in a *germinal epithelium*, usually a single layer of cuboidal or low columnar cells. It contains a *medulla* and *cortex*. The medulla consists of fibroelastic connective tissue and an extensive network of nerves and spiral blood vessels which enter the ovary at the *hilus*. It also houses large polyhedral intestitial cells which are more prominent in the ovaries of rodents and carnivores than in those of primates and ungulates. The cortex contains immature, developing, and graffian follicles, atretic follicles, and corpora lutea. It is the site of both egg and hormone production. It also contains fibroblasts, collagen,

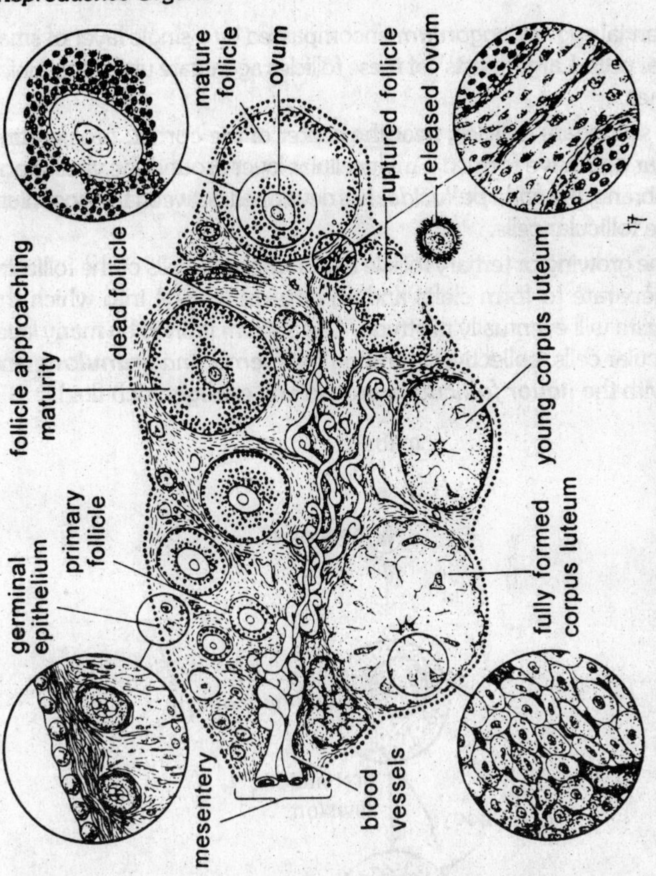

Fig. 11.8. Diagram of the stages in the development of an egg. follicle and corpus luteum in a mammalian ovary. Successive stages are depicted clockwise, beginning at the mesentery. The insets show the cellular structure of the successive stages.

and reticular fibers, blood vessels, lymphatics, nerves, and smooth muscle cells. The connective tissue cells and fibers which form the tunica *albuginea* near the surface of the ovary (just below the germinal epithelium) are arranged somewhat more compactly and regularly than those in the interior.

A. Development of Ovarian Follicles

The germinal epithelium of the female lies on the surface of the ovary. Its stage of development at birth is still a matter of controversy: the consensus is that all oogonia become primary follicles during fetal early neonatal life.

The reach maturity, these follicles than pass through successive developmental stages: from primary to secondary, to growing (tertiary), and finally to mature (graafian) follicles. The primary follicle consists of a "potential egg" or *oogonium* encompassed by a single layer of small follicular cells. Large "nests" of these follicles aggregate under the tunica albuginea.

In secondary follicles, near the center of the cortex, proliferating follicular cells have formed a multicellular layer around the vitellus and a membrane, the *zona pellucida*, has developed between the oogonium and the follicular cells.

The growing or tertiary follicle arises when the cells of the follicular layer separate to form clefts and an *anturm* (cavity) into which the oogonium will eventually protrude. The antrum is lined by many layer of follicular cells, collectively forming the *membrana grunulosa*, and filled with the *liquor folliculi*, a protein and estrogen rich fluid.

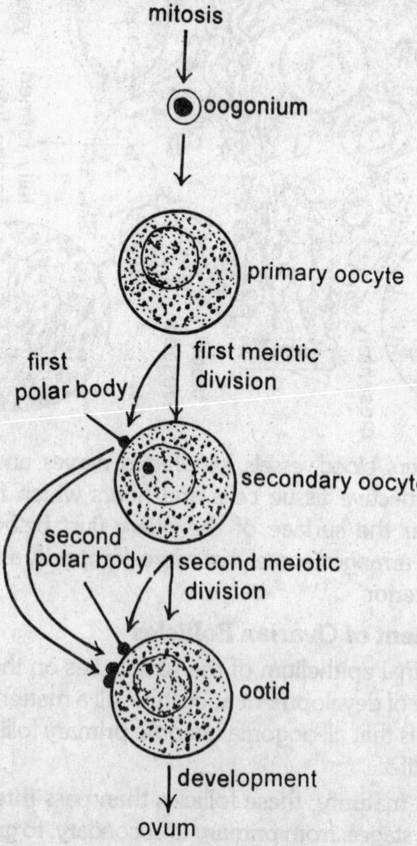

Fig. 11.9. Figure showing oogenesis.

Because of its size, the mature or graafian protrudes from the ovarian surface like a "blister". Two layers of stromal cells condense around the membrane granulosa in the cortex to form an additional coat, the *theca folliculi*, which canbe subdivided into an inner vascular *theca interna* and a peripheral fibrous *theca externa*. Estrogens are probably secreted by cells of the theca interna directly into the antrum through a basement membrane, the *membrana propria*, which separates the theca interna from the membrana granulosa.

The egg-containing follicle reaches its maximal size just before ovulation. A watery follicular fluid is released at this time and the thecal blood vessels which enclose the follicle collapse.

The number of graafian follicles which develop per estrous cycle depends on hereditary and environmental factors. In most primates, 1 or 2 eggs are released at each ovulation. In rodents, 4 to 14 eggs are released at ovulation. Polyovular follicles occur in a number of species, however. They are especially frequent in the opossum (*Didelphis virginiana*) and the striped skunk (*Mephitis mephitis*). If 1 ovary is removed, the remaining one undergoes compensatory hypertrophy.

Hormonal Control of Follicular Development

Pituitary gonadotropins regulate both the rate at which ovarian follicle develop and the number which mature per estrous cycle. If hemicastrated mice and treated with small doses of PMSG, they produce nearly twice as many mature ova as control animals. One would expect this ovulatory compensation because the single ovary is exposed to twice the normal amounts of FSH. However, the dose-response line is steeper for normal females than for hemicastrated ones and at high PMSH doses, the single ovary sheds no more eggs than does each ovary of the control female.

Many follicles grow during the first stages of the estrous cycle, but few mature completely. Perhaps less FSH is required to initiate the growth of small follicles than to maintain larger follicles and bring them to ovulatory size, since the number of follicles which mature can be greatly increased (superovualtion) by injecting animals with large doses of gonadotropins. Why then do so many follicles begin to grow? They may supply substances, like estrogen, which are essential for the ovulation of other larger follicles.

B. Formation of Eggs

The consecutive stages of egg development are (1) the proliferation of the oogonium; (2) the growth and maturation of the oocyte; (3) the formation of the ovarian follicle containing the mature egg; and (4) the release of the egg (ovulation). These processes occur cyclically in nonpregnant animals during the sexual season.

1. Proliferation of the Oogonium

The precursor cells of female gametes, called gonocytes, probably originate outside the gonad from extra-embryonic endoderm and then migrate into the embryonic ovary to differentiate into oogonia. Oogonia proliferate before or shortly after birth. In the rabbit and hamster, oogenesis continues into the postnatal period, whereas the oogonial divisions of the rat and mouse and completed several days before birth. In the guinea pig, neo-oogenesis ends when the animal is about 50 days of age. Hence, most, if not all, definitive ova are formed by birth, and de novo oocytogenesis does not generally occur in postnatal life.

2. Growth of the Oocyte

The growth of the oocyte is characterized by (a) the accumulation of yolk in the cytoplasm, (b) the development of a zona pellucida, and (c) the mitotic proliferation of the follicular epithelium and adjacent tissue. Initially, the ovum and ovarian follicle grow explosively. Later, when the antrum forms, the oocyte does not enlarge further, but the ovarian follicle, responding to pituitary hormones, continues to do so very rapidly.

3. Maturation of the Oocyte

The oocyte matures during the second phase of follicular growth. The nucleus, which entered the prophase of the first meiotic division during the first phase of growth, now undergoes a meiotic or reduction division. Nucleoli and nuclear membranes disappear and the chromosomes condense. The *centrosome* divides into 2 *centrioles* which move to the opposite poles of the cell. *Spindle fibers* appear joining the centrioles with the chromosomes. Diploid pairs of the latter move to the center of the cell and arrange themselves in one plane, the *equatorial plate* of the spindle (metaphase 1). They then separate, one member of each pair going to each pole of the cell. A membrane forms at the equatorial plate and divides the original cell into 2 daughter cells.

The primary oocyte undergoes 2 such meiotic divisions. The 2 daughter cells which arise from the first division contain one half (n) of the chromosome complement (2n) of the parent cell. Unlike the meiotic division process which occurs in spermatogenesis, 1 of the daughter oocytes acquires almost all of the cytoplasm and is called the *secondary oocyte;* the other, much smaller cell becomes the *first polar body.* During the second maturation division, the secondary oocyte divides

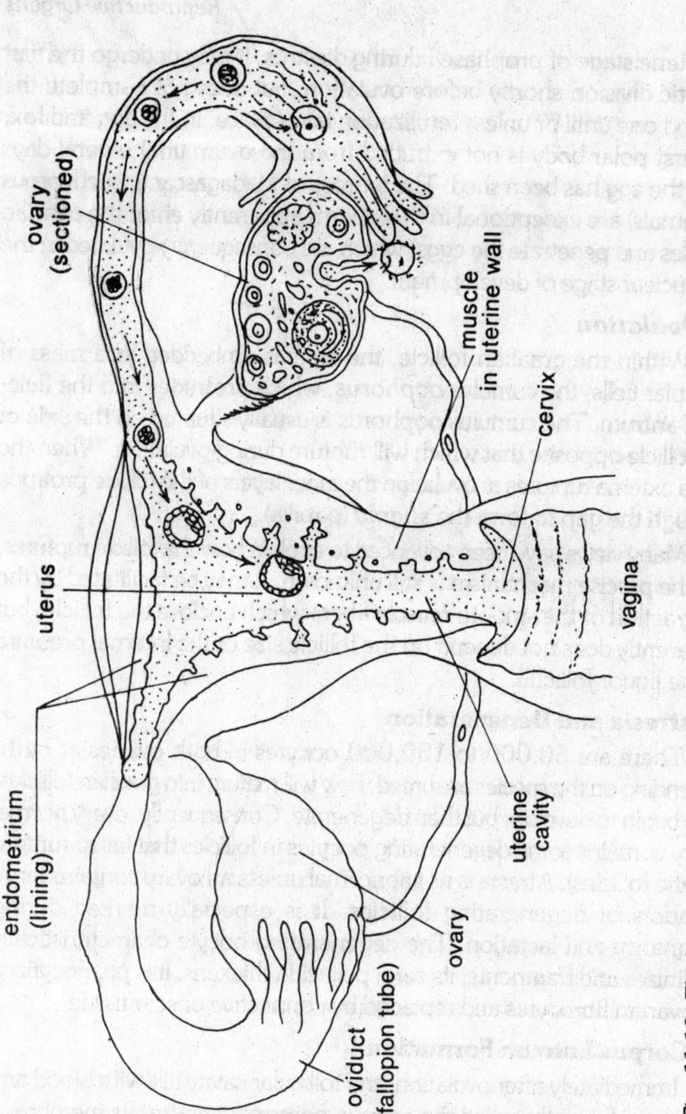

Fig. 11.10. The human female reproductive tract with the uterus and one oviduct opened up to show progress of the developing embryo at various stages. The embryo normally becomes implanted in the lining of the uterus, as shown here. Implantation in other areas causes pregnancy complications.

into an *ootid* (with n number of chromosomes) and a *second polar body* (also with n number of chromosomes). The polar bodies are extruded into the perivitelline space and degenerate.

The time at which the 2 reduction divisions occur is not necessar coincident with ovulation. The oocyte is usually in the pachytene diplotene stage of prophase I during diestrus. It may undergo the first meiotic division shortly before ovulation, but does not complete the second one until or unless fertilization takes place. In the dog and fox, the first polar body is not extruded from the ovum until several days after the egg has been shed. The tenrees of Madagascar (insectivorous mammals) are exceptional in that sperm apparently enter the ovarian follicles and penetrate the eggs, which are consequently ovulated in the pronuclear stage of development.

4. Ovulation

Within the graafian follicle, the eggs is embedded in a mass of follicular cells, the cumulus *oophorus*, which protrudes into the fluid-filled antrum. The cumulus oophorus is usually situated on the side of the follicle opposite that which will rupture during ovulation. When the theca externa ruptures at ovulation the inner layers of the follicle protrude through the gap to form the *stigma* (papilla).

Many views have been advanced to explain how the follicle ruptures, but the precise mechanism is still unknown. It may be facilitated by the contraction of the smooth muscle fibers which enclose the follicle, but apparently does not depend on the follicle size or the internal pressure of the liquor folliculi.

C. Atresia and Degeneration

There are 60,000 to 100,000 oocytes in both ovaries at birth, depending on the species are breed. Few will mature into graafian follicles; may begin to develop, but then degenerate. Consequently, every normal ovary contains some degenerating oocytes in follicles that fail to rupture (atretic follicles). Atresia is not abnormal unless an ovary contains large numbers of degenerating follicles. It is especially marked during pregnancy and lactation. The degenerating oocyte characteristically hyalinizes and fragments; its zona pellucida thickens. It is phagocytized by ovarian fibrocytes and replaced by connective or scar tissue.

D. Corpus Luteum Formation

Immediately after ovulation, the follicular cavity fills with blood and lymph and is aptly called the *corpus hemorrhagicum*. Its membrana granulosa plicates and proliferates cords of hypertrophic granulosa cells which bridge the cavity. Cells and blood vessels of the theca interna then invade this mass, forming a *corpus luteum* with columns of lutein cells interlaced with blood vessels and connective tissue. Its shape and colour vary with the species and the reproductive cycle. It grows very rapidly for slightly longer than half of the estrous cycle. If fertilization does not occur, it then regresses and is replaced by connective tissue and is consequently called the *corpus albicans*. After 2 or 3 estrous cycles, a barely visible scar is all that remains of it.

The mammalian corpus luteum is a short-lived endocrine gland which synthesizes and releases progesterone. No one has yet determined how the luteal phase of ovarian activity, i.e., the endocrine activity of the corpora lutea is regulated. The nongravid uterus is apparently involved, since hysterectomy prolongs the life of the corpus luteum in rodents such as guinea pigs or pregnant or pseudopregnant rats, and in

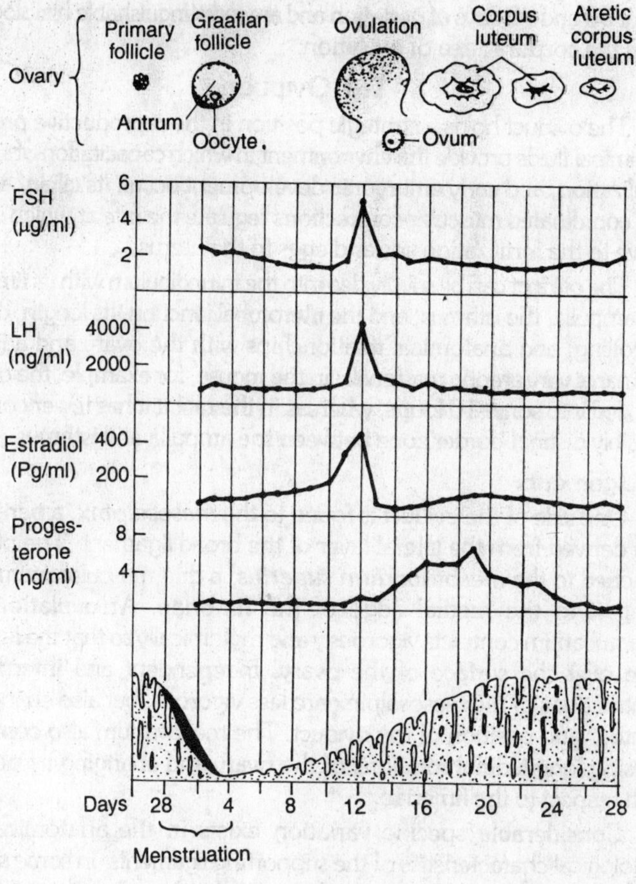

Fig. 11.11. Diagrammatic representation of the human menstural cycle, showing the relative concentrations of FSH, LH, estradiol, and progesterone and the corresponding states of the ovary and the endometrium.

ungulates. It does not, however, affect the corpora lutea of the unmated rat. In the rabbit, a reflex ovulator, the effect of hysterectomy upon the corpora lutea depends upon the time when it is done. There is some evidence that LH and the uterus dually control the luteal phase of the estrous cycle.

If pregnancy occurs, the corpus luteum remains functional and is known as the *corpus luteum verum;* it may be larger than the *corpus luteum spurium* (false yellow body) of the estrous cycle. In the chinchilla, corpora lutea persist throughout gestation and many accessory ones are also present. The latter probably arise from follicular luteinization at about 50 and 90 days of gestation and are indistinguishable histologically from the corpora lutea of ovulation.

The Oviduct

The oviduct holds a strategic position in the reproductive process. Its luminal fluids provide the environment in which capacitation of sperm, fertilization, and early embryonic development occur. Its ciliary activity and coordinated muscular contractions regulate the rate at which sperm move to the fertilization site and eggs to the uterus.

The oviduct can be subdivided into the infundibulum with its fimbriae, the ampulla, the isthmus, and the uterotubal junction. Its length, degree of coiling, and anatomical relationships with the ovary and adjacent ligaments vary among mammals. In the mouse, for example, the oviduct is coiled into some 10 loops, whereas in the rabbit it has fewer coils and a visibly distinct border zone between the ampulla and isthmus.

A. Ligaments

One side of the oviduct is found to the *mesosalpinx*, a peritoneal fold derived from the lateral layer of the broad ligament. The other is attached to the *mesotubarium superius,* a thin muscular membrane tying it to the medial edge of the fimbriae. At ovulation, the mesotubarium contracts vigorously and rhythmically so that the fimbriae slide over the surface of the ovary. Independent and intermittent contractions of the mesosalpinx are less vigorous, but also change the contour and position of the oviduct. The mesovarium also contracts, causing longitudinal rotations of the ovary and changing its position with respect to the fimbriae.

Considerable specific variation exists in the anatomical and histological characteristics of the supporting ligaments. In some species at least, the ligaments are particularly well developed and seem to have major physiological significance in moving, flexing, and closing the uterotubal junction.

B. Ovarian Bursa

The ovary and oviduct are closely associated with each other anatomically. In Lagomorpha, for example, the ovary lies in an open pouchlike *ovarian bursa* consisting of a think peritoneal fold of the mesosalpinx which is attached to the upper oviduct. In Mustelidae and Muridae, the ovary lies in a closed fluid-filled *periovarial sac* into which the eggs are shed. Large numbers of sperm reach the ovarian bursa of the rat, ferret, and bat.

C. Fimbriae and Infundibulum

The end of the oviduct adjacent to the ovary is expanded into a funnel-like structure called the *infundibulum,* whose size varies with species and age. The opening of the infundibulum (*ostium abdominal*) is in the center of a fringe of irregular processes, the *fimbriae*. Animals with a well developed ovarian bursa (mink, mice, rat, and dogs) have poorly developed fimbriae, whereas animals with only slightly developed ovarian bursae (rabbits, primates, and ungulates) have well developed fimbriae. These anatomical arrangements assure close association of the oviduct with the ovarian surface.

Specific variation also occur in the anatomy of the fimbriae and infundibulum and their relationship with the ovary. In the rat and mouse, for example, the infundibulum occupies only a small area of the periovarial sac. Its fimbriae makes very limited contact with the surface of the ovary. On the other hand, the expansive fimbriae of the guinea pig and rabbit almost completely enclose the ovary during ovulation. The anatomical association of fimbriae and ovary in primates is intermediate to these 2 extremes.

The appearance and contractability of the fimbriae also vary with the stage of sexual cycle. In the rabbit, for example, the fimbriae appear shrunken and cover only part of the ovarian surface until ovulation, when they expand into a bursa which enclose the entire ovary. The fimbriae of the rat are narrow during diestrus, but more expansive and vigorously active during estrus.

The fimbriae contain erectile tissue, a network of smooth muscle fibers and several large blood vessels, especially veins, which become engorged with blood at ovulation. The fimbrial surfaces are also covered by innumerable cilia which beat actively in the direction of the oviductal ostium. The vigorous peristaltic activity of the oviduct not only assumes that the ciliated fimbriae contact all surfaces of the ovary, but also creates a negative pressure which, together with the directional beat of the fimbrial cilia, vaccums the egg into the ostium.

D. Ampulla and Isthmus

The *ampulla*, which accounts for about half of the length of the oviduct, merges with a constricted segment of the oviduct called the *isthmus*. The significance of the ampullary-isthmic junction is not fully understood. The isthmus has a narrower lumen, fewer mucosal folds and ciliated cells, and a thicker muscularis than the ampulla. It is adrenergically innervated and apparently acts as a sphincter in several species (rabbit, cat, dog, rat, and man).

1. Mucosa and Submucosa

The tunica mucosa lines large primary and small secondary folds of the oviductal wall. The latter are particularly complex in the ampulla. The mucosa consists of epithelial cells. The underlying submucosa of smooth muscle fibers and connective tissue are permeated by fine blood and lymph vessels. The epithelium contains ciliated and nonciliated cells, together with "pig cells" (Stifzellen or intercalary cells) which are presumably depleted secretory cells. At the ovarian end of the oviduct, the wall is lined mostly with ciliated columnar cells; such cells are less common at the uterine end of the oviduct.

Epithelial folding is almost nonexistent in the mouse and rat, but extensive in the rabbit and primates. The number of ciliated cells in the mucosa also varies: they are rare in the oviducts of the mouse and rat, found in small numbers in the guinea pig, and relatively numerous in the rabbit and primates.

Ciliated Cells. The ciliated cells of the oviductal mucosa have slender motile cilia (*kinocilia*) that extend into the lumen. In the rhesus monkey, cilia are 5 to 6 μ in length and 0.2 to 0.3 μ in diameter. Those of the rabbit beat some 1200 times each minute. The frequency of the beat increases by 25 percent after ovulation and facilitates the movement of the egg toward the uterus. Their primary function is probably to keep the canal free of choking debris, as is true of cilia in the respiratory tract. Their activity, coupled with oviductal contractions, keeps oviductal eggs in the constant motion which is essential for bringing egg and sperm together (fertilization) and preventing oviductal implantation.

Ciliation of the oviduct is hormonally controlled in primates. Cilia disappear almost completely after hypophysectomy and develop in response to exogenous estrogens. Progesterone, on the other hand, antagonizes the estrogen-driven growth of cilia.

Secretory Cells and Fluid. The secretory cells of the oviductal mucosa are nonciliated and characteristically contain secretory granules whose size and number very widely among species and during different phases of the estrous cycle. They are, for example, large and numerous in the rabbit, but small and sparse in primates. In general, they become increasingly abundant as one moves from the infundibulum toward the isthmus; this is also true of the nonciliated cells which produce them. Like cilia, the presence and number of secretory granules depend on ovarian activity. Ovariectomy causes them to disappear in all regions of the oviduct except the isthmus. They reappear if the animal is subsequently treated with estrogen.

The metabolic activity of the epithelium of the oviduct appears to be most pronounced in the vicinity of the egg. In the rabbit, for example, the uptake the radioactive sulfate is always highest in the segment which contains the egg as it passes along toward the uterus.

By means of extra- and intra-abdominal devices which are used to cannulate the oviduct and collect its fluids, it has been possible to show that ovarian hormones also regulate the secretory activity of the epithelium of the oviduct. The latter exhibit predictable variation during the estrous cycle: secretory fluid volume is low during the luteal phase, increases the onsets of estrus, reaches a maximum 1 day later, and then declines again. The secretions are composed predominantly of mucoproteins and mucopolysaccharides.

Cyclical Changes. During the sexual cycle, there are cyclical changes in the size of the epithelial cells, the activity of secretory cells, the number of activity of cillia, and the size of subepithelial capillaries. The width of the oviductal lumen also varies and is related to contractions of the wall muscle and the fluid content of the wall and the lumen. In the mouse and rat, for example, oviductal loops which contain eggs are distended with fluid, whereas neighboring empty loops are smaller are have occluded lumina.

2. Muscularis and Serosa

The tunica muscularis consists of an inner circular layer and an outer longitudinal layer of smooth (unstriated) muscle. Muscle also extends these layers into the connective tissue of the mucosal folds, permitting coordinated contractions of the entire wall. The thickness of the musculature increases as one advances from the ovarian to the uterine end of the oviduct. In the mouse and rat, the isthmus contains the typical outer longitudinal and inner circular layers of muscle; however, in the guinea pig the outer layer is circular and the inner one longitudinal. The serosa consists of connective tissue and is enveloped externally by the same mesothelial coat which lines other visceral organs and the peritoneal cavity.

3. Contractions of the Oviduct

The contractile activities of the oviduct are independent of those which occur in the uterus. They appear to occur in specific peristaltic and antiperistaltic patterns which propel the egg at a definite rate and in a defined rotational manner. The latter is reflected in the remarkably uniform mucinous coats or rabbit ova.

In part, these patterns of contraction arise because of the architecture of the oviduct. The ampulla, for example, has a wide lumen surrounded by a wall with a think layer of circular muscle and a mucous membrane with highly branched folds. The folds are laced with numerous capillaries and muscle fibers (erectile tissue). The circular muscle occurs in spiral sheets of different shapes and sizes which contract in a segmented sequential fashion. In areas where the muscle has relaxed, the wall becomes engorged with blood and may expand to twice its normal size. Transport of an ovum through the ampulla depends mainly on the rate amplitude of these segmenting contractions, which originate in the infundibulum and pass progressively toward the isthmus.

Norepinephrine and acetylcholine may also initiate oviductal contractions. The oviduct and the uterine horns have opposite sensitivities to these 2 compounds; the ampulla is more sensitive to acetylcholine and less sensitive to norepinephrine than the uterus.

The contractile activities of the fimbriae, oviduct, and oviductal ligaments are also partly coordinated by the ratio of estrogen to progesterone in the blood. The oviduct is most efficient at engulfing ovulated eggs during estrus, less efficient during other periods of the sexual cycle. In some species, neurohormonal mechanisms also increase the muscular activity of the fimbriae at copulation, but in ways which are not yet well understood.

E. Uterotubal Junction

The uterotubal junction is the area where the oviduct meets the uterine cornu. In primates, it has an extramural and intramural portion. Its mucosa is thrown into longitudinal folds, which may be sparse or numerous, high or low, broad or narrow, and may also show a variety of branching, according to the species. The uterotubal junction is a papilla (*colliculus tubaris*) in the rat, an elevated site in the dog, and a rosette-like structure in the rabbit, but has no projections in primates.

Other anatomical and histological differences have also been observed in the uterotubal junctions of different species. These include differences in (a) the degree of flexure and the angle at which the oviduct and the uterine cornu meet; (b) the narrowing of the lumen in the caudal isthmus; (c) blood and lymph supplies; (d) the number and morphology of ciliated cells; (e) the cytology of mucosal secretory cells and uterine glands; (f) ciliary activity during stages of the reproductive cycle; (g) the relative thickness of the circular and longitudinal bands which make up the tunica muscularis; and (h) the blending of the oviductal musculature with the myometrium.

12

Endocrinology

The Discovery of Hormones

Endocrinology, the study of hormone-producing tissues and their secretions, has become a major branch of biology only in the past century, but the actions of some endocrine glands have been widely known for many years. For example, castration (removal of the testes) of young male vertebrates has been practiced for several thousand years to prevent certain male animals from breeding or to produce animals that are more docile and that develop better meat. In fact, Aristotle, in the fourth century B.C., wrote an accurate, detailed description of the effects of castration on birds and humans.

However, scientific experimentation on the functioning of endocrine glands did not begin until 1849, when A.A. Berthold transplanted testes back into the bodies of young castrated male chickens. Berthold found that this treatment restored the secondary sex characteristics normally seen in roosters. The males with transplanted testes developed combs, gained the ability to crow, and began to exhibit the generally pugnacious behaviour expected of roosters. These treated males looked and acted like roosters even when the transplanted testes were far from their normal location in the body and had none of the normal nerve connections. Berthold concluded that testes must release some substance that causes the development of male characteristics, and that their action does not depend directly on nerve connections, as was earlier believed. It was latter learned that the testes secrete the hormone *testosterone,* which is responsible for development of many male characteristics.

Over the next few years, similar gland removal and replacement experiments helped to identify other hormone-producing structures.

Correlation of disease symptoms with degenerated glands found during autopsies helped establish still more structures as endocrine glands. For example, in 1855 Thomas Addison described a set of symptoms—including poor appetite, low blood pressure, muscular weakness, and a characteristic skin discoloration—associated with the deterioration of

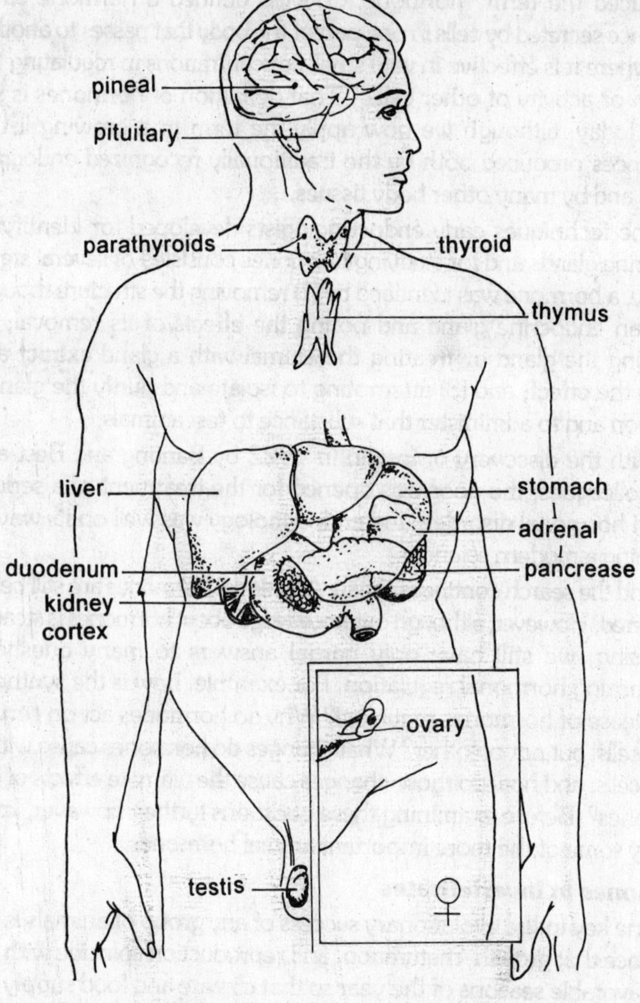

Fig. 12.1. The approximate locations of the endocrine glands in man. Although the pineal body, thymus and stomach are shown, the hormones they may secrete have not been completely characterized.

the outer portion (cortex) of the adrenal glands. The syndrome is still known today as Addison's disease.

By 1904, the English physiologists Starling and Bayliss (who discovered the hormone secretion, which stimulates the release of pancreatic juice) were writing about an "endocrine system" and had introduced the term "hormone." Bayliss defined a hormone as "a substance secreted by cells in one part of the body that passes to another part, where it is effective in very small concentrations in regulating the growth or activity of other cells." That definition of hormones is still useful today, although we now apply the term to a growing list of substances produced both by the traditionally recognized endocrine glands and by many other body tissues.

The techniques early endocrinologists developed for identifying endocrine glands and for studying hormones consisted of several steps. Usually, a hormone was identified by (1) removing the structure thought to be an endocrine gland and noting the effects of its removal; (2) replacing the gland or treating the animal with a gland extract and noting the effect; and (3) attempting to isolate and purify the gland's secretion and to administer that substance to test animals.

With the discovery of insulin in 1922 by Banting and Best and their colleagues, the door was opened for the treatment of a serious human hormonal disorder, and endocrinology was well on its way to becoming a modern science.

And the search continues today. Additional hormones are still being discovered. However, although our knowledge about hormones is steadily increasing, we still have only partial answers to many questions concenrning hormonal regulation. For example, how is the synthesis and release of hormones regulated? Why do hormones act on certain target cells, but not on other? What changes do hormones cause within target cells, and how do those changes cause the ultimate effects of the hormones? Before examining these questions further, however, let us identify some of the more important animal hormones.

Hormones in Invertebrates

One key to the evolutionary success of any group of animals is the assurance that growth, maturation, and reproduction coincide with the most favorable seasons of the year so that climate and food supply are optimal. It seems likely that the hormones regulating growth and reproduction were among the first hormones to appear during the course of animal evolution.

The first hormones were probably neurosecretions, and most of the substances functioning as hormones in modern invertebrate animals are also neurosecretions. Only in a few of the more complex invertebrates (such as molluscs, arthropods, and echinoderms) have non-neurosecretory hormones been identified.

The principal hormones involved in the regulation of insect growth and maturation are undoubtedly the most thoroughly studied set of invertebrate hormones.

Hormones and Insect Development

Insects (and all other arthropods) have a rigid outer body covering called the *exoskeleton*. Because exoskeletons do not expand to allow for enlargement, insect growth is associated with the periodic shedding

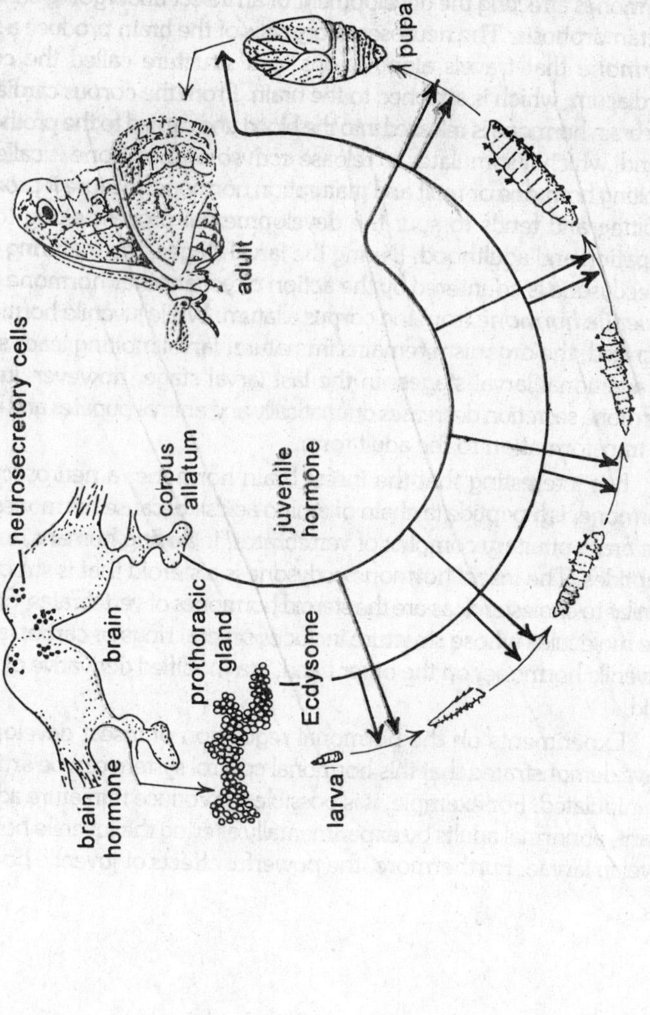

Fig. 12.2. *Developmental hormones in insects.*

of the exoskeleton in a process known as *molting* (*ecdysis*). Molting is followed by the expansion and hardening of a new, soft, larger exoskeleton that has development includes a series of molts and stages between molts, which are known as *instars*.

In insect development, *Metamorphosis* is the change in body form that converts an immature individual into an adult. Metamorphosis is either *complete* (*abrupt*), involving *larval, pupal,* and *adult* (*imago*) stages, or *incomplete* (*gradual*), involving a series of *nymphs*. Nymphs generally resemble adults and become somewhat more adultlike at each molt, until a fully developed adult emerges from the final molt.

Several hormones regulate insect growth, molting, and metamorphosis; figure summarizes the interactions of the principal hormones affecting the development of an insect undergoing complete metamorphosis. The neurosecretory cells of the brain produce a *brain hormone* that travels along axons to a structure called the corpus cardiacum, which is attached to the brain. From the corpus cardiacum, the brain hormone is released into the blood and carried to the prothoracic gland, which it stimulates to release *ecdysone*. Ecdysone is called the molting hormone or molt and maturation hormone because it promotes molting and tends to spur the developmental process along toward pupation and adulthood. During the larval stages, the maturing effect of ecdysone is countered by the action of yet another hormone called *Juvenile hormone* from the corpus allatum. While juvenile hormone is secreted, the organism remains immature; larval molting leads simply to additional larval stages. In the last larval stage, however, juvenile hormone secretion decreases dramatically and animal pupates and begins its transformation to the adult form.

It is interesting that the insect brain hormone, a neurosecretory hormone, is a peptide (a chain of amino acids), because hormones from the brain-pituitary complex of vertebrates, including humans, are also peptides. The insect hormone ecdysone is a steroid that is structurally similar to cholesterol, as are the steroid hormones of vertebrates. Steroids are molecules whose structure includes several rings of carbon atoms. Juvenile hormone, on the other hand, is a modified derivative of a fatty acid.

Experiments on the hormonal regulation of insect development have demonstrated that this hormonal control system can be artificially manipulated. For example, it is possible to produce miniature adults or giant, abnormal adults by experimentally altering the juvenile hormone level in larvae. Furthermore, the powerful effects of juvenile hormone on insect development make this hormone an interesting subject for insecticide research. If larval insects or the food they consume could be sprayed with juvenile hormone, the larvae might develop abnormally and die, or fail to reach an appropriate stage of their life cycle in time to survive the winter.

Research on juvenile hormones as insecticides seems well worth the effort because the hormones are effective at very low concentrations and are naturally occurring, biodegradable compounds that should not pose the threat of cumulative toxicity that many insecticides do. Because the juvenile hormones of different species differ, it might be possible to limit the effect to harmful insects without affecting beneficial insects in the same environment. It also seems unlikely that the insects would develop resistance to juvenile hormone insecticides because their own development depends on sensitive, specific cellular responses to very small amounts of juvenile hormones.

From an evolutionary point of view, it is interesting that some plants produce compounds similar or identical to ecdysone or juvenile hormones. This apparently protects the plants by disrupting the development of insects that eat them.

Vertebrate Glands and Hormones

More is known about mammalian hormones than about the hormones of other vertebrates, and much of the study of mammalian hormones has focused on hormonal regulation in the human body. The locations of some of the major human endocrine glands: the hypothalamus, the pituitary, the thyroid, the parathyroids, the pancreas, the kidneys, the adrenals, the ovaries, and the testes. Although we do not discuss them here, the pineal body and possibly the thymus, as well as well as some tissues in the stomach, intestine, and other organs, also function as hormone sources in mammals.

As we discuss vertebrate hormones, we will explore three main topics: (1) the hormones of the pituitary gland and the relationship between the pituitary and the hypothalamus, (2) the hormones regulating metabolism and development, and (3) the hormones regulating ionic balance in the body. Grouping hormones in this way helps emphasize the role of hormonal regulation in the maintenance of homeostasis through the coordination of specialized body functions.

Immunoassay techniques that measure the secretory products of the pituitary and advances in the neuroradiologic evaluation of the sella turcica have improved the diagnosis of pituitary disorders. Trans-sphenoidal microsurgery and bromocriptine now allow earlier and more effective therapy. This chapter describes the clincial evaluation of the anterior pituitary (adenohypophysis), including its anatomy, embryology, and histology; the physiologic action and secretion of its 6 major hormones; and disorders of the gland with respect to neuroradiologic findings and secretory hypo- and hyperfunction.

ANATOMY

Gross Anatomy

The pituitary gland lies at the base of the skull in a portion of the sphenoid bone called the sella turica ("Turkish saddle"). The anterior portion, the tuberculum sellae turcicae, is flanked by posterior projections of the sphenoid wings, the anterior clinoid processes; the dorsum sellae forms the posterior wall, and its upper corners project into the posterior clinoid processes. The gland is surrounded by dura, and the roof is formed by a reflection of the dura attached to the clinoid processes, the diaphragma sellae. In normal individuals the arachnoid membrane and, therefore, cerebrospinal fluid are prevented from entering the sella turcica by the diaphragma sellae. The pituitary stalk and its blood vessels pass through an opening in this diaphragm. The lateral walls of the gland are in direct apposition to the cavernous sinus. The optic chiasm lies 5-10 mm above the diaphragma sellae and anterior to the stalk.

The size of the pituitary gland, of which the anterior lobe constitutes two-thirds, varies considerably. It measures approximately 15 x 10 x 6 mm and weighs 500-900 mg; it may double in size during pregnancy. Since the sella turcica tends to conform to the shape and size of the gland, there is considerable variability in the contour of this body structure.

Blood Supply

The anterior pituitary is the most richly vascularized of all mammalian tissues, receiving 0.8 mL/g/min from a portal circulation connecting the median eminence of the hypothalamus and the anterior pituitary. This arterial blood is supplied by the internal carotid arteries via middle and inferior hypothyseal arteries. The arteries form a capillary network in the median eminence of the hypothalamus that recombines in long portal veins draining down the pituitary stalk to the anterior lobe, where they break up into another capillary network and re-form into venous channels. The posterior pituitary is supplied directly from branches of the middle and inferior hypophyseal arteries.

Venous drainage of the pituitary, the route through which anterior pituitary hormones reach the systemic circulation, is variable, but venous

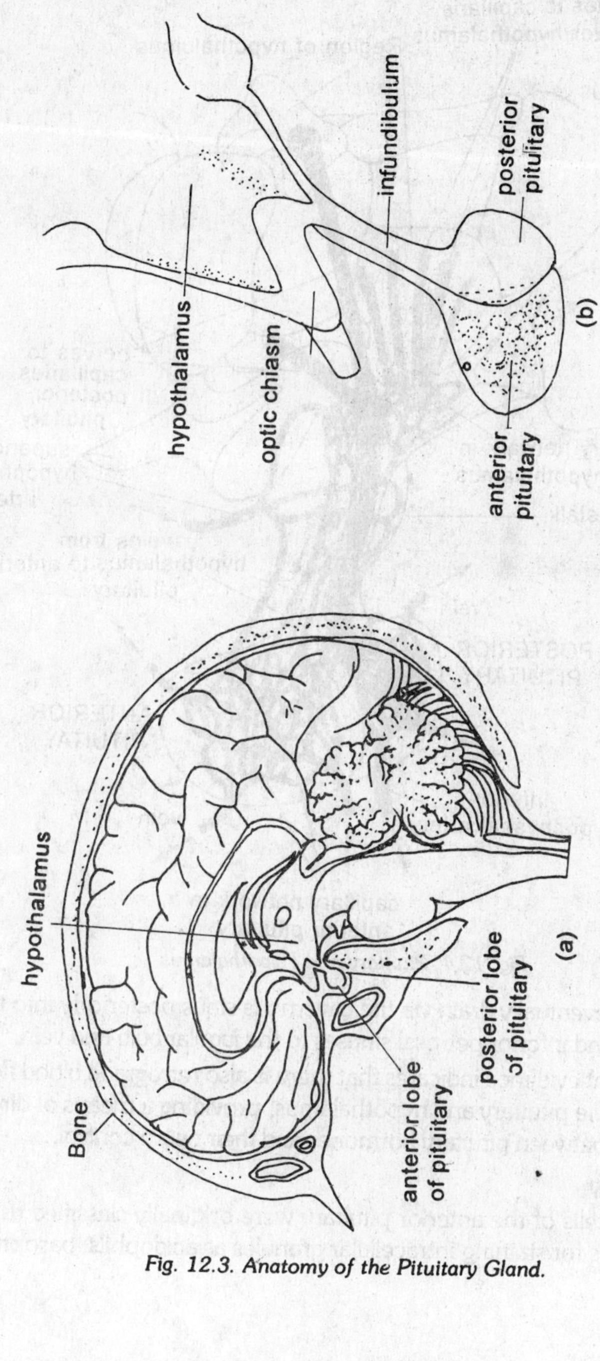

Fig. 12.3. Anatomy of the Pituitary Gland.

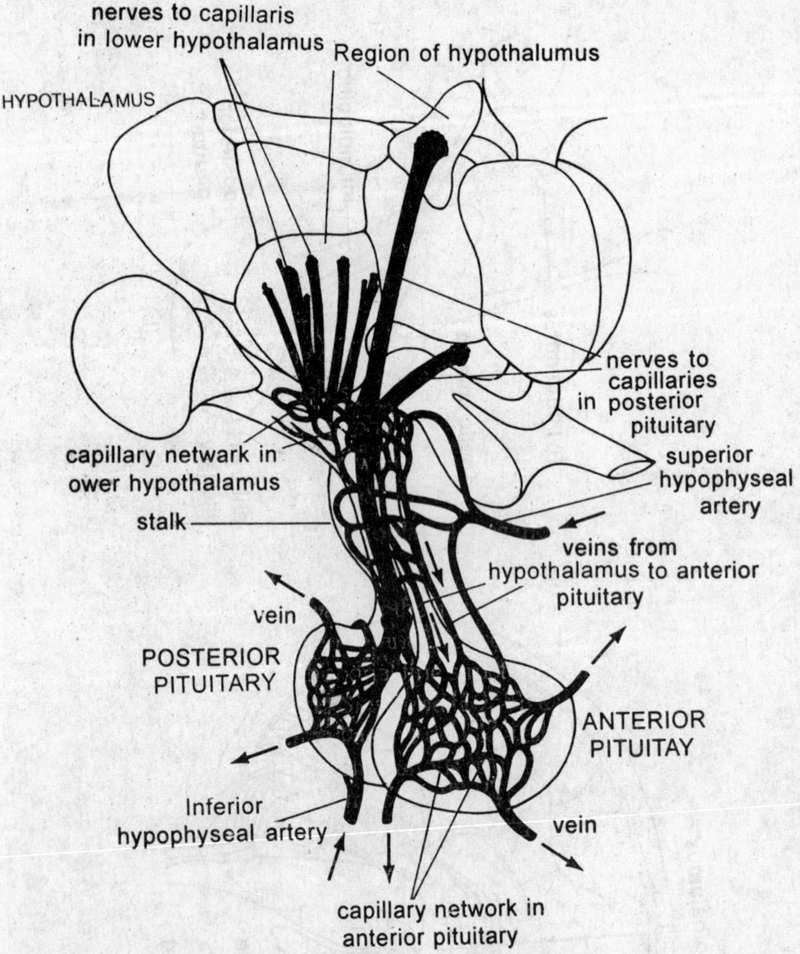

Fig. 12.4. Pituitary and Hypothalamus.

channels eventually drain via the cavernous sinus posteriorly into the superior and inferior petrosal sinuses to the jugular bulb and vein.

Recent evidence indicates that there is also retrograde blood flow between the pituitary and hypothalamus, providing a means of direct feedback between pituitary hormones and their neuro control.

Histology

The cells of the anterior pituitary were originally classified using techniques for staining intracellular granules as acidophils, basophils,

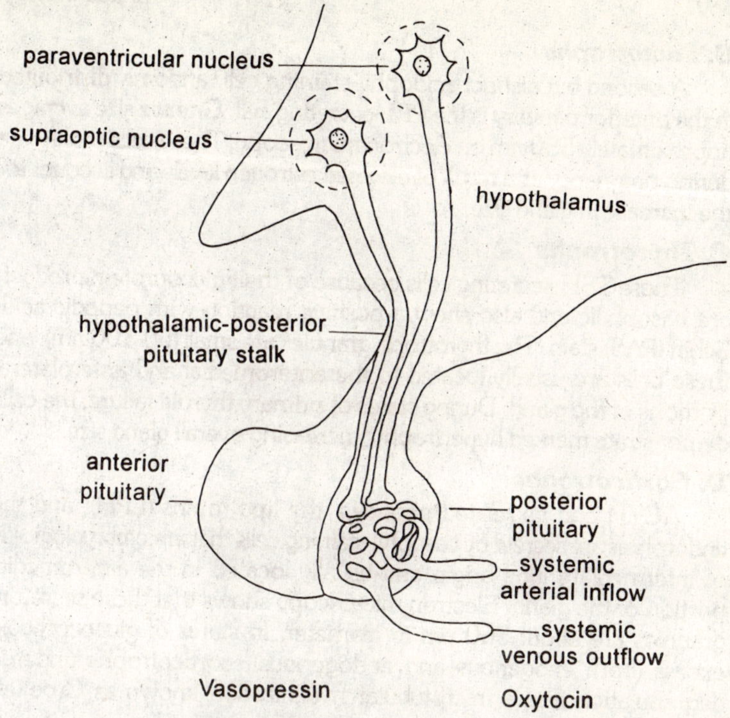

Fig. 12.5. Relationship of Hypothalamus and Posterior Pituitary.

and chromophobe cells. Immunocytochemical and electron microscopic techniques now permit classification of cells by their specific secretory products: somatotrophs (growth hormone [GH]-secreting cells), lactotrophs (prolactin [PRL]-secreting cells), thyrotrophs (thyrotropin [TSH]-secreting cells), corticotrophs (cells secreting adrenocorticotrophic hormone (corticotropin, [ACTH] and related peptides), and gonadotrophs (luteinizing hormone [LH]-and follicle-stimulating hormone (FSH]-secreting cells). The specificity of the antibody used determines the ability of immunocytochemical techniques of differentiate the various cells types.

A. Somatotrophs

The GH-secreting cells are identified as acidophilic in standard hematoxylin and eosin preparations and are usually located in the lateral portions of the anterior lobe. Granule size by electron microscopy is 300-400 nm in diameter. These cells account for about 4-10% of the net weight of the anterior pituitary.

B. Lactotrophs

A second but distinct acidophil-staining cell randomly distributed in the anterior pituitary is the PRL-secreting cell. Granule size averages approximately 550 nm on electron microscopy. These cells proliferate during pregnancy as a result of elevated estrogen levels and account for the increase in gland size.

C. Thyrotrophs

These TSH-secreting cells because of their glycoprotein product, are basophilic and also show a positive reaction with periodic acid-Schiff (PAS) stain. The thyrotroph granules are small (50-100 nm), and these cells are usually located in the anteromedial and anterolateral portions of the gland. During states of primary thyroid failure, the cells demonstrate marked hypertrophy, increasing overall gland size.

D. Corticotrophs

ACTH and its related peptides, the lipotropins (LPH), and the endorphins are secretd by basophil-staining cells that are embryologically of intermediate lobe origin and usually located in the anteromedial portion of the gland. Electron microscopy shows that these secretory granules are about 360 nm in diameter. In states of glucocorticoid excess (both exogenous and endogenous), corticotrophs undergo degranulation and a microtubular hyalinization known as Crooke's hyaline degeneration.

E. Gonadotrophs

Both LH and FSH originate from these basophil-staining cells, whose secretory granules are about 200 nm in a diameter. These cells are located in the lateral portion of the gland. They become hypertrophied and cause the gland to enlarge during states of primary gonadal failure such as Klinefelter's syndrome and turner's syndrome.

F. Other Cell Types

Despite immunocytochemical staining with antibodies directed against all of the known anterior pituitary hormones, some cells remain unstained. These are chromophobes by conventional staining methods, but electron microscopy has identified secretory granules in many of them. It is not certain whether they represent undifferentiated primitive secretory cells or whether they produce an as yet unidentified hormone, such as adrenal androgen-stimulating hormone or ovarian growth factor.

Growth Hormone

Regulation of Secretion

The factors controlling the secretion of GH by the pituitary are unusually complex. Although it was known that lesions placed in the rat hypothalamus impaired growth and caused degranulation of acidophils in the adenohypophysis the mechanism was not understood until hypothalamic hormonal control of adenohypophyseal function was recognized. The secretion of GH by the pituitary is increased by extracts from the hypothalamus; this observation was made first *in vitro* and later *in vivo.* Electrical stimulation of the ventromedial nucleus of the hypothalamus resulting in secretion of GH by the andenohypophysis has established this nucleus as one area that secretes GRH. Stravation causes a reduction of GH levels in the pituitary and GRH levels in the hypothalamus. Insulin causes a discharge of pituitary GH and

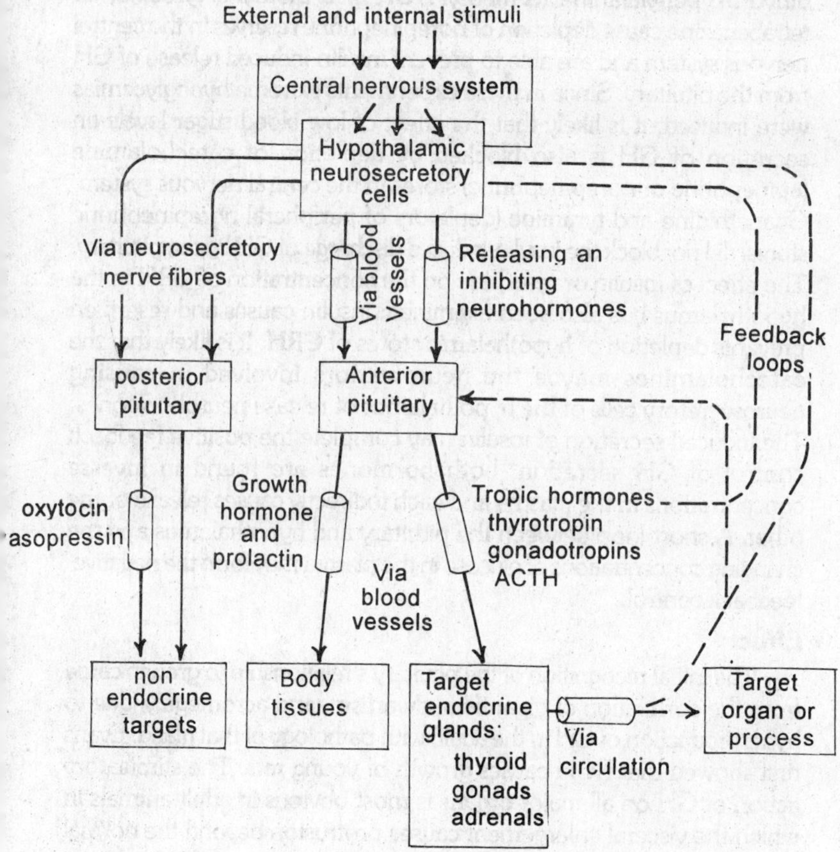

Fig. 12.6. Summary of some of the hormonal interactions mediated by the hypothalamus and the pituitary gland in mammals.

administration of glucose blocks this discharge α-Deoxy-D-glucose stimulates GH secretion and hyperglycemia; it causes a cellular glucopenia when converted to α-2-Deoxy-D-glucose stimulates GH secretion and hyperglycemia; it causes a cellular glucopenia when converted to a-2-deoxy-D-glucose-D-phosphate, which partially inhibits phosphoglucose isomerase and glucose-6-phosphate dehydrogenase. Lesions in the hypothalamus block the increase in plasma GH levels that would otherwise be induced by insulin.

Insulin may induce GH release by causing a secretion of norepinephrine or L-dopamine. Drugs such as reserpine, α-methyl-dihydroxypenylalanine (α-methyl-DOPA), α-methyl-α-tyrosine, or tetrabenazine cause depletion of norepinephrine reserves in the central nervous system and are able to prevent insulin-induced release of GH from the pituitary. Since in these experiments extreme hypoglycemias were induced, it is likely that the effect of low blood sugar levels on secretion of GH is also blocked by depletion of catecholamine (epinephrine or norepinephrine) stores in the central nervous system. Guanethidine and tyramine (depletors of peripheral norepinephrine stores) did not block the insulin-induced discharge of GH by the pituitary. The effect of insulin or reserpine on the concentration of GRH in the hypothalamus has also been examined: insulin causes and reserpien prevents depletion of hypothalamic stores of GRH. It is likely that the catecholamines maybe the neurohumors involved in causing neurosecretory cells of the hypothalamus ot release neurohormones. The induced secretion of insulin may complete the positive feedback control of GH secretion; both hormones are found in inverse concentrations in the plasma and each indirectly causes release of the other. A short loop between the pituitary and hypothalamus and the changing concentrations of glucose in the plasma may form the negative-feedback control.

Effect

The initial recognition of the pituitary's relationship to growth came from the correlation of gigantism, dwarfism, and acromegaly (due to hyperproduction of GH in the adult) with pathology of that gland. Evans first showed that BGH causes growth of young rats. The stimulatory action of GH on all major organs is most obvious in adult animals in which the visceral enlargement causes protrusion beyond the normal confines. Associated with growth is a markedlypositive nitrogen balance; GH causes a reduction in plasma amino nitrogen, increases amino acid transport across cell membranes, and increases protein synthesis.

Growth hormone causes mobilization of nonesterified fatty acids from fat deposits while it also inhibits glucose utilization by muscle tissues and decreases the sensitivity of hypophysectomized animals to insulin. The first of these actions is the so-called ketogenic effect and the latter two are diabetogenic actions. Greater utilization of fat is also reflected by the respiratory quotient.

The development of sensitive radioimmunochemical assay techniques for highly purified pituitary hormones has made it possible to determine blood concentrations of these hormones in humans. The plasma concentration of GH in fasting human subjects is 1 to 2 ng/ml plasma; after exercise it rises to 17 to 53 ng/ml, reaching the highest point in the first hour of a 2-hr exercise period. Plasma free fatty acid concentrations also rise, while the blood glucose level remains constant. These changes are prevented if the subject ingests glucose. By making depot fat available, GH may play a role in proving cellular fuel, since fatty acids are metabolized only after they are mobilized from depots.

Current studies on GH action point to a role in the transcription or translation steps leading to protein biosynthesis. Widnell and Tata have shown an *in vivo* increase in liver ribonucleic acid (RNA) polymerase within 24-hr after GH treatment. Growth hormone has also been reported to stimulate transfer RNA (tRNA) and messenger RNA (mRNA) formation. Injection of GH into hypophysectomized rats can partially restore the lost ability of ribosomes to incorporate amino acids into proteins.

Thyroid-Stimulating Hormone

Regulation of Secretion

Adenohypophyseal secretion of TSH is under dual control. There is evidence that thyroid hormone acts directly on the pituitary. Thyroid gland implants into the adenohypophyseal area of thyroidectomized rats retard the appearance of thyroidectomy cells among the thyrotrophs closest to the implants. Thyroxine (T_4) reduces the rate of ^{131}I release from the thyroids of hypophysectomized rabbits with implants of pituitary tisue in the anterior chamber of the eye. A dose of thyroxine so low that it causes no systemic effect, depress ^{131}I rate of release from rabbit thyroid when instilled into the pituitary, but is ineffectual when injected into the hypothalamus. In organ culture of pituitaries, thyroxine reduces the secretion of TSH.

The existence of a hypothalamic feedback pathway is suggested by the isolation from the hypothalamus of neurohormone (TRH) which induces the release of TSH. Other less direct evidence is that the thyroids of goitrogen (compounds causing enlargement of the thyroid) treated rats hypertrophy more than do the glands of rats given the same goitrogen but bearing bilateral hypothalamic lesions. In the latter case, the destroyed link in the chain is assumed to be the source of a substance regulating TSH secretion by the pituitary. The role of thyroid hormones in the regulation of TRH secretion is not fully established; a number of different stimuli are probably involved and the thyroid hormone may act through these stimuli by affecting caloric homeostasis.

Effects

Thyroid-stimulating hormone acts on two different sites: (1) the thyroid, and (2) the fat deposits, an extra-target site. In the latter, TSH induces release of fatty acids. Physiological levels of the hormone are sufficient for this lipolytic effect to be exerted.

The various actions of TSH on the thyroid have not been clearly differentiated. The first effect of TSH is an increased blood flow through the gland, and the second is an increase in the rate of breakdown of the colloid thyroglobulin. This proteolytic effect results in an increased thyroid hormone discharge and a reduced amount of colloid in the follicles. Simultaneously there is an increased biosynthesis of thyroid hormone, resulting in an increased rate of ^{131}I accumulation bythe gland. A late consequence of TSH stimulation is an increase size of the follicle cells and, when the stimulus is long-acting and intense, an increased number of cells and follicles.

LONG-ACTING-THYROID STIMULATOR

In 1956 Adams and Purves described in the plasma of hyperthyroid patients a stimulatory substance which induced the discharge of organically bound ^{131}I from rat thyroid glands. Response to a dose of TSH required 2 to 3 hr. while response to serum from hyperthyroid patients required 12 to 16 hr. Long-acting-thyroid stimulator (LATS) has been extracted from this serum and has been characterized as a 7-Svedberg (7-S)... α-globulin. It seems to be a specific α-globulin rather than an α-globulin-TSH complex. The molecule has been hydrolyzed to A and B chains with the A chain showing LATS activity. This substance is not present in the pituitary. LATS loses activity after incubation with minced thyroid tissues. Thus LATS may be autoantibody against a thyroid components.

EXOPHTHALMOS-PRODUCING SUBSTANCE

Surgical intervention in the clinical condition of Graves disease is frequently followed by an exophthalmos (bulging of the eyes). Guinea pigs also show an exophthalmos after thyroidectomy. Although crude TSH preparations may induce exophthalmos, highly purified preparations do not. Many pituitary extracts and various subfractions have been examined; some have high concentrations of exophthalmos-producing substance and low levels of TSH activity, whereas others show the opposite condition. The consensus is that TSH and another hormone are involved; however, the exophthalmos factor has not yet been characterized.

ADRENOCORTICOTROPHIN

Regulation of Secretion

A number of diverse factors affect the adenohypophyseal secretion of ACTH. ACTH simultaneously inhibits the cellular secretion of CRH via a short negative feedback loop and stimulates its target cells in the adrenal cortex. Furthermore, the corticalsteroids in pharmacological doses may modify the activity of the adrenocortical cells and in smaller dosage the cells that secrete CRH. Fig. summarizes the presently conceived pattern or regulation of ACTH secretion. Stress (any of a large variety of noxious stimuli) may provoke an increased secretion of ACTH, acting on the hypothalamus via central nervous system connections. Steroids implanted in the midbrain usually cause a reduction in adrenocortical secretion of corticosterone. This probably induces diffusely distributed steroid receptors to discharge, inhibiting the hypothalamic release of CRH. Hypothalamic extracts containing CRH have been shown to regulate ACTH secretion by the pituitary, and ACTH implanted into the hypothalamus has been shown to reduce the ACTH content of the adenohypophysis, probably due to the inhibition of neurosecretory-cell secretion of CRH. Glucocorticoids may have a direct action on the adenohypophyseal secretion of ACTH: they depress oxygen consumption of pituitary cultures and, when implanted into the vicinity of the pituitary, prevent the adrenal hypertrophy that occurs after unilateral adrenalectomy. Despite these indications of direct action of cortical steroids on the pituitary, the consensus is that the primary site of feedback action is the hypothalamus and perhaps other nervous areas. Vasopressin has also been shown to stimulate ACTH secretion but is not believed to be physiologically important.

Effects

Adrenocorticotrophin acts upon three sites. The targets that it primarily influences are the two innermost zones of cells in the adrenal cortex (the zona fasciculata and zona reticularis). These cell layers constitute the site of glucocorticoid synthesis and secretion. The zona glomerulosa (the outer cell layer of the cortex), which secretes aldosterone, is minimally affected by ACTH. The hormone also affects fat depots, where it exerts a lipolytic action in a manner similar to GH, TSH, and catecholamines. It stimulates melanocytes, causing darkening of skin in a manner similar to melanocyte-stimulating hormone (MSH).

The adrenal cortex is affected by ACTH in several ways: (1) It causes a reduction in the ascorbic acid content. (2) It concurrently converts cholesterol to gluco-corticoids and increases their secretion rate. (3) It raises the level of metabolic activity of adrenal tissue through an increase in oxygen consumption and glucose utilization. (4) It stimulates cell division of the inner two layers and causes the cortex to grow. (5) It increases adenylcyclase activity and causes c-AMP synthesis with enchanced steroidogenesis proportionate to the ACTH dosage. Protein-synthesis inhibitors block the respone to c-AMP but not the effect of ACTH on c-AMP concentration and glycogenolysis. The increased steroidogenesis may be due to increased conversion of cholesterol to pregnenolone.

PROLACTIN

Regulation of Secretion

There is evidence that the hypothalamus continually inhibits prolactin secretion by the adenohypophysis. Transplanted pituitary tissue continues to produce prolactin while the secretions of other hormones cease. Furthermore, cultured pituitary tissues release less prolactin to the medium when grown in a flask containing hypothalamic tissue than when grown alone; under similar conditions, more GH is formed and released when hypothalamic tissue is present. The hypothalamus is the source of a continuously secreted substance that inhibits prolactin release from the adenohypophysis. This substance is called prolactin-release-inhibiting hormone (PIH).

Secretion of PIH is decreased in the lactating animal; termination of lactation has been associated with increased secretion of PIH. Therefore there should be an inverse relationship between amounts of prolactin and PIH secreted. Such a relationship has been observed and correlated with frequency of suckling by rat pups. As time of weaning approaches, the pups suckle less frequently and the prolactin content of the pituitary rises, reflecting the increased secretion of PIH.

The corpus luteum in the rat fails to function in the absence of prolactin, but this luteotrophic action of the hormone is seen only in rodents. Maintenance of a functional corpus leuteum in rats is assumed, therefore, to be evidence of prolactin secretion. Estrogens, progesterone, and testosterone cause corpora lutea to continue functioning. Estrogen induces prolactin secretion by the pituitary; testosterone probably has dual effect: (1) causing secretion of estrogen by the ovary, and (2) blocking neurosecretory-cell secretion of PIH. The progresterone effect is an example of positive-feedback action, the pituitary production of prolactin is unaffected by large doses of the steroid.

Effects

It is apparent in nonmammalian vertebrates that prolactin has functions other than those of a lactogenic hormone. In pigeons, prolactin causes the crop sac to secrete "crop milk." In tadpoles the hormone inhibits tail resorption and urea excretion, and promotes growth after hypophysectomy. Prolactin stimulates a water drive in the land form of a newt (the red eft), and in some teleosts cause melanogenesis. In some euryhaline fish transferred to fresh water after hypophysectomy, administration of prolactin makes survival possible.

GONADOTROPHINS

Regulation of Secretion

It is difficult to discuss one of the pituitary gonadotrophins, FSH or LH, without simultaneously considering the other, since functionally they complement each other. Spermatogenesis is possible only if both are secreted together, since FSH is needed in the intermediate stages of sperm development and interstitial-cell-stimulating hormone (ICSH) is needed to stimulate the androgen secretion necessary for sperm maturation. In the female, ovulation occur only if both FSH and LH are sequentially available.

The negative-feedback loop illustrates our current ideas on the regulation of gonadotrophin release in the male. The testosterone levels in the blood are thought to regulate activity of the neurosecretory cells that secrete the gonadotrophin-releasing hormones, FRH and LRH. Although the testes of some animals are rich sources of estrogens (for example, bull, stallion), the role of estrogens in male reproductive physiology and biochemistry is unknown. There is some evidence that the sex steroids directly affect the adenohypophysis; however, these pathways are not illustrated here.

Regulation of gonadotrophin secretion in the female is much complicated than in the male. The secretion of both FSH and LH occurs at a constant basal level. In the human an intense surge of LH and FSH occurs on approximately the fourteenth day of the menstrual cycle.

Rats maintained in animal rooms with light: dark cycles of 14 hr : 10 hr and with the lights going on at 5.00 AM and off at 7.00 PM have 4 day cycle. In such rats FSH is probably released simultaneously with LH on the day of proestrus between 2.00 and 4.00 PM. Hilliard has shown that in the rabbit a positive feedback system dependent on the LH stimulated secretion of 20α-hydroxypregn-4-en-3-one supports the secretion of LH until ovulation occurs.

Effects of Follicle-stimulating Hormone

The role of FSH is males in unclear since FSH given in small doses to hypophysectomized rats does not maintain either the germinal epithelium or testicular weight. However, FSH and LH adminstrered together in ratios of 4:1 to 400:1 do maintain testicular function. The combined action of these hromones probably involves stimulation of androgen secretion as final maturation of the sperm requires the presence of androgens.

In females, FSH acts on growing follicles to stimulate their maturation. The transition from an immature oocyte surrounded by a single layer of granulosa cells to a mature structure involves stimulation of mitotic activity and consequent growth of three different cell types: (1) connective tissue cells that form the outer layers of follicles; (2) theca internal cells in the middle layer; and (3) granulosa cells in the innermost layer.

Effects of Luteinizing Hormone

In males, ICSH causes the interstitial cells to secrete testosterone. There is little evidence to indicate cyclic release of ICSH from the adenohypophysis, except in seasonal breeders, for example, deer. In animals such as rabbit, dog, rat, and man, there is continuous stimulation of the interstitial cells.

In females, ovulation may either occur spontaneously as part of a cycle, or as a reflex in response to a stimulus (most commonly coitus). Rats, dogs, sheep, and most primates are spontaneous ovulators, whereas cats, ferrets, and rabbits are reflexive. The action of LH is some easily understood in the latter group. The initial response of a rabbit of LH occurs within seconds of the secretion of the hormone and constitutes a hyperemic reaction, perhaps due to a local release of histamine. The next response appears within minutes and is manifested by discharge of a progestin (20α-hydroxypregn-4-en-3-one) from the interstitial cells. The mature follicle is noticeably enlarged 6 hr after coitus and ruptures after 10 to 11 hr. Following the rupture, the granulosa cells hypertrophy and increase their secretory activities and the entire follicular space becomes filled with large lipid-containing cells, becoming a corpus luteum. Stimulation of steroid biosynthesis by ICSH (LH) in interstitial cells and the corpus luteum may involve the generation of c-AMP as a "second messenger" in a manner similar to the action of other hormones such as ACTH, TSH, vasopressin, and norepinephrine upon their targets.

The Lipotrophins

In 1936 Best and Campbell described a fat-mobilizing effect of anterior pituitary extracts. Subsequent preparations having specific lipotrophic effects were obtained, and in 1965 and 1966 Li and co-workers isolated to lipolytic polypeptides and determined their structures: one, with 90 amino acids and a molecular weight of 9,500 was designated β-lipotrophin (β-LPH); the second, with 58 amino acids and a molecular weight of 5,810, was named γ-liptrophin (γ-LPH). The amino acid sequences are shown in Fig. The γ-LPH molecule is represented in its entirety in the β-LPH molecule. These lipolytic polypeptides. ACTH, and MSH all share a common heptapeptide core, which exhibits MSH-type activity.

In vitro MSH activity of β-LPH is about twice as great as that of γ-LPH: the former has an activity of about 1.9×10^{11} units/mole and the latter about 0.9×10^{11} units/mole. The same two-to-one relationship of the two liptrophins is true for in vitro lipolytic activity in rabbit adipose tissue. In the rat, sheep γ-LPH is inactive while β-LPH is active at concentrations 10 to 50 times higher than in the rabbit by the lipolytic assay performed on adipose tissue of the two species.

Little is known about theregulation of secretion of the lipotrophic hormones or the role they play in the vertebrate organism.

Melanocyte-Stimulating Hormone

Regulation of Secretion

The dispersion of pigment in the frog skin is caused by MSH; concentration of pigment may be controlled by melatonin, produced by the pineal gland. Figure illustrates the presently conceived regulation of MSH secretion as being dependent on the release of an MSH-release-inhibiting hormone (MIH), produced in the supraoptic region of the hypothalamus. Melatonin may act on the hypothalamus to stimulate secretion of MIH and directly on the melanophore to cause pigment concentration around the nucleus and blanching of the skin.

Effects

Melanin-containing cells are generally of two types. In cold-blooded vertebrates, melanin is contained in *melanophores*. Skin colour is determined by the distribution of pigment in the cytoplasm; the skin is pale when the pigment is concentrated near the nucleus and dark when widely dispersed through the cytoplasm of the cell. In birds and mammals, melanin is contained in *melanocytes*. These cells synthesize and store their pigment, and can contribute it to adjacent cells and structures. The α-MSH causes increased synthesis of melanin in both melanophores and melanocytes and dispersion of melanin in melanophores. The role of α-MSH and β-MSH in mammals is not clear. In cold-blooded vertebrates—fish, amphibia, and reptiles—α-MSH serves a protective function; the animal becomes pale on a light background and dark on a dark background. Adrenocorticotrophin can also cause darkening of the skin (bronzing of the skin in Addison's disease) in a manner similar to that of α-MSH.

The α-MSH exerts a thyrotrophic action on thyroid tissue and both α- and β-MSH cause an increased lipolysis of adipose tissue in vitro. These effects may not be physiological important.

The Neurohypophyseal Hormones

The secretion associated with the neurohypothysis and produced in the hypothalamus. Seven natural cyclic nonapeptides have been identified in vertebrate neurohypophyseal extracts: oxytocin, vasotocin, mesotocin, ichthyotocin (isotocin), glumitocin, arginine vasopressin, and lysine vasopressin. In reptiles, birds, and mammals, these secretions originate in the supraoptic and paraventricular nuclei, while in fish and amphibia they originate in the preoptic nuclei. The hormones are transmitted through the axons of the neurons that compose these nuclei to the capillary beds in the neurohypophysis. There the hormones are stored in granules accumulated at the axon terminals; they must be released from these granules (neurophysins) before they can be secreted. The neurophysins may be involved in transport and storage of the hormones. Three hormone-binding proteins, neurophysin I, II and III, have been isolated. Neurophysin III appears in smaller amounts than either I or II. The mechanism that causes hormone release is not known; the axon terminals may abut on either the basement membranes of capillaries or cells surrounding the capillaries. In mammals, each hormone is secreted independently from its own hypothalamic nucleus; the supraoptic nucleus has been associated with vasopressin secretion and the paraventricular nucleus with oxytocin secretion.

Regulation of Vasopressin Secretion

Regulation of vasopressin (antidiuretic hormone) secretion may be initiated by a number of different pathways converging on the hypothalamus. Alteration of plasma osmotic pressure is probably the primary regulatory mechanism; however, alteration of blood volume or blood pressure, which activates volume receptors in the right atrium or baroceptors of the carotid arteries, also contributes to the control of vasopressin secretion. An increase in blood volume and consequent distention of the right atrium causes a diuresis, while hemorrhage, when accompanied by a reduced blood volume and pressure, causes water retention. Extrinsic factors such as pain, which affect the nervous system, induce an emotional stress inhibiting the secretion of vasopressin.

Regulation of Oxytocin Secretion

Although oxytocin has no demonstrated role in males, it has a well-established role in females. In pregnant animals parturition can be induced and brought to completion with oxytocin. During parturition the plasma oxytocin in cows increases from 55 μU/ml (microunits per milliliter) in the first stage of labor to between 10 to 20 times this amount during the second stage, and falls to control levels 2 hr after delivery. In second-stage labor in women, there is a threefold increase in plasma oxytocin concentration (from 50 to 150 μU/ml). Oxytocin release is inhibited by progesterone in rats, rabbits, pigs, and cows, but not in sheep or women.

Oxytocin also is important in the lactating female since in its absence no milk is releated from the mammary gland. Ejection of milk occurs after administration of oxytocin or after the application of an electrical stimulus to the paraventricular nucleus, infundibulum, or any point on the hypothalamohypophyseal tract. Secretion of oxytocin can be prevented by anesthetizing the suckled teat, cutting the innervation of the gland, placing lesions in the paraventricular nucleus, or sectioning the pituitary stalk.

Effects of Vasopressin and Oxytocin

These hormones are capable of causing similar responses of different magnitudes. In physiological doses, vasopressin exerts an antidiuretic effect; in larger doses, it causes contraction of smooth muscle, especially in the blood vessels. At present, oxytocin is thought to be important only in the pregnant and lactating female, where it causes contraction of the uterine muscle and ejection of milk. The mechanism by which it acts on smooth muscle cells is not known; however, the speed of this response is thought to be due to a membrane-dependent phenomenon.

Vasopressin manifests its action largely by affecting water and urea permeability of renal nephrons, and frog and bladders. The absence of vasopressin in mammals results in diabetes insipidus, a condition in which large volumes of water are consumed and then lost in the urine. The most acceptable hypothesis to explain the action of vasopressin (and perhaps oxytocin) was proposed by Orloff and Handler, who suggested that the hormone triggers the activity of the adenyl cyclase system in cell membranes. This proposal is based on the observation that vasopressin increases concentrations of c-AMP in susceptible tissues. (c-AMP induces effects similar to vasopressin as does theophylline, which protects c-AMP against destruction by phosphodiesterase).

Thyroid Gland

The thyroid gland secretes two major iodine-containing hormones, thyroxine (or tetraiodothyronine, T_4) and triiodothyronine (T_3), both of which control and regulate the overall metabolic rate of the body cells and tissues. The principal function of the thyroid gland is to sustain an optimum metabolic level that is essential to the normal function of all the tissues of the body. The general functions of the thyroid hormones are to stimulate oxygen consumption by the cells, to participate in the regulation of carbohydrate and fat metabolism, and to exert important effects in the growth and maturation of the individual.

Thyroid secretion is regulated by thyrotropic (or thyroid stimulating) hormone of the adenohypophysis, as discussed earlier. In turn, the secretion of thyrotropic hormone is regulated partly by a direct negative feedback mechanism produced by elevated circulating thyroid hormone levels which act partly upon the anterior pituitary directly, and partly by a neural mechanism which operates through the hypothalamo-hypophyseal complex and whose effects are mediated via the thyrotropic hormone releasing factor.

Thus, homeostatic regulation of thyroid secretion is achieved in such a way that the metabolic requirements of the individual are adapted to alterations in the internal as well as external environments. In addition to thyroxine and triiodothyronine, the thyroid gland also secretes a third hormone known as calcitonin (or thyrocalcitonin, which reduces the plasma level of calcium and phosphate upon injection into experimental animals. Thus, calcitonin counteracts the hypercalcemia produced by excess parathyroid hormone secretion.

Functional Morphology of the Thyroid Gland

Gross Anatomy

The thyroid gland is located in the neck, anterior to the trachea. In normal adult humans this organ weighs roughly 32 gm (between 25 and 40 gm), and it consists of two lateral lobes which are bridged by a

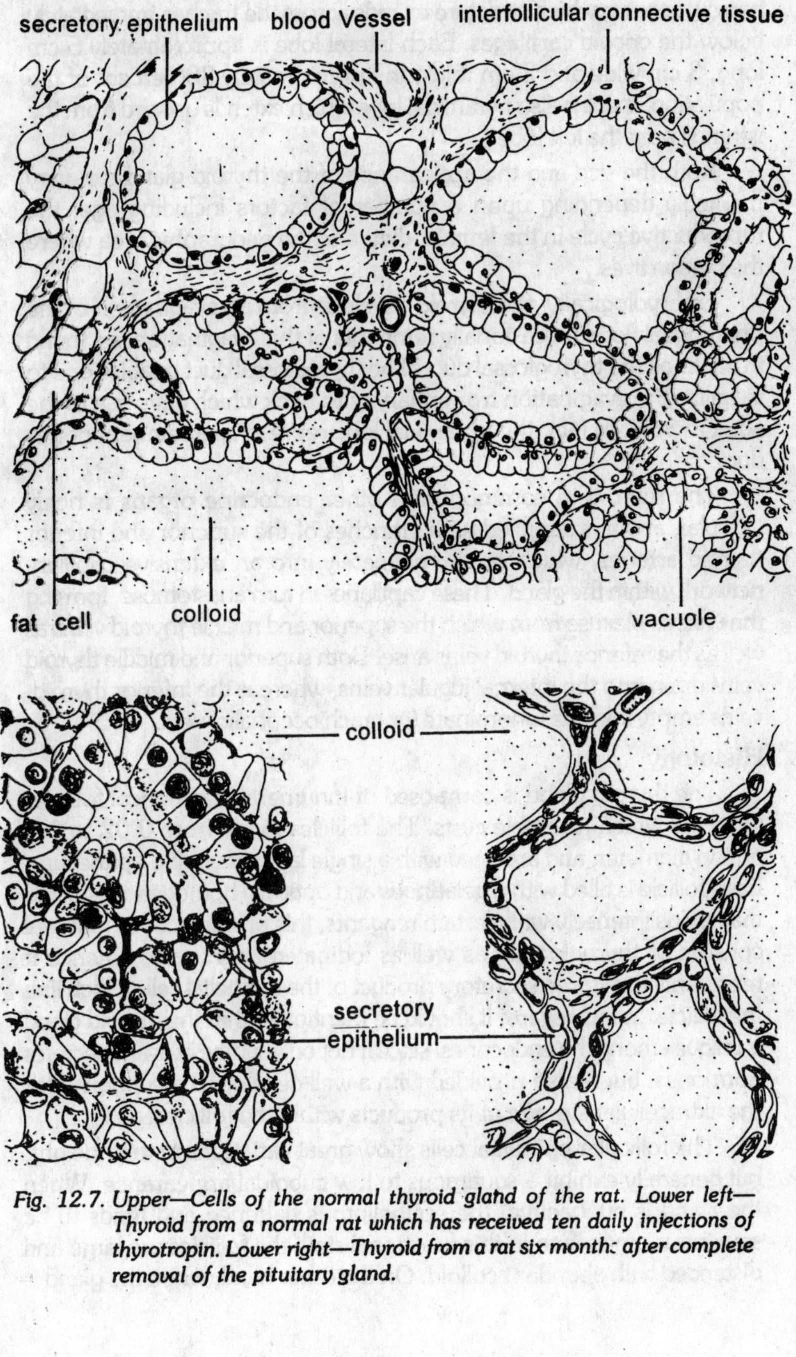

Fig. 12.7. Upper—Cells of the normal thyroid gland of the rat. Lower left—Thyroid from a normal rat which has received ten daily injections of thyrotropin. Lower right—Thyroid from a rat six months after complete removal of the pituitary gland.

narrow isthmus which structure exends across the trachea immediately below the cricoid cartilages. Each lateral lobe is approximately 5 cm long, 3 cm wide and 2 cm thick. In approximately 30 percent of the population, there is also pyramidal lobe which extends upward from the isthmus near the left lobe.

Both the size and the appearance of the thyroid gland can vary markedly depending upon a number of factors including age, the reproductive cycle in the female, diet, and the geographic area where the person lives.

Embryologically, the thyroid develops from an evagination of the pharyngeal floor. Occasionally remnants of this evagination are found in adults as the thyroglossal duct. The thyroglossal duct enables one to visualize the evagination from pharynx to neck which gave rise to the adult thyroid. In fact, the pyramidal lobe represents a remnant of this duct.

The thyroid in common with other endocrine organs is richly vascular, and it is supplied with branches of the superior and inferior thyroid arteries, which ramify ultimately into an extensive capillary network within the gland. These capillaries in turn anastomose, forming the venous plexuse from which the superior and middle thyroid veins as well as the inferior thyroid veins arise. Both superior and middle thyroid veins drain into the internal jugular veins, whereas the inferior thyroid, veins empty into the innominate (or brachiocephalic) vein.

Histology

The thyroid gland is composed of innumerable spherical follicles (or acini) which resemble cysts. The follicles range from 0.02 to 0.9 mm in diameter, and are lined with a single layer of simple epithelium. Each follicle is filled with a gelatinous and optically homogenous colloid that stains intensely with certain reagents; this proteinaceous material consists of thyroglobulin as well as iodinated amino acids, hence it represents the stored secretory product of the epithelial cells which line the follicle. In this regard, it should be mentioned that the thyroid gland is unique among the endocrines, since it not only synthesizes and secretes hormones, but is also provided with a well-developed mechanism for the extracellular storage of its products within the follicles.

The follicular epithelial cells show great variations in their height, but generally exhibit a squamous to low cuboidal appearance. When the gland is hypoactive, the epithelium is flattened and tends to be squamous, and when in this functional state the follicles are large and distended with abundant colloid. On the other hand, when the gland is hyperactive, the appearance of the cells ranges from a cuboidal toward a columnar appearance, and the follicles are much smaller with correspondingly less colloid. There are many exceptions to these general statements, however, so that an accurate evaluation of the functional state of the thyroid gland solely from microscopic study is difficult.

The nuclei of the glandular (or principal) cells of the follicles are centrally located, spherical, and contain little chromatin. One or several nucleoli may be present. The basophilic cytoplasm contains numerous thin rodlike mitochondria as well as an abundance of fat droplets and a highly developed endoplasmic reticulum, a feature common to most glandular cells.

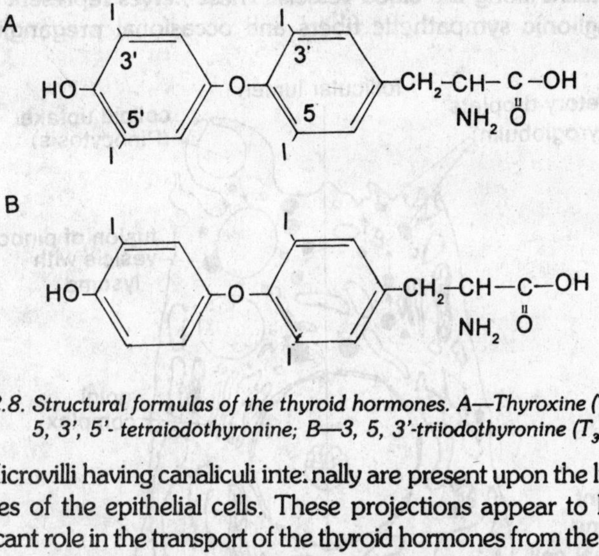

Fig. 12.8. Structural formulas of the thyroid hormones. A—Thyroxine (T_4) or 3, 5, 3', 5'- tetraiodothyronine; B—3, 5, 3'-triiodothyronine (T_3).

Microvilli having canaliculi internally are present upon the luminal surfaces of the epithelial cells. These projections appear to have a significant role in the transport of the thyroid hormones from the follicle to the capillary blood. Lysosomes having a diameter of about 0.5 μ are abundant in the cytoplasm.

In addition to the principal cells of the follicles, there is present in the thyroid gland a much smaller population of larger cells that commonly are known as parafollicular cells, clear (C) cells, or "mitochondria-rich" cells. These structures stain less deeply in routine histologic preparations than do the principal cells, hence the name clear (or C) cells is applied to those elements. The parafollicular cells are found interspersed among the principal cells of the follicles as well as in an inter-follicular location. It has been demonstrated that the parafollicular cells have a quite different biochemical structure than the principal cells. For example, the enzyme patterns are different in both cell types. The parafollicular cells have been shown to be associated with synthesis of the hormone thyrocalcitonin whereas the principal cells elaborate the other thyroid hormones, viz., thyroxine and triiodothyronine. The parafollicular cells of the thyroid gland also are similar to the calcitonin-secreting cells of the parathyroid glands.

Surrounding each follicle is an exceedingly thin basement membrane or lamina which is about 500Å thick, as well as a delicate reticular fiber network. Each follicle is also invested with a plexus of fenestrated capillaries, and between the capillary networks of adjacent follicles are found the blind-ending terminations of lymphatic vessels. Nerve fibers are abundant along the blood vessels. These nerves represent both postganglionic sympathetic fibers and occasional preganglionic

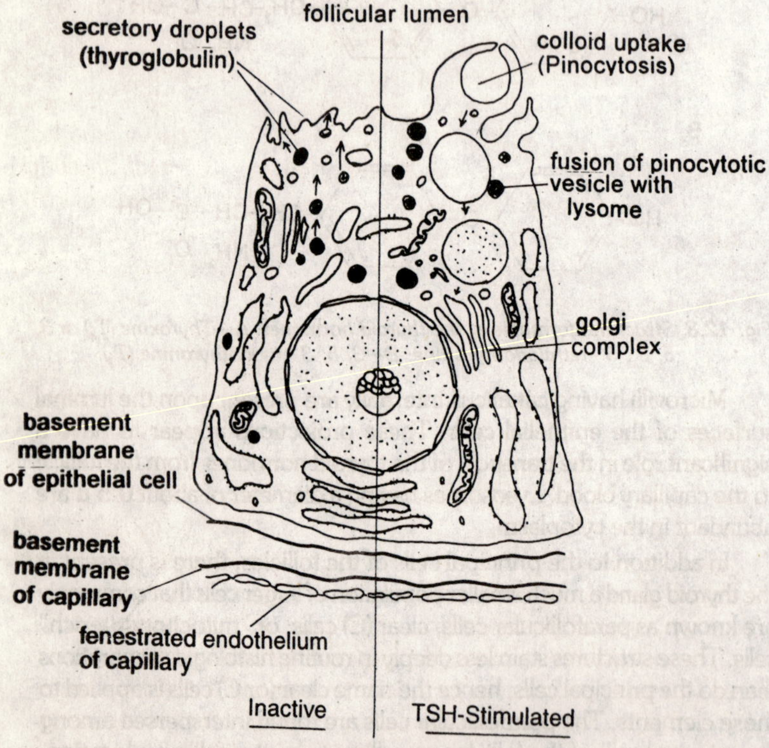

Fig. 12.9. Ultrastructure of a thyroid cell under normal conditions (left side) and following prolonged administration of thyroid stimulating hormones.

parasympathetic fibers also are found. The nerves that innervate the thyroid apparently serve a vasomotor function, as extrinsic denervation of the thyroid by transplantation to an ectopic locus in the body does not affect the secretion of hormones by this organ.

Metabolic Effects of Thyroid Hormone

Calorigenesis

Homoiotherms. In warm-blooded animals the most characteristic effect of thyroid hormone is to increase the energy production and oxygen consumption of most normal tissues. The basal metabolic rate (BMR) falls rapidly after thyroidectomy or hypophysectomy and may be elevated above the normal level by administering thyroid hormone. The increased metabolism of hyperthyroid animals is reflected in the accelerated respiration of its excised tissues; tissues taken from thyroidectomized animals respire at a comparatively low rate. After administration of thyroid hormone to the animal, the respiration of surviving liver, kidney, skeletal muscle, cardiac muscle, gastric mucosa, and diaphragm is strikingly increased. Several workers have found that the respiration of surviving brain, spleen, and testis is not enhanced by hormone administered to the animal and that the respiration of thyroid tissue is decreased by the same procedure. A number of enzyme systems, notably cytochrome oxidase, cytochrome c, and succinoxidase, increase in the tissues after thyroxine administration to the animal and are diminished by thyroidectomy or thiouracil feeding.

Calorigenesis in the nornal organism is regulated by a balance of hormonal factors, thyroid hormone not being the only hormone involved. Heat production in the rat declines more abruptly after hypophysectomy than after thyroidectomy. Thyroxine and triiodothyronine elevated the BMR more effectively in thyroidectomized animals than in hypophysectomized animals. Doses of TSH sufficient to causes thyroid hypertrophy in hypophysectomized rats increase the BMR but do not restore it to normal. It is apparent that the anterior pituitary affects calorigenesis in some manner in addition to its production of TSH. Coticotrophin (ACTH), acting to promote the secretion of adrenal steroids, appears to be an important factor. Certain adrenal cortical hormones (and ACTH) are calorigenic in the absence of the pituitary, of the thyroid, or of both glands. There is a gradual decline in ACTH production after thyroidectomy, and the adrenal cortices undergo considerable involution. When thyroxine is administered to the thyroidectomized aniamal it not only restores the functional capacity of the adrenal cortex but also promotes the release of somatotrophin (STH) by the anterior pituitary and intensifies the action of this hormone on the peripheral tissues. The decline in growth rate that follows thyroidectomy of the young rat parallels the decline in BMR. Hypophysectomized rats stop growing almost immediately, whereas thyroidectomized animals are capable of growing at a reduced rate. It is probable that the immediate loss of STH after hypophysectomy accounts, in part at least, for the more rapid decline in BMR and growth than occurs after thyroidectomy.

Experiments on the rat provide evidence that STH augments the action of exogenous TSH on the thyroid gland of the hypophysectomized rat, both hormones increasing calorigenesis more effectively than when TSH is given alone. STH also potentiates the calorigenic action of thyroxine in hypophysectomized rats, indicating that the calorigenic effect of STH is exerted in part at the level of the peripheral tissues.

Epinephrine, a hormone of the adrenal medulla, also a calorigenic effect. It has been proposed that an important role of thyroxine in cold acclimatization of the rat is to potentiate the calorigenic action of endogeneous epinephrine. Observations of this type provide an excellent example of the interaction of multiple hormones in conditioning metabolic processes.

Poikilotherms

While thyroid hromones are identifiable in the thyroids of all vertebrate classes, it has been difficult to demonstrate calorigenic effects in the poikilotherms, or cold-blooded vertebrates. Both positive and negative results have been reported with fishes and amphibians. Numerous observations of fishes seem to imply that the piscine thyroid is more involved in the metabolism of salts and water than in oxidative metabolism. The thyroid gland of the goldfish (*Carassius auratus*) appears to be very inactive throughout the year and at no season does it undergo hypertrophy after the administration of thiourea and other goitrogens. It does, however, respond strongly to the administration of TSH. It has been suggested that the inactive thyroid in this species may enable it to tolerate an exceptionally wide range of temperatures (0 to 41°C). The thyroid of *Triturus viridesceus,* an aquatic urodele, is more inactive than that of *Desmognathus fuscus,* a terrestrial urodele, and the thyroids of the two species respond differently to goitrogenic agents. The thyroidal differences may correlate with the temperature ranges of the two amphibians. Thyroidectomy of the teleost parrot fish does not adiminish oxygen consumption, nor does the administration of thyroxine elevate it.

While birds and mammals can maintain their body temperatures in the cold, the reptile can only demonstrate a temperature sensitivity. There is evidence that the turtle, a poikilotherm, is sensitive to temperature changes, although it cannot regulate its body temperature. Exposing turtles to cold stress produces eosinopenia, whereas epinephrine injections cause eosinophilia. Cold stress appears to activate nervous and hormonal mechanisms, but the animal fails to maintain a fixed body temperature; this may indicate that the stress hormones cannot stimulate heat production or that the cannot animal reduce heat loss at low temperatures.

There are strong indications that environmental temperature may be a very important factor in conditioning thyroidal responses in the poikilotherms. If the lizard, *Anolis carolinensis,* is maintained at room temperature (20 to 24°C), thyroidectomy, or the administration of thyroxine or TSH, has no effect on oxygen consumption. When these lizards are maintained at a constant temperature of 30°C, thyroidectomy reduces oxygen consumption, and thyroxine or TSH elevates it.

Hibernation

In warm-blooded animals that are active throughout the year, the thyroids are generally more active during the cold months than during the summer. However, in hibernating mammals, the thyroid tends to be inactive during the winter and reaches a high peak of activity in the spring as the animals emerge from hibernation. Most observations indicate that the reverse is true among hibernating poikilotherms. In hibernating frogs and toads, for example, the thyroid is moderately active during the winter and is most inactive during the summer.

Electrolytes and Water

In hypothyroidism there is an extracellular retention of sodium and chloride and water, the blood volume being considerably reduced. When thyroid hormone is administered under these conditions, it causes diuresis and urinary loss of sodium. The plasma volume is elevated. Administration of the hormone to normal subjects also produces excessive water loss through the kidney, but under these conditions the urine is especially rich in potassium. This suggests that there has been a mobilization of intracellular fluid. When small amounts of thyroxine are given to young growing animals, they enhance calcium retention. This is probably a secondary effect resulting from the protein anabolic action of the hormone, which facilitates the deposition of new bone matrix. In hyperthyroid states there is increased mobilization of calcium from the skeleton and increased loss through the urine and feces. The calcium concentration of the blood is not appreciably changed.

Protein Metabolism

One of the most characteristic effect of cretinism and myxedema in human subjects is the deposition of a mucoprotein substance in the skin and other tissues. This protein is thought to be the same as that which constitutes the intercellular cement normally found between tissue cells. It is similar in composition to the mucin in which the fetal tissues are embedded. Thyroid hormones relieve the excessive accumulation of mucoprotein in the tissue of hypothyroid subjects.

Thyroid hormone may produce either a protein anabolic or a protein catabolic effect. Human cretinism may be simulated experimentally by removing the thyroid gland of the young animal or by feeding antithyroid drugs. Cretinic rats may be produced by feeding thiouracil to the pregnant mother and continuing it during the course of lactation. At the time of weaning (22 days) the cretinic rats weigh about 19 gm. as compared with 42 gm. for normal controls. The subnormal growth of these animals is due partly to the voluntary reduction of food intake consequent upon a low BMR, but it also suggests a deficiency in protein synthesis. The carcass of young thyroidectomized rats contains less protein and more fat than normal controls. When such animals are force-fed to cause a weight gain, there is a further reduction in carcass protein and a further increase in carcass fat. Although thyroid deficiency favors increased nitrogen excretion, thyroidectomized rats are able to maintain a positive nitrogen balance. Profound shifts in protein stores occur in the thyroidectomized animal. The total plasma protein of the rat increases as much as 30 percent after thyroidectomy, whereas the liver and kidneys are small and deficient in protein. The deficiencies of hypothyroidism can be corrected by a adequate doses of thyroid hormone. Nitrogen retention is obtained and growth occurs at a normal rate; liver enlargement and the elevated plasma globulin are corrected.

A protein anabolic effect of thyroxine has been demonstrated in the young hypophysectomized rat. The growth of nearly all tisues may be stimulated by thyroxine in the absence of pituitary hormones. The protein content of the pelt, including hair, is greatly increased. Physiologic doses of thyroxine must be employed in order to demonstrate protein anabolic effects; toxic amounts of the hormone do not stimulate growth.

Fat Metabolism

Hypothyroidism is generally associated with a rise in serum cholesterol and phospholipid, which may be restored to normal levels by administering thyroxine. Since thyroid hormone increases the oxidative processes of the body, the carbohydrate stores may be depleted. If this occurs, fat is mobilized from the tissue deposits to the liver, and the rate of ketone body formation is augmented. Depot fat is usually not increased appreciably during thyroid deficiency because of reduced caloric intake.

Carbohydrate Metabolism

Thyroid hormone accelerates the rate of absorption of monosaccharides from the alimentary tract. Liver glycogen stores are diminished as a consequence of hepatic glycogenolysis, and the blood sugar levels tend to rise. The hyperglycemic effect of thyroid hormone is partially offset by the simultaneous increase in the oxidation of sugar by the tissues. Hypothyroidism increases hepatic glycogen without altering the fat content. The administration of thyroid hormone can produce diabetes mellitus in animals after the removal of a portion of the pancreas, but it does not produce diabetes in animals having an intact and healthy pancreas. Insulin requirements generally increase in hyperthyroidism and decrease in hypothyroidism. The administration of thyroxine or triidothyronine to the rat increases the rate of degradation of insulin-1, and removal of the thyroid decreases it. Pancreatic diabetes in the cat and dog is not ameliorated by thyroidectomy, as it is by hypophysectomy or adrenalectomy.

Effects of Thyroid on Reproduction

There is ample evidence that normal reproductive functions depend upon thyroid activity, though there appear to be many species and age differences. Testicular functions are more easily impaired in young animals than in adults. Histologic defects have been observed in the testes of cretinic rats, produced by feeding thiouracil to the mohter during pregnancy and lactation. At 40 days of age the testes from the cretinic animals weigh about 112 mg. as against 893 mg. for the controls. The cretinic tests contain spermatozoa, but the infantile nature of the sex accessory organs indicates that androgens are not being secreted at this age. Hypothyroidism appears to delay sexual development in the young male and toxic levels of thyroxine to impair reproductive functions. However, moderate amounts of thyroid hormone have been reported to stimulate spermatogenesis in the rabbit, mouse, ram, and human subject, even when their is no demonstrable thyroid deficiency.

Thyroid deficiencies or excesses in the female may impair the ovaries and cause the cycles to cease or to become irregular. Reproduction in most species may occur after removal of the thyroid, but fecundity is usually subnormal. The litter size of hypothyroid rats is reduced and many of the young die because of insufficient lactation by the mother. There is evidence that thyorid deficiency may render the ovary particularly susceptible to cyst formation, a disorder frequently associated with infertility in man and domestic animals. At 10 days of age the ovaries of cretinic rats do not contain vesicular follicles and lack lipid and cholesterol. When equine gonadotrophin is administered to the cretinic animals, the ovaries respond by forming large follicles without luteinization; normal ovaries respond to this gonadotrophin by forming both follicles and corpora lutea. Thus hypothyroidism in the brings about some change in the ovary that favors the development of cystic follicles.

Little is known about the exact manner in which thyroid hormones influence reproductive processes. Some of the impairments may be consequent upon disturbances in protein metabolism; others may be attributable to pituitary malfunctions. The anterior pituitary secretes large amounts of TSH after thyroidectomy, and under these conditions the gland enlarges and vacuoles appear in the basophils (thyrotrophs). The vacuolated basophils are generally regarded as active rather than degenerating elements. Thyroxine not only reinstates growth in the thyroidless rat but also repairs the pituitary defects. In the hypophysectomized animal, thyroxine increases the rate of oxygen consumption, accelerates the heart rate, etc., but it does not restore normal body growth, adrenocortical growth, or reproductive development. These facts suggest that thyroid hormones exert some control over the release from the hypophysis of somatotrophin, corticotrophin, and gonadotrophins. There is no direct evidence indicating whether these defects of the thyroidectomized animals are due to reduced ability of the pituitary to secrete gonadotrophins, ACTH, and STH or to an inability of the target organs to respond normally in the absence of thyroid hormone.

Growth and Differentiation of Tissues

Most vertebrates cannot attain normal adult form and dimensions in the absence of thyroid secretions. There is evidence that thyroid hormone and somatotrophin act synergistically in promoting normal skeletal growth. The arrest of endochondral ossification that occurs in the young hypophysectomized rat may be prevented by the administration of thyroxine. In the absence of pituitary hormones, thyroxine permits the erosion of cartilage and its replacement by bone to continue, but chondrogenesis itself is not maintained.

Thyroxine has been shown to have a direct effect on skeletal tissue of the chick grown *in vitro*. In these experiments it was found that the same concentration of hormone may be stimulatory or toxic depending upon the stage of differentiation of the tissue and the particular bone rudiment treated.

Thyroid hormone is essential for the metamorphosis of amphibian larvae, and this response has provided a useful method for bioassaying the hormone. The capacity of thyroid hormone to promote tissue differentiation is not directly related to its calorigenic action. Dinitrophenol increases the metabolic rate, but it does not produce this morphogenetic effect. Furthermore, certain stages of amphibian metamorphosis are promoted by thyroid hormone without being accompanied by an elevated metabolic rate.

Effects on Cardiovascular and Nervous Systems

In hypothyrodism the heart enlarges, and its rate and amplitude are diminished. Under these conditions the mass movement of blood is less than normal. The isolated perfused heart increases in rate and stroke volume when thyroxine is added to the perfusion medium. In hyperthyroidism there is typically a tachycardia, increased stroke volume, peripheral vasodilation, and an increase in pulse pressure.

Numerous controlled experiments have demonstrated that thyroid hormone sensitizes the nervous system. There is good correlation between the basal metabolic rate and the rate of emanation of alpha waves from the brain. That a functional impairment of the nervous system results from hypothyroidism is indicated by the lowered resistance to narcotics, the increased thresholds to light and sound stimulations, the diminished alpha rhythms, and the decreased speed of reaction to electric shocks.

In hyperthyroid patients, as well as in rats receiving thyroxine injections, the BMR is elevated but the oxygen consumption of the brain remains normal. Dinitrophenol elevates the BMR but has no effect on brain excitability. These facts indicate that the increased excitability of the central nervous system after thyroid hormone administration is separable from the metabolic effects of the hormone. Changes in electrolyte distribution brought about by the hormone appear to be associated with brain excitability. Triiodothyronine is about five times more effective than thyroxine in increasing brain excitability in the rat. However, the effects of triiodothyronine are more transient than those of thyroxine.

Hormones and Molting

In mammals without seasonal molts (e.g., mouse and man), the replacement of hair is an autonomous property of the hair follicle, but endocrine factors influence the cycle to some extent. Thyroidectomy of the mouse retards hair replacement, whereas adrenalectomy or ganadectoxy accelerate it. Seasonal molting is characteristic of many mammals (e.g., ferret), and in these species, neuroendocrine mechanism play a very important role in the mediation of environmental stimuli.

Molting has been extensively studied in birds, and it is very complex process involving the interaction of genetic and neuroendocrine factors. There are many species differences, and not all areas of the body respond in the same way. Throxine acts to promote molting, though in some species it is not absolutely essential; thyroidectomy has the reverse effect. Estrogens and androgens counteract the molt inducing effects of thyroxine; progesterone encourages molting; especially in those species that breed throughout the year. Prolactin also causes molting in some birds. Seasonal molting is apparently controlled by external factors, acting *via* the hypothalamus, to induce variations in the endocrine system.

Molting in amphibians is a dual process involving (a) an inherent tendency of the skin to stratify and to keratinize and (b) the periodic removal of the slough which is regulated by neuroendocrine mechanisms. The specific hormonal controls differ in the anurans and urodeles.

Toads (Bufonidae) molt at intervals varying from one to two weeks, each animal maintaining its own individual rhythm rather precisely under uniform environmental conditions. Normal molting is abolished by hypophysectomy and is not restored by the administration of thyroxine. Adrenocortical steroids, on the other hand, induce molting in hypophysectomized toads. Molting can be produced in toads carrying ectopic autotransplants of the anterior pituitary by administering vasopressin. The neurohypophysial principle apparently stimulates the graft to release ACTH, which causes the adrenocortical tissue to discharge its steroids. Autonomous cyclic processes seem to be inherent in the skin, but these the sensitive to varying levels of hormones in the circulation. The hormones of the steroidogenic component of the adrenal, rather than those of the thyroid, seem to be the major ones affecting the anuran integument.

Among urodeles, the thyroid hormones seem to perform a more important role in the molting process. Thyroidectomy or hypophysectomy prevents molting in urodeles; after these operations, the keratinized layers accumulate but are not sloughed. TSH induces shedding when administered to hypophysectomized urodeles; exogenous thyroxine is effective both in hypophysectomized and thyroidectomized animals. Prolactin has been reported to produce shedding in hypophysectomized urodeles. The implantation of thyroxine pellets into the skin of thyroidectomized newts produces localized molting. This indicates that thyroxine acts directly upon the skin in these amphibians.

Role of Thyroxine in Amphibian Metamorphosis

The induction of amphibian metamorphosis is one of the most dramatic effects of thyroid hormone, illustrating its role in both differentiation and development. Typical of amphibia, embryonic development results in a tadpole larval form that differs greatly from the adult forg in both structure and habitat. This transition of the tadpole to the frog during a discrete period of postembryonic change is known as metamorphosis; this process is quite distinct from postmetamorphic growth and maturation. Amphibian metamorphosis has become a model system for the study of comparative biochemistry and cellular differentiation, especially, since it illustrates a series of remarkable biochemical and structural adaptation. In 1912, F.G. Gudernatsch showed that feeding thyroid glands to tadpoles initiated metamorphosis; this effect was eventually traced to T_4. Later B.M. Allen demonstrated the pituitary control over the secretion of thyroid hormone. More recently, W. Etkin and A. Voitkevitch showed that pituitary activity is in turn dependent on signals from the hypothalamus in the brain. If any of these tissues—thyroid, pituitary, or hypothalamus—is destroyed, normal metamorphosis is arrested. Metamorphosis can be induced by T_3, T_4, orTSH. These hormones can even induce similar transformations in other amphibian species—for example, the Mexican axolotl—which does not metamorphose under normal conditions.

The extensive anatomical transformations during metamorphosis were described by biologists even before 1900. These changes reflect many millions of years of evolution. The frog tadpole resembles a fish, with a streamlined body and tail, gills, lidless, eyes, thin skin, and a mouth suited to eating aquatic plants; within a period of several months it loses its fish like characteristics and changes to a young frog, a versatile land animal with lungs, eyelids, thick skin, large limbs with powerful muscles, and a mouth and tongue adapted for capturing insects. Biochemical studies on the role of thyroid hormones in this process have been conducted only within the last 15 yr. Of the numerous biochemical systems known to be extensively modified salientian metamorphosis, only several of the more dramatic molecular changes are summarized below:

Change in Blood Proteins

An extensive adaptive change in homeostatic mechanisms that is mediated through the blood precedes the shift in environment. During metamorphosis, the amounts of serum albumin, copper protein, ceruloplasmin, and numerous other proteins greatly increase as a result of extensive changes in liver function. Tadpole hemoglobins are replaced by frog hemoglobins with a considerably lower oxygen-blinding capacity, more suited to the new aerobic environment. Comparisons of these proteins reveal a number of significant differences in their chemistry. For example, in contrast to human fetal and adult hemoglobins, the tadpole and frog appear to have no hemoglobin polypeptide chains in common. Tadpole hemoglobins are unique in having no sulfhydryl groups, whereas frog hemoglobins have eight per molecule. Also, the site of reticulocyte formation shifts from the kidney (and possibly the liver) in the tadpole to the spleen in the frog.

Increase in Tail Hydrolases

During metamorphosis, disappearance of the tadpole tail is accompanied by large increases in concentrations of numerous hydrolytic enzymes, including cathepsins, deoxyribonuclease (DNase), ribonuclease (RNase), β-glucuronidase, phosphatases, and so forth. While other proteins disappear, these enzymes accumulate and produce tissue resorption.

The Shift from Ammonia to Urea

A remarkable shift in the nitrogen-excretion pattern of the amphibian takes place during metamorphosis. Over 90 percent of the tadpole's nitrogen-excretion product is ammonia, characteristic of aquatic forms; this shifts to 90 percent urea after metamorphosis. Accompanying this shift from ammonotelism to ureotelism is a major increase in enzymes of the Krebs ornithine-urea cycle. One of the enzymes of this cycle, argininosuccinate synthetase, has been identified as the rate-limiting enzyme in the biosynthesis of arginine and, in turn, urea.

Basic Initiating Mechanism in the Liver

Many of the changes mentioned above occur in the tadpole liver (such as those in the serum protein and the urea-cycle enzymes), reflecting a major metabolic reorganization of this tissue during metamorphosis. In recent years, the liver has been utilized in the search for the basic mechanism involved in the T_4 and T_3 effect on metamorphosis. As a result of the studies of Cohen, Frieden, and Tata, and their associates, it has been shown that T_3 and T_4 can induce an increased RNA biosynthesis prior to many of the changes in protein metabolism cited above. Thus, in common with the mechanisms cited for the effects of thyroid hormones on liver tissue in the growing rat, RNA-and protein-biosynthetic mechanisms have been shown to be stimulated by thyroid hormones in the metamorphosis amphibian.

Neoteny and the Thyroid

It is well known that certain urodeles fail to metamorphose and retain the larval body form throughout life. The animals become sexually mature and are able to reproduce without adults ever appearing. Neoteny was first described in the American axolotl, *Ambystoma tigrinun,* in which it was noted that member of this species inhabiting high mountain lakes failed to metamorphose, whereas specimens of the same species living at lower altitudes metamorphosed as do most other amphibians. Axolotls can be induced to metamorphose by treating them with thyroxine or various iodine compounds. Inorganic iodine is also effective if it is implanted under the skin or into the peritoneal cavity so that it is rapidly absorbed. These observations suggest that the failure of metamorphosis in the axolotl is due to the inability of the thyroid to secrete hormone or to a decreased sensitivity of the tissues to the hormone.

It is reasonably certain that the Mexican axolotl, *Ambystoma mexicanum,* never undergoes metamorphosis in its natural habitat. When pituitary glands of *A. tigrinum,* a readily metamorphosing form, are transplanted into *A. mexicanum* larvae, metamorphosis is induced. On the other hand, pituitaries of *A. mexicanum* do not promote metamorphosis when transplanted to the larvae of *A. grinum.* This indicates that the pituitary gland of the Mexican axolotl is defective and apparently does not release the proper thyrotrophin.

The perennibranchiate salamanders (e.g., *Necturus, Amphiuma*) are not known to undergo metamorphosis. They retain gills, tail fins, and other larval traits throughout their lives. Even large doses of thyroid hormone or other iodine compounds fail to produce any appreciable morphologic change in these species. The indications are that the perennibranchiate amphibians retain larval characters chiefly because their tissues have not developed a responsiveness to thyroid hormone. Studies have shown that the thyroids of *Necturus* and *Amphiuma* are very slow in accumulating radioiodine and in synthesizing thyroid hormone. In salamanders, as well as in certain fishes and reptiles, great variations are observed in the capacity of the thyroid to bind iodine and synthesize hormone, complex factors such as seasonal changes and the accompanying fluctuations in endocrine balance undoubtedly being involved.

Metamorphosis in Fishes

It is well known that cyclostomes and certain teleost fish undergo metamorphic changes during their life histories, but the hormonal mechanisms involved are not so clear as in amphibians. Metamorphosis of the lamprey is striking, but it has not been possible to influence this process through the administration of exogenous thyroxine to the larvae. There are some reasons for believing that adrenocortical hormones, rather than thyroid hormones, may be involved in the metamorphic process of cyclostomes.

While metamorphosis in teleost fishes is not so impressive as in amphibians, it does occur in certain species. In the salmon, for example, the freshwater parr transforms into the migratory, marine smolt. This involves a silvering of the integument, due to the deposition of guanine, and certain physiologic adjustments which equip the smolt for life in the sea. These transformations are accompanied by increased metabolic rate and hyperactivity of the thyroid. These facts imply that thyroid hormones are important in bringing about these changes in salmonids, though it is probable that other hormones play some part.

Antithyroid Agents (Goitrogens)

A number of chemical agents can interfere, in one way or another, with the thyroid mechanisms involved in the synthesis of hormones. These antithyroid substances have the common property of causing a fall in the blood level of thyroxine, and this leads to an augmented output of thyrotrophin (TSH) by the anterior pituitary. The latter hormone causes hypertrophy of thyroid tissue and, if there is conspicuous enlargement of the thyroid it may be termed *goiter*. Hence these agents are sometimes called goitrogens.

Parathyroid Glands

The first anatomic description of the parathyroids was provided by Sandstrom. He dissected such domesticated mammals as the dog, cat, rabbit, and ox; it is apparent now that the parathyroids lying some distance from the thyroid lobes escaped his attention. Glay (1891) discovered that the rabbit possesses an external pair of parathyroids, in addition to the pair that is closely applied to the thyroid lobes, and demonstrated that total thyroidectomy (external parathyroids remaining intact) did not result in tetany. This was the first clear demonstration that the thyroid and parathyroids have separate and different functions.

Almost all the information available on parathyroid functions: is confined to a limited number of eutherian mammals. There has been a conspicuous lack of comparative research on nonmammalian species. This may be due in part to the impression that parathyroids are not very important in poikilothermic vertebrates since they seem to be absent in fishes, the awareness that only mammalian extracts are available, or to surgical difficulties attendant upon the of small organ whose positions may be variable.

Anatomy of the Parathyroid Glands

The human parathyroids are flattened, ovate, or pyriform bodies situated on the posterior surfaces of the lateral lobes of the thyroid, near their mesial edges. There are usually four glands in man, a superior and an inferior pair. The number and location of the glands are quite variable in all vertebrates that have them. Occasionally, some of them may be buried deeply in the substance of the thyroid or thymus, and this association stems from the close origin of these structures in the embryo. Though one or both of the inferior parathyroids may be located some distance caudad of the thyroid, they usually follow the branches of the inferior thyroid arteries. Parathyroid rests may be scattered widely in the fat and connective tissue of the neck or in the mediastinum. In an adult they average from 6 to 8 mm in length and from 3 to 4 mm in breadth.

The blood supply is chiefly from the inferior thyroid arteries, but a certain amount of blood may be received through anastomosing branches of the superior thyroid arteries. Veins from the parathyroids enter those that drain the thyroid gland. Although the vascular supply is rich, the nervous connections are very meager and are probably vasomotor. Secretory nerve fibers have not been demonstrated, and transplanted glands appear to function normally in the absence of the usual nervous connections. The parathyroids undergo no conspicuous change after hypophysectomy and do not appear to be controlled by any of the pituitary trophic hormones.

Each parathyroid is surrounded by a connective tissue capsule from which septa extend into the substance of the gland and imperfectly divide it into lobules. The substance of the gland consists of densely packed masses and cords of epithelial cells between which are interspersed numerous small blood vessels.

In glands from adult human beings, principal cells and oxyphile cells are distinguishable. The principal cells are the more numerous and probably are the source of the parathyroid hormone. The oxyphile cells of the human gland appear between the fourth and seventh years and are said to increase in number after puberty. The parathyroids of the horse are cytologically more comparable to human glands than those of other mammals that have been studied. Follicle-like groups of cells enclosing colloid are found infrequently in the parathyroids; these resemble thyroid follicles but the colloid does not contain iodine.

Seasonal changes have been described in the parathyroids of amphibians. The summer glands of *Rana pipiens* are composed of spindle-shaped cells arranged in whorls; during the winter, the cells seem to lose their functional integrity and undergo a type of reticular degeneration. The glands of the bullfrog (*R. catesbeianna*) undergo extreme cytolysis during the winter, and replacement occurs in the spring from the few viable cells that remain below the fibrous capsule. While dense capillary networks are present about the periphery of the glands, internal blood vessels are not conspicuous. The parathyroid cells of all vertebrates contain relatively few secretory granules, indicating that hormone storage is meager.

Comparative Anatomy

Parathyroid glands have been identified in *cyclostomes* and teleost fishes, and appear not to develop in *noetenious* amphibians. Two pairs of parathyroid are typically found in amphibians, they are small rounded bodies lying posterior to the thyroids in close association with branches of the external jugular vein. Since there are spatially separated from the thyroids, they may be removed without interfering with the thyroid glands. Most reptiles have two pairs of parathyroids, though as mentioned earlier, three pairs or one pair may be found in certain species. Birds often have two separate parathyroids, localid cauded of each thyroid lobe in other species, the parathyroids on each side tend to fuse. When the parathyroids of the chicken are approached by ventral dissection, each seems to be double structure; dorsal dissection, however, reveals that they are fused to form a single structure on each side.

One or two pairs of parathyroids may be present in mammals, and at least one pair is commonly associated with or embedded within the thyroid tissue. In certain species (e.g., rabbit, sheep, and ox) some or all of the parathyroids may be spatilaly sparated from the thyroid lobes. Four parathyroids are present in man, and in many other species such as the dog, cat, rabbit, horse, guinea pig, and opossum. Only two parathyroids are normally present in rat, mouse, pig, seal, shrew, and certain other species. Aberrant parathyroid rests are difficult to detect anatomically, and one cannot be certain about the frequency of their occurrence.

Mineral Metabolism

The parathyroid hormone (PTH) is primarily involved in the metabolism of calcium and phosphorus, the principal mineral constitutents of the skeleton. Over 99 percent of the total body content of calcium and about 90 percent of the total body content of phosphorus are found in the skeleton. Therefore it is not surprising to find that PTH modifies the osseous system. Without attempting to give a detailed and exhaustive treatment of the subject, some aspects of calcium and phosphorus metabolism in relation to the skeletal system are presented as an aid to understanding parathyroid physiology.

Calcium Metabolism

The human body contains more calcium than any other cation. A human adult of average size contains 1200 to 1400 gm of calcium: most of this is found in the skeleton; about 12 gm is present in the soft tissue and is chiefly intracellular; less than 1 gm is found in blood and extracellular fluid. Although the concentration of calcium ions in the

Fig. 12.10. Homogeneous osteoporosis produced in the dog by the surgical removal of the stomach.

body fluids is low, it performs a vital role in neuromuscular excitability, membrane permeability, enzyme activity, the clotting of blood, and the responses of the organism to fluctuations in acid-base balance. The normal organism carefully regulates the amount of calcium in the blood, and even minor variations in its concentration may produce profound effects on neuromuscular irritability. Tetany, or tonicspasms of the skeletal muscles, occurs when the Ca^{++} in the body fluids falls too low. The magnitude of contraction of the isolated frog's heart is directly proportional to the amount of Ca^{++} present, up to a certain concentration. In the absence of Ca^{++} the heart remains in diastole, but if the concentration is too great it goes into a state of sustained contraction

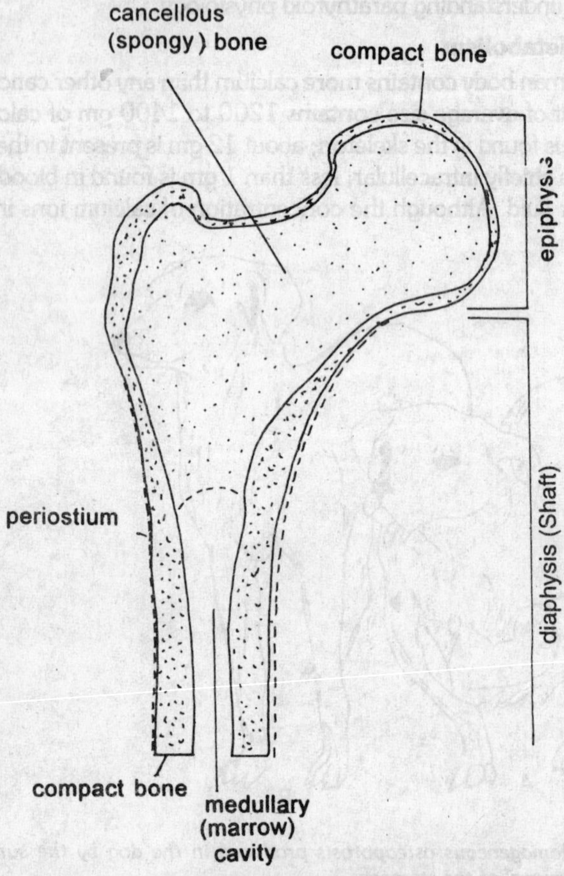

Fig. 12.11. Gross anatomy of bone.

or rigor. Potassium and magnesium ions antagonize the effects of Ca^{++} on contraction of the heart.

By using radicalcium (Ca^{45}) and radiophosphorus (P^{32}) it is possible to follow the movement of these elements from the blood into the tissues. Intravenous Ca^{45} leaves the blood very rapidly and is deposited in the skeleton. Long-term studies on the retention of Ca^{45} indicate that the turnover of total bone calcium is very slow.

Calcium is absorbed largely from the upper part of the small intestine, and factors that hasten the passage of intestinal contents through this part of the tract reduce the amount of calcium that can be absorbed. Since hydrochloric acid has a solubilizing effect on dietary calcium, the efficiency of calcium absorption may be influenced by the reservoir function of the stomach and its frequency of emptying. Vitamin D promotes the intestinal absorption of calcium. The presence of alkali and an excess of fat interfere with calcium absorption. A high calcium to phosphorus ratio in the diet impedes the absorption of calcium and raises the requirement for vitamin D. An excess of dietary phosphate has no striking effect on the absorption of calcium. Factors that reduce calcium absorption also impair phosphate absorption. Since parathyroidectomy reduces the intestinal absorption of calcium, it is probable that PTH has some effect on calcium metabolism at this level.

The human erythrocyte does not contain calcium. The normal level of the serum calcium is about 10 mg per 100 ml with a range from 9 to 11 mg. per 100 ml; in infants and during early childhood the range is slightly higher. Approximately 55 to 60 percent of the calcium in the serum in ultrafiltrable and diffusible; the remaining 40 to 45 percent is bound to serum proteins, largely to the albumin fraction. The diffusible fraction of serum calcium is generally considered to be ionized, although some of the diffusible calcium may be in the nonionized form. Citrate forms a nonionized compound with diffusible calcium. If the citrate levels of the serum are high, all of the serum calcium may become diffusible although not ionized. The serum appears to be supersaturated with calcium and phosphate, more being present than can be dissolved in water at the same pH. A partial precipitation of serum calcium occurs when the serum is shaken with species of cartilage. An excess of phosphate in the serum biologic activity, resides in a much smaller portion of the molecule.

Metabolism of Parathyroid Hormone

In man, the half-life of parathyroid hormone in the circulation has been estimated at about 20 to 30 minutes. Following parathyroidectomy in calcium-depleted rats, the plasma calcium level commences to fall within an hour and this steady decline continues until death ensues usually within 12 hours. Data such as these are indicative of a continuous and closely-regulated secretion of parathyroid hormone in the normal individual.

Regulation of Parathyroid Hormone Secretion

The mechanism whereby parathyroid hormone secretion is controlled resides in a simple negative feedback loop. The concentration of ionized serum calcium directly affects the rate of parathyroid hormone secretion, so that an inverse relationship exists between these two factors. When a hypercalcemia is present, regardless of its cause, parathyroid secretion is inhibited, so that the excess calcium is deposited in the bones. Conversely, a hypocalcemia stimulates parathyroid hormone secretion; thereby calcium is mobilized from the bones. These responses to serum ionized calcium are relatively rapid in onset, and the net result is an exquisitely delicate homeostatic mechanism for the regulation of the circulating calcium ion level.

An increase in the plasma phosphate level will also increase the rate of parathormone secretion. This effect is indirect, since an elevated phosphate concentration reduces the plasma calcium level, and this lowered calcium level in turn acts as the direct stimulus on the parathyroid glands to enhance their secretion.

Adrenal Gland

The adrenal glands were first described in man by Eustachius in 1563. Other anatomists, dissecting poorly preserved cadavers, identified these organs along the anterior borders of the kidneys and were impressed by the fact that they were generally filled with fluid. Not appreciating the effects of post mortem decay, the term suprarenal capsules was used to describe them. Curvier recognized that each gland consists of an inner and outer region, now referred to as the medulla and cortex, respectively. The first hint that these glands might be of functional significance came from Thomas Addison's description of a clinical condition resulting from their deterioration, a syndrome now bearing his name.

Anatomy of Adrenals

This gland is a compound structure consisting of an outer cortex and an inner medulla. The hormones of the cortex are steroids, whereas those of the medulla are amines. The two components of the organ originate from different embryonic primordia. The cortex is derived from mesoderm in close association with the developing gonads. The medulla is ectodermal since it differentiates from neural crest cells along with the sympathetic ganglia. The medullary cells are modified ganglion cells and remain in intimate contact with the preganglionic fibers of the sympathetic system. Secretion of the medulla is regulated very largely by means of thse nerves. The cortex, on the other hand, resembles the anterior hypophysis in being practically devoid of secretory nerve terminals.

Fig. 12.12. Homogeneous osteoporosis produced in the dog by the surgical removal of the stomach.

Like other endocrine glands, the adrenals receive a rich blood supply. The human organs are flattened bodies situated in the retroperitoneal tissue along the cranial ends of the kidneys. They vary considerably in shape, but are usually described as triangular or cresentric accessory deponts of cortical tissue are frequently found in many mammalian species, most commonly in the perirenal fat and along the path of descent of the gonads.

During the third month of intrauterine life, the human adrenal reaches its maximum relative size and exceeds that of the kidney. This large size of the festal adrenal is due to the presence of a thick boundary zone between the definitive cortex.

Histology

The cortex forms the major portion of the total mass of tissue found in the adrenal gland, and in adults three concentric zones are

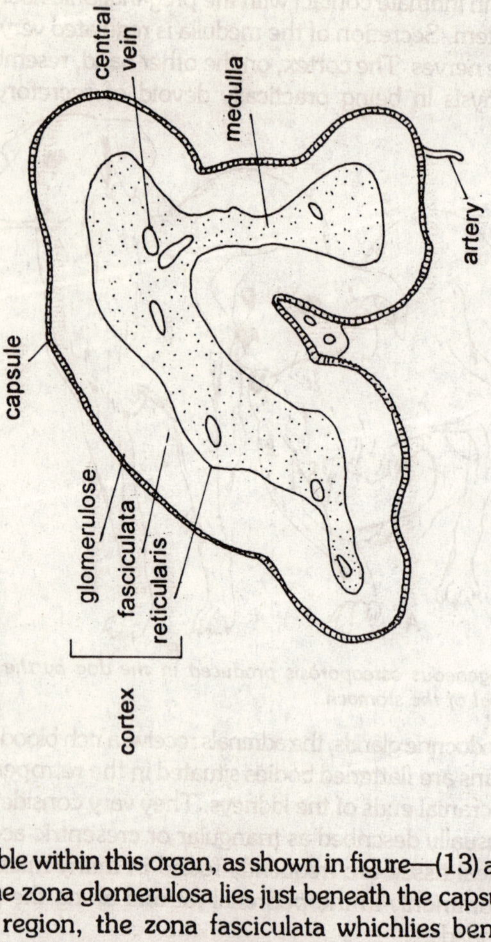

Fig. 12.13. Cross anatomy of the adrenal gland.

discernible within this organ, as shown in figure—(13) a thin outermost layer, the zona glomerulosa lies just beneath the capsule; (14) a thick middle region, the zona fasciculata whichlies beneath the zona glomerulosa; (15) a relatively thick inner layer, the zona reticularis which lies adjacent to the medulla. The transition between one cortical zone and the next is gradual rather than sharply delineated.

Zona Glomerulosa

The zona glomerulosa comprises about 15 percent of the total cortical mass. The cells of the zona glomerulosa are arranged in tight groups and pillars consisting of large, columnar epithelial cells that are continuous with the cells of the zona fasiculata. The individual pillars of glomerulosa cells are separated by vascular sinuses.

The spherical nuclei of the glomerulosa cells generally contain one or more nucleoli, and stain deeply; the Golgi apparatus is juxtanuclear. The cytoplasm of glomerulosa cells is more sparse than in the cells of the other two cortical zones, and is usually acidophilic, although clumps of basophilic material are often present. Filamentous mitochondria are abundant.

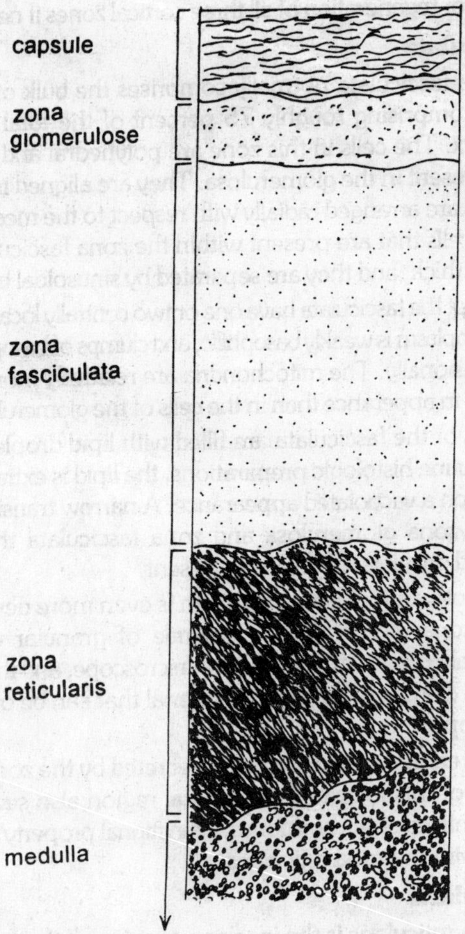

Fig. 12.14. Microscopic anatomy of the adrenal gland.

The cytoplasm also contains a well-developed smooth endoplasmic reticulum that is present as an anastomosing network of tubules, as well as abundant free ribosomes in the matrix. Some rough endoplasmic reticulum is also present in these cells.

The glucocorticoid hormone corticosterone is secreted by all three layers of the adrenal cortex. However, the enzymatic machinery necessary for the synthesis of the mineralocorticoid hormone aldesterone is restricted to cells of the zona glomerulosa. Furthermore, the cells of the glomerulosa that are located adjacent to, and just beneath, the capsule are able to provide cells for regeneration of all three cortical zones if necessary.

Zona Fasciculata

The zona fasciculata normally comprises the bulk of the cortical tissue, thus comprising roughly 75 percent of the total mass of the adrenal cortex. The cells in this zone are polyhedral and much larger than those present in the glomerulosa. They are aligned in long cords, and the cords are arranged radially with respect to the medulla. Usually the cords of cells that are present within the zona fasciculata are only one cell layer thick, and they are separated by sinusoidal blood vessels.

The cells of the fasciculata have one or two centrally located, spherical nuclei, the cytoplasm is weakly basophilic, and clumps of basophilic material are seen occasionally. The mitochondria are relatively more scarce and more variable in apperance than in the cells of the glomerulosa.

The cells of the fasciculata are filled with lipid droplets normally, but during routine histologic preparations, the lipid is extracted, so that the cells take on a vacuolated appearance. A narrow transitional region between the zona glomerulosa and zona fasciculata that lacks the abundant lipid droplets may also be present.

The smooth endoplasmic reticulum is even more developed than that in the zona glomerulosa. Cisternae of granular endoplasmic reticulum are also seen with the electron microscope, and these cisternae represent the clumps of basophilic material that can be observed with light microscopy.

As noted earlier, corticosterone is secreted by the zona fasciculata and the cells of this intermediate cortical region also secrete cortisol and small quantities of sex hormones, a functional property that is shared in common with the zona reticularis.

Zona Reticularis

The zona reticularis is the innermost region of the adrenal cortex and comprises roughly 10 percent of the total cortical tissue mass. The cells of the zona reticularis are arranged in an anastomosing network, rather than dispersed in an ordered parallel array of cords such as are found in the zona fasciculata. The transition between these two inner zones is gradual, and the cells in both regions are quite similar morphologically, save for the fact that lipid droplets are more scanty in the cytoplasm of the reticularis cells. The nongranular endoplasmic reticulum is abundant in reticularis cells like in the cells of the fasciculata.

"Light" and "dark" staining nuclei are common in cells of the zona reticularis, and this effect possibly may represent degenerative changes that have taken place in some of the cells, rather than an artifact that was induced during preparation of the tissue for histologic examination.

The cells of the zona reticularis normally secrete corticosterone and cortisols as well as minute quantities of sex hormone (androgens and estrogens).

Cortisol and Androgen Biosynthesis

Steroidogenesis

The major hormones secreted by the adrenal cortex are cortisol, the androgens, and aldosterone. The carbon atoms is the steroid molecule are numbered as shown in figure.

A. Zones and Sterodogenesis

Because of enzymatic differences between the zona glomerulosa and the inner 2 zones, the adrenal cortex functions as 2 separate units, with differing regulation and secretory products. Thus, the zona glomerulosa, which produces aldosterone, lacks 17 α-hydroxyprogesterone, which are the precursors of cortisol and the adrenal androgens. The synthesis of aldosterone by this zone is primarily regulated by the reninangiotensin system and by potassium.

The zona fasciculata and zona reticularis produce cortisol, androgens, and small amounts of estrogens. These zones, primarily regulated by ACTH, do not contain the enzymatic system necessary to dehydrogenate 18-hydroxycorticosterone and thus do not synthesize aldosterone.

B. Cholesterol Uptake and Synthesis

Synthesis of cortisol and the androgens by the zonae fasciculata and reticularis begins with cholesterol, as does the synthesis of all steroid hormones. Plasma lipoproteins are the major soruce of adrenal cholesterol, though synthesis within the gland from acetate also occurs. A small pool of free cholesterol within the adrenal is available for rapid synthesis of steroids when the adrenal is stimulated. When stimulation

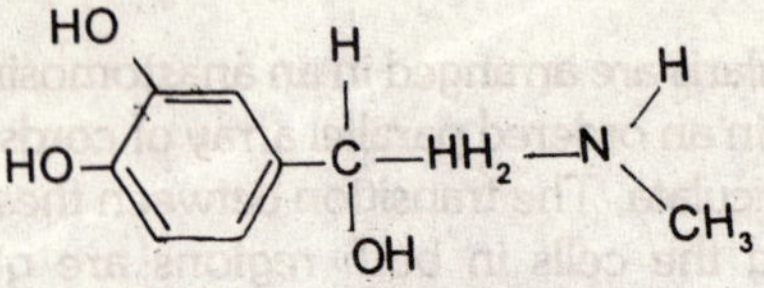

Epinephrine

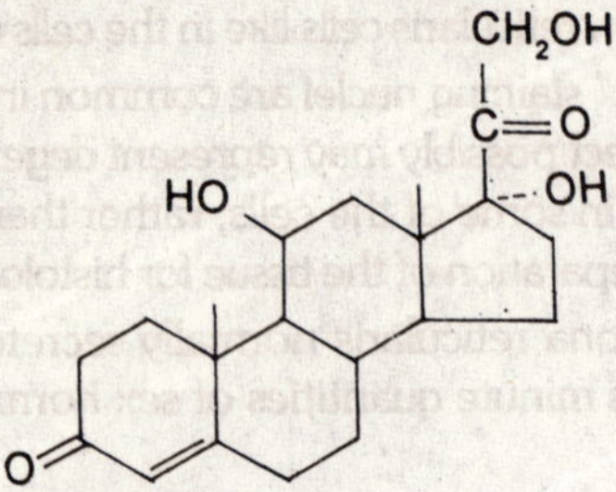

Cortisol (hydrocortisone)

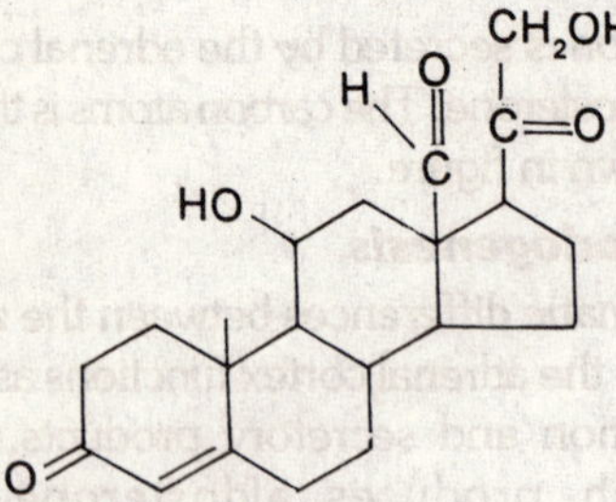

Aldosterone

Fig. 12.15. Structure of adrenal hormones.

occurs, there is also increased hydrolysis of stored cholesterol esters to free cholesterol, increased uptake from plasma lipoproteins, and increased cholesterol synthesis within the gland.

C. Cholesterol Metabolism

The conversion of cholesterol to pregnenolone is the rate-limiting step in adrenal steroidogenesis and the major site of ACTH action on the adrenal. This step occurs in the mitochondria and is mediated by cholesterol 20, 22-hydroxylase: 20, 22-desmolase enzyme complex, involving 2 hydroxylations and the side-chain cleavage of cholesterol. Pregnenolone is then transported outside the mitochondria before further steroid synthesis occurs.

D. Synthesis of Cortisol

Cortisol synthesis proceeds by the 17 α-hydroxylation of pregnenolone within the smooth endoplasmic reticulum to form 17 α-hydroxypregnenolone. This steroid is then converted to 17 α-hydroxyprogesterone after conversion of its 5, 6 double bond to a 4, 5 double bond by the 3 β-hydroxysteroid dehydrogenase: Δ^5-oxosteroid isomerase enzyme complex, which is also located within the smooth endoplasmic reticulum. An alternative but apparently less important pathway in the zonae fasciculata and reticularis is from pregnenolone→ progesterone→ 17 α-hydroxyprogesterone.

The next step, which is again microsomal, involves the 21-hydroxylation of 17 α-hydroxyprogesterone to form 11-deoxycortisol; this compound is further hydroxylated within mitochondria by 11 β-hydroxylase to form cortisol. The zone fasciculata and reticularis also produce 11-deoxycorticosterone (DOC), 18-hydroxydeoxycorticosterone, corticosterone, and 18-hydroxycorticosterone. However, as noted above, the absence of mitochondrial 18- hydroxysteroid dehydrogenase prevents production of aldosterone by these zones of the adrenal cortex.

E. Synthesis of Androgens

The production of adrenal androgens from pregnenolone and progesterone requires prior 17 α-hydroxylation and thus does not occur in the zona glomerulosa. The major quantitative production of androgens is by conversion of 17 α-hydroxypregnenolone to the 19-carbon compounds (C19 steroids) DHEA and its sulfate conjugate DHEA sulfate. Thus, 17 α- hydroxypregnenolone undergoes removal of its 2-carbon side chain at the C17 position by microsomal 17, 20-desmolase, yielding DHEA with a keto group at C17. DHEA is then converted to DHEA sulfate by a reversible adrenal sulfokinase. The other major adrenal androgen, androstenedione, isproduced form 17 α-hydroxyprogesterone by 17, 20-desmolase and to a lesser extent from DHEA. Androstenedione can be converted to testosterone, though adrenal secretion of this hormone is minimal. The adrenal androgens, DHEA, DHEA sulfate, and androstenedione, have minimal intrinsic androgenic activity, and they contribute to androgenicity by their peripheral conversion to the more potent androgens testosterone and dihydrotestosterone. Although DHEA and DHEA sulfate are secreted in greater quantities, androstenedione is qualitatively more important, since it is more readily peripherally converted to testosterone.

Regulation of Secretion

A Secretion of CRF and ACTH

ACTH is the trophic hormone of the zonae fasciculata and reticularis and the major regulator of cortisol and androgen production by the adrenal cortex. ACTH in turn is regulated by the hypothalamus and central nervous system via neurotransmitters and corticotropin releasing factor (CRF).

B. ACTH Effects on the Adrenal Cortex

The delivery of ACTH to the adrenal cortex leads to the rapid synthesis and secretion of steroids; plasma levels of these hormones rise within minutes after ACTH administration. Chronic stimulation increases protein, RNA, and DNA synthesis leading to adrenocortical hyperplasia and hypertrophy; conversely, ACTH deficiency results in decreased steroidogenesis and is accompanied by adrenocortical atrophy, decreased gland weight, and decreased protein and nucleic acid content.

C. ACTH and Steroidogenesis

ACTH binds to high-affinity plasma membrane receptors of adrenocortical cells, thereby activating adenylate cyclase and increasing cAMP, which in turn activates intracellular phosphoprotein kinases. This process stimulates the rate-limiting step of cholesterol to Δ^5-pregnenolone conversion and initiates steroidogenesis. ACTH may also have other actions, although these are less well documented. The exact mechanisms of ACTH stimulation of the desmolase enzyme complex are unknown, as are their relative imporance; however, ACTH has a number of effects, including increased free cholesterol formation as a consequence of increased cholesterol esterase activity and decreased cholesteryl ester synthetase; increased lipoprotein uptake by the adrenal cortex; increased content of certain phospholipids, which may increase cholesterol sidechain cleavage; and increased binding of cholesterol to the cytochrome P-450 cholesterol cleavage enzyme in mitochondria.

D. Neuroendocrine Control

Cortisol secretion is closely regulated by ACTH, and plasma cortisol levels parallel those of ACTH. There are 3 mechanisms of neuroendocrine control: (1) episodic secretion and the circadian rhythm of ACTH, (2) stress responsiveness of the hypothalamic-pituitary adrenal axis, and (3) feedback inhibition by cortisol of ACTH secretion.

1. Circadian rhythm

Circadian rhythm is superimposed on episodic secretion; it is the result of central nervous system events that regulate both the number and magnitude of CRF and ACTH secretory episodes. Cortisol secretion is low in the late evening and continues to decline in the first several hours of sleep, at which time plasma cortisol levels may be undetectable. During the third and fifth hours of sleep, there is an increase in the secretion; but the major secretory episodes begin in the sixth to eighth hours of sleep and then begin to decline as wakefulness occurs. About half of the total daily cortisol output is secreted during this period. Cortisol secretion then gradually declines during the day, with fewer secretory episodes of decreased magnitude, however, there is increased cortisol secretion in response to eating.

2. Alterations in the circadian rhythm

Although this general pattern is consistent, there is considerable intra- and interindividual variability, and the circadian rhythm may be altered by changes in sleep pattern, light-dark exposure, and feeding times. The rhythm is also changed by (1) physical stresses such as major illness, surgery or trauma, or starvation; (2) psychologic stesss, including severe anxiety, endogenous depression, and the manic phase of manic-depressive psychosis; (3) central nervous system and pituitary disorders; (4) Cushing's syndrome; (5) liver disease and other conditions that affect cortisol metabolism; (6) chronic renal failure; and (7) alcoholism. Cyproheptadine inhibits the circadian rhythm, possibly by its antiserotoninergic effects, whereas other drugs usually have no effect.

3. Stress Responsiveness

Plasma ACTH and cortisol secretion are also characteristically responsive to physical stress. Thus, plasma ACTH and cortisol are secreted within minutes following the onset of stresses such as surgery and hypoglycemia, and these responses abolish circadian periodicity if the stress is prolonged. Stress responses originate in the central nervous system and increase hypothalamic CRF and thus pituitary ACTH secretion. Stress responsiveness of plasma ACTH and cortisol is abolished by prior high-dose glucocorticoid administration and in spontaneous Cushing's syndrome, conversely, the responsiveness of ACTH secretion is enhanced following adrenalectomy.

4. Feedback Inhibition

The third major regulator of ACTH and cortisol secretion is that of feedback inhibition by glucocorticoids of CRF, ACTH, and cortisol secretion. Glucocorticoid feedback inhibition occurs at both the pituitary and hypothalamus and involves 2 distinct mechanisms—fast and delayed feedback inhibition.

Fast feedback inhibition of ACTH secretion is rate-dependent, i.e., it depends on the rate of increase of the glucocorticoid but not the dose administered. This phase is rapid, and basal and stimulated ACTH secretion both diminish within minutes after the plasma glucocorticoid level increases. This fast feedback phase is transient and lasts less than 10 minutes, suggesting that this effect is not mediated via cytosolic glucocorticoid receptors but rather via actions on the membrane.

Delayed feedback inhibition after the initial rate-dependent effects of glucocorticoids further suppresses CRF and ACTH secretion by mechanisms that are both time- and dose-dependent. Thus, with continued glucocorticoid administration, ACTH levels continue to decrease and become unresponsive to stimulation. The ultimate effect of prolonged glucocorticoid administration is suppression of CRF and ACTH release and atrophy of the zonae fasciculata and reticularis as consequence of ACTH deficiency. The suppressed hypothalamic-pituitary-adrenal axis fails to respond to stress and stimulation. Delayed feedback appears to act via the classic glucocorticoid receptor, thus reducing synthesis of the messenger RNA for the precursor to ACTH.

5. Regulation of Androgen Production

Adrenal androgen production in adults is also regulated by ACTH; both DHEA and androstenedione exhibit circadian periodicity in concert with ACTH and cortisol. In addition, plasma concentrations of DHEA and androstenedione increase rapidly with ACTH administration and are suppressed by glucocorticoid administration, confirming the role of endogenous ACTH in their secretion. DHEA sulfate, because of its prolonged metabolic clearance rate, does not exhibit a diurnal rhythm. Thus, adrenal androgen secretion is regulated by ACTH, and, in general, the secretion of these hormones parallels that of cortisol.

Glucocorticoids

General Consideration

Although glucocorticoids were originally so called because of their influence on glucose metabolism, they are currently defined as steroids that exert their effects by binding to specific cytosolic receptors which mediate the actions of these hormones. These glucocorticoid receptors are present in virtually all tissues, and glucocorticoid-receptor interaction is responsible for most of the known effects of these steroids. Alternations in the structure of the glucocorticoids have led to the development of synthetic compounds with greater glucocorticoid activity. The increased activity of these compounds is due to increased affinity for the glucocorticoid receptors and delayed plasma clearance, which increases tissue exposure. In addition, many of these synthetic glucocorticoids hae negligible mineralocorticoid effects and thus do not result in sodium retention, hypertension, and hypokalemia. This secretion described the molecular mechanisms of glucocorticoid action and the effects on individual metabolic functions and tissues.

Molecular Mechanisms

A. Glucocorticoid Receptors

Glucocorticoid action is initiated by entry of the steroid into the cell and binding to the cytosolic glucocorticoid receptor proteins. After binding, activated hormone-receptor complexes enter the nucleus and interact with nuclear chromatin acceptor sites. These events cause the expression of specific genes and the transcription of specific mRNAs. The resulting proteins affect the glucocroticoid response, which may be inhibitory or stimulatory depending on the specific tissue affected. Although glucocorticoid receptors are similar in many tissues, the proteins synthesized vary widely and are the result of expression of specific genes in different cell types. The mechanisms of this specific regulation are unknown.

B. Other Mechanisms

Although interaction of glucocorticoids with cytosolic receptors and their subsequent stimulation of gene expression are responsible for most glucocorticoid effects, other effects may occur through different mechanisms. The most significant example is that of glucocorticoid-induced fast feedback inhibition of ACTH secretion. This effect occurs within minutes of glucocorticoid administration, and its rapidity suggests that it is not due to RNA and protein synthesis but rather to glucocorticoid-induced changes in secretory function or cell membranes.

Glucocorticoid Agonists and Antagonists

The study of glucocorticoid receptors has led to the definition of glucocorticoid agonists and antagonists. These studies have also identified a number of steroids with mixed effects termed partial against, partial antagonist, or partial agonist-partial antagonist.

A. Agonists

In humans, cortisol, synthetic glucocorticoids (e.g., prednisolone, dexamethasone), corticosterone, and aldosterone are glucocorticoid agonists. The synthetic glucocorticoids have substantially higher affinity for the glucocorticoid receptor, and these have greater glucocorticoid activity than cortisol when present in equimolar concentrations. Corticosterone and aldosterone have substantial affinity for the glucocorticoid receptor; however, their plasma concentrations are normally much lower than of cortisol, and thus these steroids do not significant physiologic glucocorticoid effects.

B. Antagonists

Glucocorticoid antagonists bind to the glucocorticoid receptors but do not elicit the nuclear events required to cause a glucocorticoid response. These steroids complete with agonist steroids such as cortisol for the receptors and thus inhibit agonist responses. Other steroids have partial agonist activity when present alone, i.e., they elicit a partial glucocorticoid response. However, in sufficient concentration, they complete with agonist steroids for the receptors and thus comptitively inhibit agonist responses; i.e., these partial agonists glucocorticoids. Steroids such as progesterone, 11-deoxycortisol, DOC, testosterone, and 17 β-estradiol have antagonist or partial agonist-partial antagonist effects; however, the physiologic role of these hormones in glucocorticoid action is probably negligible, because they circulate in low concentrations.

Intermediary Metabolism

Glucocorticoid in general inhibit DNA synthesis. In addition, in most tissues they inhibit RNA and protein synthesis and accelerate protein catabolism. These actions provide substrate for intermediary metabolism; however, accelerated catabolism also accounts for the deleterious effects of glucocorticoids on muscle, bone, connective tissue, and lymphatic tissue. In contrast, RNA and protein synthesis in liver is stimulated.

A. Hepatic Glucose Metabolism

Glucocorticoids increase hepatic gluconeogenesis by stimulating the gluconeogenic enzymes phosphoenolpyruvate carboxykinase and glucose-6-phosphatase. They also have a permissive effect in that they increase hepatic responsiveness to the gluconeogenic hormones (glucagon, catecholamines), and they also increase the release of substrates from peripheral tissues, particularly muscle. This latter effect may be enhanced by the glucocorticoid-induced reduction in peripheral amino acid uptake and protein synthesis. Glucocorticoids also increase glycerol and free fatty acid release by lipolysis and increase muscle lactate release. These steroids also enhance hepatic glucogen synthesis and storage by stimulating glycogen synthetase activity and to a lesser extent by inhibiting glycogen breakdown. These effects are insulin-dependent.

B. Peripheral Glucose Metabolism

Glucocorticoids also alter carbohydrate metabolism by inhibiting peripheral glucose uptake in muscle and adipose tissue. This effect and the others described above may result in increased insulin secretion in states of chronic glucocorticoid excess.

C. Effects of Adipose Tissue

In adipose tissue, the predominant effect is increased lipolysis with release of glycerol and free fatty acids. This is partially due to direct stimulation of lipolysis by glucocorticoids, but it is also contributed to by decreased glucose uptake and enhancement by glucocorticoids of the effects of lipolytic hormones. Although glucocorticoids are lipolytic, increased fat deposition is a classic manifestation of glucocorticoid excess. This paradox may be explained by the increased appetite caused by high levels of these steroids and by the lipogenic effects of the hyperinsulinemia that occurs in this state. The reason for abnormal (central) fat deposition is states of cortisol excess is unknown.

D. Summary

The effects of the glucocorticoids on intermediary metabolism can be summarized as follows: (1) Effects are minimal in the fed state. However, during fasting, glucocorticoids contribute to the maintenance of plasma glucose levels by increasing glucogeneogenesis, glycogen deposition, and the peripheral release of substrate. (2) Hepatic glucose production is enhanced, as is hepatic RNA and protein synthesis. (3) The effects on muscle are catabolic; i.e., decreased glucose uptake and metabolism, decreased protein synthesis, and increased release of amino acids. (4) In adipose tissue, lipolysis is stimulated. (5) In glucocorticoid deficiency, hypoglycemia may result, whereas in states of glucocorticoid excess there may be hyperglycemia, hyperinsulinemia, muscle wasting, and weight gain with abnormal fat distribution.

Functional Morphology of the Adrenal Medulla

Gross Anatomy

The adrenal medulla in adult humans and other mammals is invested closely by the cortical tissue (interrenal tissue). However, in other vertebrates the two structures either may be entirely separate or else interrelated in a number of ways. In a transacted fresh adrenal gland, the medullary tissue appears grayish in sharp contrast to the yellow cortex.

Histology

Under the light microscope, the boundary between the zona reticularis and the medulla generally is irregular and columns of cortical cells may extend for some distance into the medullary tissue.

The medulla is comprised of irregularly shaped epithelioid cells that are arranged in cords or clusters, and that are surrounded by capillaries and venules.

If adrenal medullary tissue is fixed in solutions containing bichromate ion, the medullary cells are seen to be filled with fine brown granules. This is the chromaffin reaction, and presumably this reaction is caused by the action of chromic ion on the catecholamines in the granules of the medullary cells so that oxidation and polymerization of the hormones epinephrine and norepinephrine within the cytoplasmic granules takes place. Similarly, the medulla is stained green following treatment with ferric chloride.

The chromaffin reaction is seen in other tissues of the body, principally in the argentaffin cells of the gastrointestinal tract and the mast cells. These cells give this reaction because of the 5-hydroxytryptamine and dopamine that they contain. The adrenal medullary chromaffin cells, however, originate from neural ectoderm in the embryo, secrete catecholamines, and are innervated by sympathetic preganglionic fibers.

Special histochemical techniques reveal that two different cell types are present in the adrenal medulla, one of which contains epinephrine, the other norepinephrine. Furthermore, electron microscopy of the adrenal medulla reveals that the cells contain abundant membrane-limited dense granules, having diameters ranging between 50 and 350 μm. The granules of certain cells exhibit a rather low electron density and are considered to contain epinephrine, whereas another cell population contains very high density granules that are believed to consist of norepinephrine.

The cisternae of the granular endoplasmic reticulum are arranged in parallel systems. The Golgi apparatus is located in a juxtanuclear position, and its cisternae contain electron dense material that is believed to represent a precursor of the hromone granules.

The secretory mechanism wherby the adrenal medullary hormones are released into the bloodstream is not understood fully. Some investigators feel that this process is similar to the secretion of exocrine products, whereas other workers believe that the chromaffin granules are secreted intact, and the catecholamine then diffuses out of these "packets" of hormone through the membrane. Alternatively, an active transport process may liberate the hormone from the granules into the cytoplasm and thence the hormone passes out of the cell.

Medullary Blood Supply and Innervation

Certain features of the adrenal cortical blood supply may be reviewed here briefly together with a consideration of the medullary blood supply. As discussed previously, the rich blood supply to the adrenal gland is provided by a number of arteries that penetrate the capsule of the organ, and give rise to a capsular plexus. This plexus in turn gives rise to a system of cortical arteries, and these vessels are distributed to the anastomosing network of sinusoids that invest the cords of cortical parenchymal cells. The sinusoids within a specific region of the cortex then converge upon a collecting vein at the corticomedullary junction, and as noted earlier there is no venous drainage from the adrenal cortex.

Some branches of the adrenal arteries, in contrast to the cortical vessels, pass through the trabeculae and then traverse the cortical tissue with little or no branching until they reach the medulla. Within the medulla, these vessels now branch repeatedly and thus form the abundant capillary network that surrounds the cords and clumps of medullary chromaffin cells. Consequently the adrenal medulla has a double (or dual) blood supply. First, blood comes from the sinusoids of the cortex tha anastomose with the medullary capillary bed at the corticomedullary junction, and second, blood also comes from the medullary arteries that pass directly from capsule to medulla.

The medullary capillaries drain into the same venous as do those of the cortex. Ultimately the numerous adrenal venules form large central veins within the medulla which anastomose and emerge from the capsule as the suprarenal vein.

The cells that line the medullary capillaries are composed of typical endothelial cells.

The abundant capsular nerve plexus contains some sympathetic ganglion cells; branches of these nerves pass through the cortex in the trabeculae and thence run directly, and almost exclusively, to the medulla. These preganglionic fibers terminate in multiple endings that surround individual medullary cells. In actual fact, therefore, the adrenal medulla may be considered from a functional standpoint to be a physiologic extension of the sympathetic nervous system.

Embryologic Considerations

As noted earlier, the adrenal cortex develops from coelomic mesoderm located on the medial side of the urogenital ridge. The adrenal medulla, on the other hand, arises from ectodermal tissue of the neural crest. This tissue also gives rise to the cells of the sympathetic ganglia. The sympathochromaffin cells of the neural crest that are destined to form the adrenal medulla migrate ventrally and penetrate the analgen of the adrenal cortex medially, so that ultimately they assume a central position within the primitive adrenal gland.

Chromaffin System (or paraganglia)

Within the body are a number of broadly scattered accumulations of cells that appear to share many common features with the adrenal medullary cells. These groups of cell collectively are termed the paraganglia or chromaffin system of the body. Included in this category are the organs of Zuckerkandl (which is located in the retroperitoneum) as well as similar cell groups found in the heart, liver, testis, and ovary. These paraganglia all contain cells that exhibit a typical chromaffin reaction described above, and these cells also stain with ferric ion. Furthermore, the paraganglia contain cells that are usually arranged in cords, and also possess a rich vascular supply. However, a definite endocrine role has not been demonstrated as yet for this group of tissues, although they are known to synthesize catecholamines.

Physiologic and Biochemical Actions of Norepinephrine and Epinephrine

The hromones secreted by the adrenal medulla exert a broad spectrum of physiologic and biochemical effects upon a wide variety of bodily processes. The several possible effects of a single catecholamine may result from the mode of administration of the hormone (i.e., endogenous or exogenous), as well as the dosage of the hormone to which the subject is exposed. These factors are important in determining the degree as well as the type of response elicited by the adrenal medullary hormones.

Physiologic Effects of the Catecholamine Hormones. Norepinephrine and epinephrine not only mimic the effects of sympathetic nervous discharge or stimulation, but also these hormones themselves stimulate the nervous system.

The catecholamines directly augment both the rate and contractile force of the isolated heart as studied in vitro. The heart in situ, however, may respond quite differently, because the various reflexes present in the body may cause secondary alterations in heart rate because of shifts in blood pressure induced by these hormones. The catecholamines also directly increase myocardial irritability, so that extrasystoles or even fibrillation may develop under certain conditions subsequent to the administration of these compounds.

Both norepinephrine and epinephrine dilate the coronary vessels. However, norepinephrine causes vasoconstriction in other organs, whereas epinephrine dilates the vascular bed in skeletal muscles and viscera when administered in moderate doses. Thus, epinephrine when injected intravenously produces a marked rise in heart rate and cardiac output concomitantly with a sharp rise in blood pressure that is due to a generalized vasoconstriction.

As noted above, norepinephrine produces vasoconstriction in most organs other than the heart; hence it too produces a marked pressor response. However, since epinephrine causes vasodilation of the vascular bed in skeletal muscles, this effect overcompensates for the vasoconstriction produced elsewhere; the net effect of an increased adrenal medullary secretion in vivo is a fall in the total peripheral resistance. The systolic blood pressure rises because of the direct effect of epinephrine in stimulating the heart rate increases the cardiac output sufficiently to overcome the concomitant vasodilatory effect of epinephrine in the skeletal muscles.

Infusion of norepinephrine into normal humans (or animals) results in an increase in both the systolic and diastolic blood pressures. The ensuing hypertension then stimulates the aortic and carotid pressoreceptors, with the result that reflex bradycardia occurs, and this reflex-induced bradycardia dominates the direct stimulatory effect of norepinephrine upon heart rate; hence this rate shows a net decrease and the cardiac output falls.

Epinephrine, on the other hand, causes an increased pulse pressure; however, the pressoreceptor stimulation evoked by this hormone is inadequate to obscure the direct effects of epinephrine upon the heart rate and cardiac output. The heart rate and cardiac output increase.

In addition to the effects exerted upon the nervous and cardiovascular systems by the adrenal medullary hromones, these compounds also directly affect the physiologic activity of smooth muscle. In this regard the effect of norepinephrine and epinephrine are quite variable, and their effects upon this kind of tissue depend upon the site in the body where the muscle is located.

Noreponephrine exerts a weakly inhibitory action upon the contractile activity of smooth muscle in the gastrointestinal tract, but does not relax the smooth musculature of the pulmonary bronchioles. As noted above, this hormone also elicits a marked pressor response because of its direct effect upon vascular smooth muscle, with an attendant rise in total peripheral resistance.

Epinephrine, on the other hand, markedly inhibits the contractile activity of the smooth musculature of the gastrointestinal tract, producing relaxation, while simultaneously stimulating contraction of the pyloric and ileocecal sphincters. Epinephrine also produces a strong dilatation of the bronchial musculature.

Note that these effects of the adrenal catecholamines strongly resemble the effects of adrenergic nerve stimulation to the same organs.

Epinephrine decreases the rate of onset of fatigue in skeletal muscle and also increases the rate and depth of respiration. Presumably these effects are due to the enhanced glycogenolysis and lypolysis produced by this hormone as well as the calorigenic action of these amines.

In adrenalectomized animals suffering from an overall adrenal insufficiency, the vascular smooth muscle becomes unresponsive to the effects of norepinephrine and epinephrine. Capillary dialation ensues and permeability of these vessels (e.g., to dyes) increases markedly in the terminal stages. Lack of response to adrenergic stimulation (hence liberated norepinephrine) possibly hinders the compensation of the vascular bed to the reduced circulating blood volume (or hypovolemia) that attends adrenal insufficiency, and this in turn favours cardiovascular collapse. Vascular reactivity in such animals is restored by glucocorticoid administration.

Metabolic Effect of the Catecholamine Hormones

Both epinephrine and norepinephrine exert certain effects upon the metabolism of carbohydrate and fats, and these actions of the catecholamines may be far more significant physiologically than their roles as endocrine extensions of the sympathetic nervous system.

Carbohydrate Metabolism

Epinephrine stimulates glycogenolysis in both liver and muscle. This results in an elevation of the blood glucose level as well as an increased production of lactic acid in muscle. Epinephrine activates phosphorylase in both of these tissues and the plasma potassium level rises coincidentally with the increased glycogenolysis.

The biochemical mechanisms whereby epinephrine mediates glycogenolysis in liver and muscle are illustrated in detail in figure. Aside from their inherent physiologic significance, the mechanisms of these reactions provide examples of the first elucidation of the action of a hormone on a fundamental cellular process at the molecular level.

An increased oxygen consumption accompanies these glycogenolytic effects; this amounts to about 20 to 40 percent above the resting level in humans. The carbon dioxide production concomitantly rises to an even greater extent than does the increase in oxygen consumption, so that the respiratory quotient also increases.

Norepinephrine, on the other hand, exerts very little effect upon carbohydrate metabolism and oxygen consumption, a fact that provides the major exception to the generalization that the metabolic actions of the two adrenal catecholamines are similar.

Lipid Metabolism

Epinephrine as well as norepinephrine are almost equally potent in their effects upon lipid metabolism. Thus, both of these catecholamines exert a profound lipid-mobilizing action, and they increase the blood level of nonesterified fatty acids by stimulating the release of free fatty acids glycerol from adipose tissue. There effect results directly from an elevated rate of lipolysis that is produced in adipose cells by these hormones, and an increased oxygen consumption (or calorigenic action) also attends these processes.

In normal subjects, the administration of epinephrine also elevates serum cholesterol and phospholipids levels. The turnover rate of the cholesterol and phospholipids, also is elevated in cardiac muscle by exogenous epinephrine.

Epinephrine directly inhibits insulin release from the pancreas, and thus serves as an emergency hormone by exerting the following metabolic actions. First, epinephrine rapidly provides a supply of fatty acids for use as fuel by skeletal muscle. Second, epinephrine mobilizes glucose by stimulating hepatic glycogenolysis as well as gluconeogenesis directly. Third, by reducing the endogenous secretion to insulin, epinephrine decreases glucose uptake by the peripheral tissues, thereby increasing the supply of this substrate that is available to the central nervous system.

Mechanism of Action of the Adrenal Medullary Hormones

The general effects of norepinephrine and epinephrine have been divided into two categories depending upon their sensitivity to certain drugs. These effects depend in turn upon the actions of these catecholamines upon postulated α and β receptors located within the effectors organs. Drugs which inhibit α receptors (such as phentolamine) prevent such actions as the pressor effects of the catecholamines, whereas inhibition of β receptors by other compounds (e.g., isoproterenol or propranolol) prevents the inotropic and chronotropic responses of the heart to the catecholamines as well as the ability of these hormones to stimulate glycogenolysis in muscle and liver and lipolysis in adipose tissue. Experimental evidence has shown that most of the β responses are due to stimulation of adenyl cyclase activity, so that an increase in the synthesis of cyclic 3', 5'-AMP results. Epinephrine, for example, produces this effect in a number of tisues, including liver, skeletal muscle, adipose tissue, and cardiac tissue. The adenyl cyclase system in turn has been shown to be localized within the cell membranes of hepatic and other tissues as discussed in the introduction of this chapter. Furthermore, the effects of norepinephrine and epinephrine upon glycogen metabolism have been shown to be mediated by a marked increase in the activity of the adenyl cyclase system for epinephrine.

On the other hand, certain α responses have been demonstrated to reside in an inhibition of cyclic AMP synthesis.

It is important to realize that in certain tissues where epinephrine can stimulate both α and β types of response, the net effect of the amine depends upon the relative quantity and/or sensitivity of each type of receptor that is present in the tissue. For example, in the pancreas the α-adrenergic response to epinephrine is predominant; therefore, cyclic AMP concentration decreases, inhibiting insulin secretion. If, however, an α adrenergic blocking agent such as phentolamine (Regitine) is present, the β effect becomes predominant, and epinephrine now stimulates cyclic AMP synthesis, and insulin secretion is stimulated rather than inhibited.

Prostaglandins

The *prostaglandins* are a family of twenty-carbon polyunsaturated fatty acids formed in the plasma membranes of cells in a wide variety of tissues. They are produced in small amounts, are extremely potent, and exist for only a short time.

The prostaglandins were first discovered in the 1930s and were named by the Swedish scientist Ulf S. von Euler. Von Euler first found prostaglandins in semen, where they are one hundred times more concentrated than they are in most body tissues, and he assumed that they are secreted by the prostate gland, hence the term *prostaglandin*. This name is somewhat of a misnomer, however, because most of the prostaglandins in semen come from the seminal vesicles. Furthermore, prostaglandins have been found in virtually every tissue of the mammalian body.

There are many different prostaglandins, and for convenience, they were orginally grouped into five series that differ slightly in molecular structure: PGA, PGB, PGC, PGE and PGF. Each series was further subdivided as additional prostaglandins were identified; for example, PGE_1 and PGE_2 are chemically similar molecules. Figure shows the structure of PGE_2, one of the better-known prostaglandins.

Prostaglandins have widespread and varied effects. Some of them are even antagonists; for example, PGE_2 lower pressure and PGF_2 raises it. However, both PGE_2 and PGF_2 stimulate the contraction of the smooth muscle of the uterus, perhaps the best-known effect of the prostaglandins. This uterine muscular response can make the careful admininstration of prostaglandins useful in aiding childbirth. Prostaglandins also are possible abortion-inducing agents, not only because they cause the uterus to contract but also because PGE_2 may cause a reduction in the secretion of progesterone, a hormone necessary for the maintenance of pregnancy. It is possible that some of the prostaglandins may prove useful for birth control in the future.

Prostaglandin research has already provided a partial explanation for the way in which aspirin acts. Aspirin has been used as a drug for several hundred years without a clear understanding of what it does; in fact, some early cynics doubted that aspirin did anything but make people think they were going to feel better. It is now known that aspirin blocks the synthesis of prostaglandins, and this discovery seems to be a step toward explaining some of the effect of this most commonly used of all drugs.

Prostaglandins (and chemically similar compounds, the thromboxanes, prostacyclins, and leukotrienes) are released from tissues in response to a variety of stimuli, including hormones. Prostaglandins circulate in the blood and act as hormones in that they can exert their effects at distant locations, but they have their principal effects within the cells in which they are produced. Since hormonal stimulation of a cell sometimes results in increased prostaglandin synthesis as well as increased synthesis of cyclic AMP, prostaglandins may act as an intermediary between some hormones and cyclic AMP, or they may function independently as second messengers. No doubt, many basic mechanisms of cell function will be clarified by research on prostaglandins and their interactions with other hormones and hormonelike substances.

PHEROMONES

In addition to the hormones secreted within an organism that affect its own physiological processes, there are other chemical regulatory compounds that are secreted by organisms into the environment. These compounds are called *pheromones*. Pheromones may evoke behavioural, developmental, or reproductive responses within a population of organisms. Therefore, pheromones are sometimes called "social hormones."

A number of phermones that affect behaviour patterns have been intensively studied. Well-known examples include trail-marking behaviour in ants and territory marking by dogs, deer, and antelope through pheromones secreted in urine or from scent glands. In these cases, the phermone proclaims to other males of the species that the territory belongs to a particular individual.

An even more intriguing example of pheromones are the *sex attractants*, which have been most extensively studied in insects. The first sex attractant identified and synthesized was bombykol, which comes from females of the silkworm moth *Bombyx mori*. Bombukol can attract male silkworm moths several kilometers away. Sex attractant pheromones have been discovered in other moths and in many other insect species as well. For example, the queen bee releases a pheromone during her nuptial flight that entices the drones to follow her and mate with her. Sex attractants have a very compelling effect on insect behaviour.

Sex attractants have also been studied in mammals. In a number of mammalian species, females emit a special odor during their period of fertility and sexual receptiveness, which is called estrus. Normally docile, well-behaved male dogs will scratch doors, break leashes, or jump fences and roam great distances to reach a female in estrus once they detect her chemical signal.

Isolation of a sex attractant from female rhesus monkeys has focused new attention on primate pheromones in general and human pheromones in particular. Whether or not sex attractants and other pheromones exist in human remains undetermined. At any rate, modern men and women seem to prefer to wash off natural body chemicals and replace them with laboratory-compounded scents. Who knows what would happen if manufactures of these scents should succeed in isolating, identifying, synthesizing, and marketing human pheromones?

Pancreas

The mammalian pancreas is a compound gland consisting of exocrine and endocrine tissues. The exocrine constituent secretes pancreatic juice which is poured into the duodenum via the pancreatic duct; the islets of Langerhans are aggregations of endocrine cells which liberate their hormones directly into the circulation. The human pancreas contains a million or more individual islets scattered widely throughout the acinar tissues; the total islet tissue constitutes only 1 or 2 percent of the pancreatic mass. The cells of the islets are arranged in irregular cords which are separated by a very rich system of capillary vessels or sinusoids. While the exocrine cells receive sympathetic and parasympathetic fibers, it is doubtful whether the nervous system performs any important role in regulating the secretory functions of the islets.

The pancreas and liver originate as outpocketings of the duodenal entoderm, at a point just caudad to the pylorus. There are generally three main pancreatic buds, one dorsal and two ventral. The dorsal primordium constricts from the duodenum and thus forms the duct of Sanatorium whch may persist in the adult as the accessory pancreatic duct, which persists in the adult as the pancreatic duct proper. The dorsal and ventral anlagen of the human embryo fuse during the seventh week to form the pancreas. In birds, the three pancreatic outgrowths do not fuse so completely, and the adult gland consists of three completely or partially separated lobes.

The endocrine cells of the islets arise from the duct epithelium and often seem to remain connected to its finer ramifications; however,

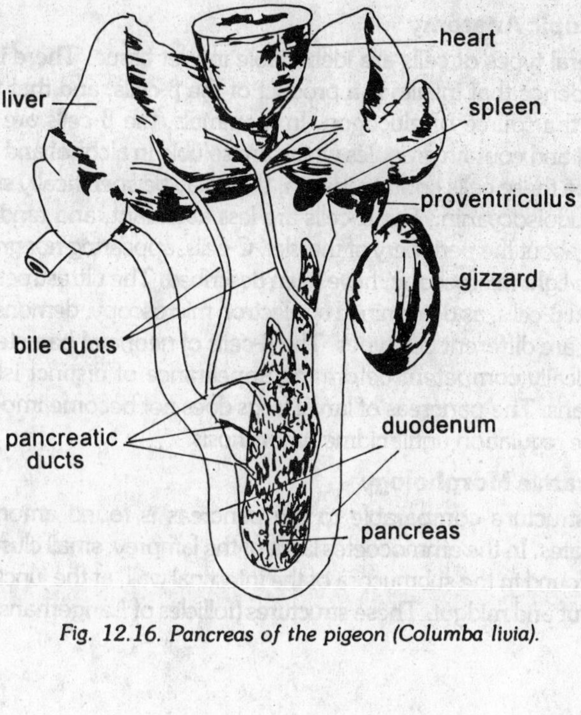

Fig. 12.16. Pancreas of the pigeon (Columba livia).

neither insulin nor glucagon is present in pancreatic juices. There is no clear evidence that acinar cells ever transform into islet tissue. Since the islets bud off fro the duct epithelium, and the duct itself is an outgrowth of the duodenal entroderm, the islets can be regarded as endocrine specializations of the duodenal mucosa. The gastrointestinal hormones, such as secretion, arise from the mucosal lining of the stomach and intestine.

Two hormones, insulin and glucagon, are known to derive from the islets of Langerhans. Both of these hormones act mainly, but not exclusively, upon the liver. Since the pancreas is drained by the hepatic portal vein, the hormones must traverse the liver before they can reach the systemic circulation. The islet hromones are enzymatically inactivated by liver tissue, and thus attain higher concentrations in the blood of the portal vein than in the systemic circulation. It is possible that large doses of exogenous hormones, added to peripheral blood, have certain effects which are pharmacologic rather than physiologic. The gastrointestinal hormones (e.g., secretin, cholecystokinin) confine their actions to the gastrointestinal tract and its appendages, and the islet hormones show this same tendency.

Anatomy of the Pancreatic Islets

Microscopic Anatomy

Several types of cells are identifiable in islet tissue. There is very good evidence that insulin is a product of the β-cells, and that the α-cells are the source of glucogon. In mammals, the β-cells are more abundant and contain granules which are soluble in alcohol and acetic acid, since these cells contain insulin, they may be specifically stained with pseudoisocyanin. The α-cells are less abundant, and tend to be arranged about the periphery of the islet. C-cells, appearing nongranular under the light microscope, have been described. The ultrastructure of the α- and β-cells, as determined by electron microscopy, demonstrates that they are different cell types. The β-cells of neonatal hamsters are biochemically competent before the appearance of distinct islets of Langerhans. The pancreas of larval frogs does not become important in glucose regulation until midmetamorphosis.

Comparative Morphology

No structure comparable to the pancreas is found among the invertebrates. In the ammocoetes larva of the lamprey, small clusters of cells are found in the submucosa of the intestinal wall, at the junction of the foregut and midgut. These structures (follicles of Langerhans) were

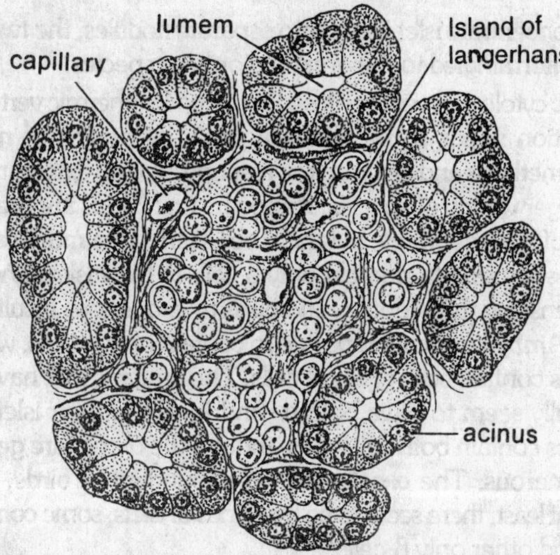

Fig. 12.17. The pancreas, showing and island of Langerhans, the cells of which produce the hormone of the pancreas, surrounded by acini, the cells of which secrete the pancreatic juice; seen in section.

described by Langerhans in 1873 and are presumed to be islet tissue. The administration of glucose to these larvae causes certain follicle cells (probably β-cells) to become vacuolated and hydropic just as in mammals. Cells appearing to correspond to the exocrine pancreas are present in the epithelium of the anterior intestine. As the larval lamprey undergoes metamorphosis the islet follicles give rise to a compact tissue at the blind end of the intestinal caecum. Encapsulated follicles of islet tissue have been found in adult hagfishes near the point where the bile duct enters the intestine. It appears that both endocrine and exocrine constituents of the pancreas are present in cyclostomes, but these remain separate from each other. These observations suggest that islet tissue arose very early in vertebrate evolution as a derivative of the intestinal tract.

Much diversity is encountered among teleost fishes, but, in some of them (e.g. *Ameiurus, Lophius, Cottus*), most of the islet tissue is concentrated into a "principal islet" which is separate from the exocrine pancreas. In the initial attempts to prepare effective extracts of mammalian pancreas, it was learned that proteolytic enzymes from the exocrine pancreas promptly destroyed the insulin. It was possible to obtain potent extracts from the principal islets of the anglerfish (*Lephius*), since these organs contain very little exocrine tissue. Although some fishes show a tendency to concentrate islet tissue into separate nodules, the two kinds of tissue are intermingled in the great majority of species.

Pancreatic cytology is very confused in the poikilothermic vertebrates and the situation needs to be reinvestigated, by means of modern cytochemical methods and electron microscopy, and by extending such studies to a larger variety of species. It appears that β cells are present in the pancreas of all vertebrates, but the α cell seem to be more variable. Some urodeles are said to possess only β cells in their islets and to be much more sensitive than reptiles and birds to exogenous insulin. The small islets of *Ambystoma* are reported to contain only α-cells, whereas the larger islets contain both types of cells. The anurans that have been studied carefully seem to have both α- and β-cells in their islets. The islets of reptiles contain both cell types, though the α-cells are generally the more numerous. The α-cells are very abundant in birds, and in some species at least, there seem to be two kinds of islets, some containing only α-cells and other only β-cells.

Pancreas yields about 10 times more glucagon than comparable quantities of mammalian gland. Exogenous insulin generally lowers the level of blood sugar but in species having a high α cell count, this treatment produces an initial hypoglycemia which is followed by a more lasting elevation of blood sugar. This probably means that the hypoglycemia produced by insulin stimulates the α-cells to release glucagon which has the reverse effect.

Chemistry of Insulin

Almost 50 years have elapsed since Banting and Best demonstrated that extract of the pancreas could cause hypoglycemia in diabetic dogs. Their success where others had failed was due to Banting's realization that the proteolytic enzymes of the pancreas destroyed the hormone involved. By ligating the pancreatic ducts of dogs, they destroyed the acinar tissue without damaging the islet tissue, which they then extracted. Abel crystallized insulin in 1926; 3 decades later, Sanger (1955) determined the complete amino acid sequence of bovine insulin. This was confirmed in 1964 through the synthesis of insulin by Katasoyannis *et al.*

Insulin derived from bovine pancreas has a molecular weight of 5,734 and is composed of two polypeptide chains designated A and B. The A chain has 21 amino acid residues, and the B chain has 30. The chains are connected to each other by disulfide bonds between cysteine residues at the A-7 and B-7 position and at the A-20 and B-19 positions. The A chain also has a disulfide ring between cysteine residues at A-6 and A-11. Structure-activity relationships indicate that the amino acid composition of this disulfide ring is not crucial to the activity of the parent molecule; in fact, it is one of the points of species variation. The ring is composed of five amino acids, as is the disulfide ring of the neurohypophyseal hormones.

Table 12.1. Species Variations of the A and B Chains of Insulin

Species	*A Chain (Disulfide Ring)*	*B Chain (C Terminal)*
Ox	6 7 8 9 10 11 -cys-cys-ala-ser-val-cys S—S	30 -ala
Sheep	-ala-gly-val	-ala
Horse	-thr-gly-ile-	-ala
Pig, dog, whale (sperm)	-thr-ser-ile-	-ala
Rabbit	-thr-ser-ile-	-ser
Human	-thr-ser-ile	-thr

The Biosynthesis of Insulin

The formation of insulin might proceed by (1) condensation of simultaneously synthesized A and B polypeptide chains, or (2) the cleavage of the insulin molecule from a macromolecule in which the polypeptide chains have been sequentially synthesized. Steiner has provided evidence that the latter hypothesis may be more correct. Slices of β-cell tumors of the human pancreas, known to cause hypoglycemia, were incubated with ^{3}H-phenylalanine or ^{3}H-leucine. Purification of the labeled protein and fractionation by gel filtration caused separation of a protein with a large molecular weight (designated component b) from a smaller protein that had a mobility identical to that of porcine or bovine insulin. Limited tryptic digestion of component b yielded a radioactive peptide which, on gel filtration, had a mobility identical to that of procrine or bovine insulin. Sulfitolysis of the cleaved protein yielded labeled sulfonated A and B chains of insulin, incubation of component b with large amounts of trypsin caused cleavage of the arginyl-glycine bond of the B chain and release of the heptapeptide B-23 to B-29. A similar approach with rat islet tissue demonstrated the presence of component b in this tissue also. During synthesis the appearance of component b precedes the apearance of insulin. Steiner has called this high-molecular-weight material proinsulin; its molecular weight was calculated as 9,082 on the basis of the complete amino acid sequence.

Chemistry of Glucagon

Murline suggested in 1924 that the initial hyperglycemia seen after administration of insulin to normal dogs was due to a hyperglycemic coldaminant, which he called glucagon. The insulin prepared by Abel caused on hyperglycemia, thereby supporting Murlin's idea. However, the commerical method of preparation in the United States resulted in a product that contained a hyperglycemic contaminant. Bouckaert and DeDuve showed in 1947 that insulin obtained from European sources lacked the hyperglycemic contaminant. In 1952, Staub used the residue from an insulin isolation to prepare a crystalline compound that caused hyperglycemia and glycogenolysis in liver tissues; in 1956, Bromer and Behrens determined the amino acid sequence of this molecule. Glucagon is a polypeptide, consisting of 29 amino acid residues, with a molecular weight of 3,485. It can be destroyed by many proteolytic enzymes, but may be freed of insulin contamination by treatment with alkali to which it is resistant.

Regulation of Insulin Secretion

A number of stimuli have been found to induce the pancreatic secretion of insulin under in vitro and in vivo conditions. The primary stimulus is blood sugar concentration. Hyperglycemia induces an increased insulin secretion and glucose is the most effective hexose in the sequence, glucose > mannose > fructose > galactose. Ingestion of food initiates a secretion of insulin. Food in the stomach and duodenum causes secretion of three gastrointestinal hormones: (1) gastrin, which increases gastric secretion of hydrochloric acid; (2) secretion, which stimulates the secretion of a watery pancreatic juice; and (3) pancreozymin, which stimulates secretion of a pancreatic juice rich in enzymes. The administration of any one of these to dogs causes in increased secretion of insulin that reaches its maximum rate within one minute. Glucagon, administered in doses of less than 10 mg, also induces secretion of insulin. However, the increase in serum insulin occurs concurrently with the increase in serum glucose. Arginine and lysine, leucine, and phenylalanine increase secretion of insulin. Both epinephrine and norepinephrine inhibit the secretion of insulin in dogs and man even when glucose is administered simultaneously with the catecholamine; epinephrine is the more effective of the two. Insulin secretion is regulated by the interaction of hormones with the β cells and by the action of blood sugar as a negative-feedback stimulus. Together these regulatory mechanisms constitute an extremely sensitive control system, resulting in the homeostasis of glucose that Cannon visualized almost 50 years ago.

Table 12.2. Effects of Gastrointestinal Hormones on Glucose and Pancreatic Secetions

Hormone Dose	*Arterial Glucose*	*Insulin Secretion*	*Glucagon Secretion*
Glucagon, 1 μg	Prompt 50 mg/ 100 ml rise: peak at 6 min.	Prompt 250% rise; peak at 6 min.	
Secretion, 20-30U (1-2 μg)	No effect	Vertical 290% rise; peak at 1 min.	None
Gastrin, 0.13-0.2U (30-34 mg)	No effect	Vertical 390% rise; peak at 1 min.	None, or very slight
Pancreozymin-cholecystokinin, 100 (17 μg)	Late rise; peak at 10 min.	Vertical 480% rise; at 1 min.	Prompt 80% rise; peak at 3 min.

Regulation of Glucagon Secretion

Secretion of glucagon by the α cells of the pancreas is stimulated by hypoglycemia. Foa demonstrated by cross-circulation experiments between the pancreatic and femoral veins that induction of hypoglycemia in a donor dog caused the appearance of a phosphorylase-activating substance in the pancreatic vein that caused hyperglycemia in the recipient dog. While blood sugar concentration is the most potent regulator of glucagon secretion, pancreozymin is also effective. Although it causes secretion of both insulin and glucagon, pancreozymin is about six times more effective in inducing insulin secretion. Glucagon-immunoreactive material has been found in extracts of the gut; these extracts have shown glucagonlike biological activity. The source of intestinal glucagon is not known.

Unlike their role in response to stress, the catecholamines are not involved in normal glucose homeostasis. Glucagon is a much more effective regulator of glucose release from the liver than is epinephrine. Sokol has emphasized that while 10^{-6} *M* glucagon induces an increase in liver phosphorylase activity and glycogenolysis, 10^{-5} *M* ephinephrine is required for an equivalent effect.

Effect of Insulin on Carbohydrate Metabolism

Insulin affects the metabolism of glucose, fatty acids, and amino acids. Its actions are so diverse that it could be considered a general metabolic hormone, modifying the utilization of nutrients in the same way that thyroid hormones in adult mammals modify the rate at which energy is released from these nutrients.

The formation of glucose-6-phosphate is key importance in glucose metabolism: synthesis of glycogen (glycogenesis); degradation to pyruvate (glycolysis); and formation of pentoses via the pentose shunt. The pool of glucose-6-phosphate is maintained by entrance of glucose into the cell and phosphorylation to glucose-6-phosphate, and by conversion of glycerol and amino acids to glucose (gluconeogenesis). The enzymes capable of performing all of these reactions are found in the liver and perhaps in the kidney. Although the enzymes involved in glycolysis are also found in muscle and adipose tissue, these tissues can neither release glucose (because they lack glucose-6-phosphate) nor form it (because they lack the transaminases necessary for gluconeogenes...

Levine and Goldstein (1949) demonstrated that insulin caused an abrupt increase of cell permeability to galactose in eviscerated, nephrectomized dogs. They later showed that this effect of insulin on cell permeability to glucose and other monosaccharides was stereospecific for D-glucose structure in the first four carbons, that is D-arabinose and D-xylose. Studies by other workers confirmed that the rate limiting step for the movement of glucose into muscles is at the cell membrane in the absence of insulin; in its presence, glucokinase activated phosphorylation becomes rate-limiting. Glucose enters liver cells freely and phosphorylation by glucokinase is the rate-limiting step in the formation of glucose-6-phosphate. Diabetes or starvation causes a reduction in glucokinase activity; recovery of activity occurs after administration of insulin or feeding respectively.

The formation of glycogen from glucose-6-phosphate requires transformation to glucose-1-phosphate, condensation with uridine diphosphate to form uridine diphosphoglucose (UDPG) under the influence of UDPG pyrophosphoerylase, and the introduction of D-glucose into the 1, 4-configuration of glycogen by glycogen synthetase. Glycogen synthetase is rate limiting and is affected by the presence of insulin; it can exist in two forms, a D form and I form, referring to its dependence on or independence of glucose-6-phosphate. In the presence of insulin the I form predominates and glycogen formation occurs in the presence of low intracellular concentrations of glucose-6-phosphate. In the diabetic liver, glucose-6-phosphate formation is impaired; since the glycogen synthetase is the D-form, little glycogen can be formed.

The rate of glycolysis is determined by the phosphofructokinase and pyruvatekinase reactions, which are metabolically irreversible and rate-limiting. The pyruvatekinase reaction is regulated by insulin. Pyruvatekinase activity is reduced in diabetes, and restored by insulin; it varies directly with the carbohydrate content of the diet.

Fasting animals and diabetic rats show decreased activities of glucose-6-phosphate and 6-phosphogluconate dehydrogenases, and a decreased oxidation of glucose through the pentose shunt. A recovery of activity occurs about 12 hr. after feeding or administration of insulin. This increase can be prevented by administration of actinomycin D or puromycin. However, since the rise in activity following administration of insulin can be prevented by imposing a fast on the animals, and since actinomycin D-treated rats are anorexic, it is not clear whether the insulin effect is due to increased availability of glucose-6-phosphate or induction of protein synthesis. The increase in fatty acid synthesis seen after insulin treatment may be due to the increased availability of NADPH that results from the initial dehydrogenation of glucose-6-phosphate in the pentose shunt.

Gluconeogenesis is inhibited by insulin and stimulated by glucocorticoids. Diabetic animals have an accelerated rate of gluconeogenesis that can be reduced by adrenalectomy or insulin. Gluconeogenesis is exaggerated in diabetic animals and is manifested by azoturia and muscle wasting; it can be relieved by insulin, hypophysectomy, or adrenalectomy. ACTH induces gluconegenesis in hypophysectomized.

Effect of Insulin on Fat Metabolism

The fat stored in adipose tissue is derived from dietary fat or synthesized from carbohydrates; it is readily mobilized by numerous hormones, among which are ACTH, TSH, GH, glucagon, epinephrine, and norepinephrine. Mobilization of fat from adipose tissue involves hydrolysis of neutral fats by a lipase and the appearance in the blood of glycerol and free fatty acids complexed with plasma albumin. The fatty acids are readily metabolized by muscle cells, and in the fasting human the high concentration of free fatty acids in the plasma is associated with a respiratory quotient (RQ) of 0.77, elevated plasma GH concentration, and low plasma glucose and insulin concentrations.

In the fasting or diabetic animals, increased mobilization of fatty acids and lipemia result from formation of lipoproteins in the liver. The increased rate of fatty acid oxidation in the liver results in the accumulation of acetyl coenzyme A (CoA). Normally, the concentration of acetyl CoA is regulated by oxidation via the Krebs cycle and by synthesis of fatty acid or cholesterol. In starvation of diabetes, the pathways for oxidation and fatty acid synthesis are inhibited, the cholesterol pathway is open, and excess cholesterol synthesis ensues. β-Hydroxy-β-methylglutaryl CoA is found in the initial stage of steroid synthesis; it is the immediate precursor of mevalonic acid and the key compound in the formation of the ketone bodies. The coenzyme is formed by the condensation of acetoacetyl CoA; its cleavage produces acetoacetic acid and acetyl CoA. Acetoacetyl CoA is formed by the condensation of two molecules of acetyl CoA in the presence of thiolase. Acetoacetyl CoA and succinic acid in the presence of acetoacetyl-succinic CoA transferase give rise to more acetoacetic acid and succinyl CoA. In the diabetic liver, oxidation of acetoacetic acid is not readily accomplished; the excess diffuses into the blood. In the presence of β-hydroxybutyrate dehydrogenase and NADH, acetoacetic acid may be reduced, giving rise to β-hydroxybutric acid. Decarboxylation of acetoacetic acid yields acetone. These three compounds-acetoacetic acid, β-hydroxybutric acid, and acetone—are the familiar ketone bodies. Their appearance in the blood and urine causes ketonemia and ketonuria, respectively. The presence of acetone in the breath or urine leads to the diagnose of ketosis. In addition to sings of deranged fat metabolism, ketone bodies cause a loss of Na^+ into the urine with consequent disturbance of acid-base balance of the blood, and a loss of water leading to dehydration.

Effect of Insulin on Protein Metabolism

Insulin increases the rate of amino acid accumulation by muscle cells, probably in a manner similar to its effect on glucose uptake. This enhanced availability of amino acids may increase the rate of their incorporation into proteins. Insulin also affects protein synthesis directly by exerting an influence on the synthesis of macro-molecules. In an in vitro system actinomycin D inhibits the biosynthetic action of insulin, but inhibition of increased amino acid transport occurs only after a long incubation period in actinomycin D.

Administration of insulin to rats causes an increased synthesis of RNA in the liver within 30 min and an increase in DNA between 24 and 36 hr. Growth of the liver is associated with the increase in DNA;

over a 3-day period, the liver increases as much as 2.5 times in net weight. Liver RNA polymerase prepared from diabetic rats treated with insulin is increased in activity and may account for the rapid synthesis of RNA. After 36 hr, RNA synthesis in the liver changes in character and reflects the need for RNA in newly-formed cells. The mechanism for activation of RNA polymerase is not known, nor is it clear that increased protein synthesis is due to activation of the sequence of events leading from transcription to translation. Wool has found that the 80 S ribosomes prepared from normal rat muscle are more active in polypeptide synthesis directed by polyuridylic acid than are similar ribosomes prepared from diabetic muscle. He suggested that decreased protein synthesis in ribosomes prepared from diabetic rat muscle is due to a defect in the 60 S subunit. The demonstration of the defect in the 60 S particle by this elegant experiment was based on the dissociation of the 80 S ribosomes into 40 S and 60 S subunits and subsequent of hybridization of these subunits.

Effect of Glucagon

Glucagon affects the kidney, causing an increased glomerular filtration rate and renal plasma flow, with a consequently increased excretion of ions. However, regulation of blood glucose is the primary role of glucagon. It can also, by virtue of its activation of adenylcyclase and consequent increase in the concentration of c-AMP cause lipolysis in adipose tissue. Glycogenolysis and hyperglycemia results from the action of glucagon and epinephrine on liver, whereas glycogenolysis and glycolysis results from the action of epinephrine on muscle. Both glucagon and epinephrine appear to activate the enzyme adenylcyclase, which converts ATP (adenosine triphosphate) to c-AMP. Unfortunately the inability to purify adenylcyclase, which appears to be membrane-bound, has interfered with an understanding of the specific mechanism involved. Acting as a "second messenger," c-AMP regulated the conversion of an inactive form of phosphorylasekinase to an active form. Glycogen phosphorylase b is transformed to glycogen phosphorylase a, through the consumption of four molecules of ATP. Glycogen phosphorylase has been crystallized and its molecular changes are at least partially understood: there is an equilibrium between phosphorylase a and phosphorylase b; phosphorylasekinase is involved in the dimerization of two molecules of phosphorylase b, each having a molecular weight of 250,000, to one molecule of phosphorylase a having a molecular weight of 50,000. Concomitantly ATP phosphorylates four serine hydroxyl groups to serine phosphate. This appears to be a principal mechanism of intracellular control of glycogen phosphorylase, but other effects have been observed. For example, phosphorylase b can be made partially active in the presence of adenosine-5'-monophosphate. There is also a specific phosphorylase phosphatase which converts phosphorylase a to phosphorylase b independently of the phosphorylasekinase reaction. In muscle, glucose-6-phosphate derived from the degradation of glycogen is oxidized principally to lactate and further oxidation products. In the liver, which contains glucose-6-phosphatase, free glucose is formed. Diffusion of this glucose into the blood causes hyperglycemia.